Studies in Big Data

Volume 98

Series Editor

Janusz Kacprzyk, Polish Academy of Sciences, Warsaw, Poland

e-mail: kacprzyk@ibspan.waw.pl

The series "Studies in Big Data" (SBD) publishes new developments and advances in the various areas of Big Data- quickly and with a high quality. The intent is to cover the theory, research, development, and applications of Big Data, as embedded in the fields of engineering, computer science, physics, economics and life sciences. The books of the series refer to the analysis and understanding of large, complex, and/or distributed data sets generated from recent digital sources coming from sensors or other physical instruments as well as simulations, crowd sourcing, social networks or other internet transactions, such as emails or video click streams and other. The series contains monographs, lecture notes and edited volumes in Big Data spanning the areas of computational intelligence including neural networks, evolutionary computation, soft computing, fuzzy systems, as well as artificial intelligence, data mining, modern statistics and Operations research, as well as self-organizing systems. Of particular value to both the contributors and the readership are the short publication timeframe and the world-wide distribution, which enable both wide and rapid dissemination of research output.

The books of this series are reviewed in a single blind peer review process.

Indexed by SCOPUS, EI Compendex, SCIMAGO and zbMATH.

All books published in the series are submitted for consideration in Web of Science.

More information about this series at https://link.springer.com/bookseries/11970

Ali Emrouznejad · Vincent Charles
Editors

Big Data and Blockchain for Service Operations Management

Springer

Editors
Ali Emrouznejad
Surrey Business School
The University of Surrey
Guildford, UK

Vincent Charles
Center for Value Chain Innovation
CENTRUM Católica Graduate Business
School, Pontifical Catholic University
of Peru
Lima, Peru

ISSN 2197-6503 ISSN 2197-6511 (electronic)
Studies in Big Data
ISBN 978-3-030-87306-6 ISBN 978-3-030-87304-2 (eBook)
https://doi.org/10.1007/978-3-030-87304-2

This Springer imprint is published by the registered company Springer Nature Switzerland AG
The registered company address is: Gewerbestrasse 11, 6330 Cham, Switzerland

Preface

Big Data is a major source of change in today's world. It is without doubt a source of immense economic and social value with the potential to impact individuals, organisations and society alike in ways that are yet to be fully explored. On the other hand, Blockchain is poised to play the role of foundation technology to store Big Data, ensuring that the data remain trustworthy, immutable and traceable. In this sense, then, Blockchain will make Big Data even more valuable. Altogether, Big Data and Blockchain are two complementary technologies that are expected to radically transform the way organisations are run in the upcoming years.

Organisations are constantly collecting a variety of data, such as standard tables, text, pictures and videos, of unprecedented sizes (millions or billions of records / variables) and from various sources, with the aim to use such data to improve their operations/services and create competitive advantage. There is a collective assumption that if organisations can learn to harness Big Data and Blockchain technologies, then their operational capabilities would be transformed. In this context, both academics and practitioners interested in Service Operations could benefit from Big Data and Blockchain technology to enhance operational performance.

The present book titled "Big Data and Blockchain for Service Operations Management" aims to provide the state-of-the-art on the use of Big Data and Blockchain to improve Service Operations in a variety of domains. Along theory and applications, the book compiles the authors' experiences so that these may be aggregated for a better understanding.

The book is well organised in fourteen chapters, contributed by authors from all around the globe: Argentina, Austria, Canada, Chile, China, Czech Republic, India, Iran, Malaysia, Peru, Portugal, Turkey, United Arab Emirates, United Kingdom and the United States.

The first eight chapters address the topic of big data and service operations management from various angles.

The chapter "Characteristics and Trends in Big Data for Service Operations Management Research: A Blend of Descriptive Statistics and Bibliometric Analysis" provides an introduction to big data for service operations management by reviewing relevant literature and highlighting developments in research and application. The analysis reveals patterns in scientific outputs and serves as a guide for global research trends in big data for service operations management. Among others, the chapter emphasises the need for research on building big data-driven analytical models which are not only explainable and interpretable, but also deployable in the Cloud.

The chapter "Strategy Formulation and Service Operations in the Big Data Age: The Essentialness of Technology, People, and Ethics" complements the chapter "Characteristics and Trends in Big Data for Service Operations Management Research: A Blend of Descriptive Statistics and Bibliometric Analysis" and explores the promise of big data in redefining strategy in service operations management by means of investigating a rich range of bibliographic material. As the authors indicate, service operations management research in the big data age implies a shift in attention from being increasingly integrative across themes to being integrative across multiple disciplines, requiring the expertise of and tuning between different actors and expertise domains.

The chapter "Modeling Big Data Enablers for Service Operations Management" identifies big data enablers in service operations management and analyses the interactions between them. The findings allow decision-makers to select the desired enablers and drop the undesired ones in the implementation of big data initiatives to improve service operations management performance. The chapter makes contributions by proposing the use of MCDM-based hierarchical models and causal diagram.

The chapter "Data Architecture for Big Data Service Operations Management (The New Vision of Data Architecture for the Future Human Society)" proposes a platform construction for data management and control called Data Architecture, which can be used in big data service operations management and provide complicated data applications with data protection in the open Internet environment.

The chapter "Big Data for Educational Service Management" presents a survey of the state-of-the-art applications of big data analytics in the field of educational services. The authors explain how big data can help in improving the overall education services, describe the challenges that institutions face while implementing big data-based solutions and suggest future research avenues on the topic.

The chapter "A Novel Big Data Approach for Text Supported Service Operations Management" presents the latest advances in artificial intelligence for text data analysis and operations management. It provides the state-of-the-art of the text processing approaches, discusses selected use-cases from the field of operations management and how the latest methods can help to solve those problems, and outlines some ideas for further improvement of the current approaches in terms of how to effectively analyse data in a multilingual environment and decrease memory demands.

The chapter "Toward a Comprehensive Framework of Social Media Analytics" proposes a practical analytics framework for gaining more actionable insights from social media content. The framework is developed based on a series of machine learning and data analysis algorithms along with the required ETL modules. Having the ability to get embedded in big data clusters, the proposed analytics engine can be utilised in analysing large social media datasets through big data analytics solutions.

The chapter "Data Mining Approach in Repair and Service Systems of Electronic Products Under Warranty" provides the only study in the literature aimed at contributing to operational processes by evaluating the only type of product in the electronic repair service sector using data mining methods relative to the type of repair. The results of the research are expected to contribute to the evaluation of the processes of firms operating in a similar field.

The role of blockchain has been expanding in recent years due to its increasing application in various domains. The second part of the book comprises six chapters that deal with the topic of blockchain in the context of big data for service operations management.

The chapter "Integrative Applications of Blockchain and Contemporary Technologies from a Big Data Perspective" introduces the concept of blockchain and highlights the integrated applications of blockchain, Internet-of-Things, fog computing and artificial intelligence. It focuses on 3D Printing based on blockchain, introduces the integrative applications of blockchain and swarm robotic systems, as well as it features the composite application geotagging and blockchain. Several managerial and policy implications of managing big data from a service operations perspective are also proffered.

The chapter "Blockchain for Disaster Management" explores the application of blockchain in disaster management to address issues related to poor coordination among responding agencies, late disaster response and inadequate distribution of resources. The authors offer a comprehensive blockchain framework for disaster management which includes governments, residents, telecommunication providers, shelter providers, food service providers, medical service providers and suppliers, transportation providers and non-governmental relief organisations.

The chapter "Blockchain Production Planning in Mass Personalized Environments" presents the tools to generate a business strategy for mass customised/personalised production in Industry 4.0 environments. To this aim, the authors propose an autonomous and decentralised blockchain-based system managed fundamentally by cyber-physical systems (CPS), which allows associating the decision-making processes concerning production planning to the CPS. The proposal is readily applicable to service operations management, given that personalised goods embody many features shared with services.

The chapter "Frontiers of Blockchain for Railways" explores blockchain-based applications in the railway ecosystem from a service operations perspective. More specifically, the authors advance a conceptual model (i.e. a specific provenance framework) to understand the contribution of blockchain technology in the Indian Railway system.

The chapter "Blockchain Interoperability Issues in Supply Chain: Exploration of Mass Adoption Procedures" provides an insight into blockchain interoperability in supply chains for mass adoption. To this end, a three-step approach is applied through conducting a literature review of blockchain technology and commonly used methodologies for blockchain interoperability, analysing four blockchain real-life use case applications in supply chains that address interoperability concerns and mass adoption, and discussing the results of the analysis based on the comments of interviewees.

Finally, the chapter "Blockchain Technology Enablers in Physical Distribution and Logistics Management" identifies the critical enablers for the adoption of blockchain technology in the logistics and physical delivery sector, which it then validates by experts for highlighting their prioritisation using an analytic hierarchy process approach. Findings show that traceability and transparency were the factors given the utmost priority; they also underline how the adoption of blockchain technology will enhance the operational efficiency in the case of service operations management.

The chapters contributed to this book should be of considerable interest and provide our readers with informative reading.

Guildford, UK
Lima, Peru
February 2022

Ali Emrouznejad
Vincent Charles

Acknowledgments

First among these are the contributing authors—without them, it was not possible to put together such a valuable book, and we are deeply grateful to them for bearing with our repeated requests for materials and revisions while providing the high-quality contributions. We are also grateful to the many reviewers for their critical review of the chapters and the insightful comments and suggestions provided. Thanks are also due to *Professor Janusz Kacprzyk*, the Editor of this Series, for supporting and encouraging us to complete this project. The editors would like to thank *Dr. Thomas Ditzinger* (Springer Senior Editor, Interdisciplinary and Applied Sciences & Engineering), *Ms. Sylvia Schneider* (Springer Project Coordinator, Production Heidelberg), *Ms. Divya Meiyazhagan* (Springer Production Editor, Project Manager), *Mr. Viju Falgon* (in the Production team) for their excellent editorial and production assistance in producing this volume.

We hope the readers will share our excitements with this important scientific contribution to the body of knowledge in Big Data and Blockchain.

Ali Emrouznejad
Vincent Charles
The Editors

Contents

Characteristics and Trends in Big Data for Service Operations Management Research: A Blend of Descriptive Statistics and Bibliometric Analysis

Vincent Charles, Tatiana Gherman, and Ali Emrouznejad

Abstract The field of service operations management has a plethora of research opportunities to capitalise on, which are nowadays heightened by the presence of big data. In this research, we review and analyse the current state-of-the-art of the literature on big data for service operations management. To this aim, we use the Scopus database and the VOSviewer visualisation software for bibliometric analysis to highlight developments in research and application. Our analysis reveals patterns in scientific outputs and serves as a guide for global research trends in big data for service operations management. Some exciting directions for the future include research on building big data-driven analytical models which are deployable in the Cloud, as well as more interdisciplinary research that integrates traditional modes of enquiry with for example, behavioural approaches, with a blend of analytical and empirical methods.

Keywords Analytics · Big data · Operations management · Services · Bibliometrics

V. Charles (✉)
CENTRUM Católica Graduate Business School, Lima, Peru
e-mail: vcharles@pucp.pe

Pontifical Catholic University of Peru, Lima, Peru

T. Gherman
Faculty of Business and Law, University of Northampton, Northampton, UK
e-mail: tatiana.gherman@northampton.ac.uk

A. Emrouznejad
Surrey Business School, The University of Surrey, Guildford, UK
e-mail: a.emrouznejad@surrey.ac.uk

A. Emrouznejad and V. Charles (eds.), *Big Data and Blockchain for Service Operations Management*, Studies in Big Data 98,
https://doi.org/10.1007/978-3-030-87304-2_1

1 Introduction

Operations management is a fundamental organisational function involved in the management of activities to produce and deliver products and services [3]. Services have long been the dominant sector of industrialised nations; nevertheless, the field of operations management has been traditionally associated with manufacturing and supply chains rather than with services [17]. And while it is true that *service operations management* (SOM) has struggled to find its fit as a distinct discipline in the literature, it is fair to say that the number of studies on the topic have increased substantially over the years. For example, a recent review study by Roth and Rosenzweig [31] concluded that, in their sample, service-oriented papers represented 58% of the publications, with the healthcare and retail sectors at the forefront. Today, big data have been accelerating this shift in research. All in all, the presence of big data has been 'pushing' organisations to review their practices and identify opportunities that would allow them to embrace data-driven decision-making processes to a greater extent.

Although there is no unique definition of big data [7, 8], it is commonly accepted that big data are "datasets that are too large for traditional data-processing systems and that therefore require new technologies" [30]. In the literature, it is common to refer to the four dimensions of big data defined by Laney [21]: volume, velocity, variety, and veracity, which are indicative of the computational complexities and technical requirements associated with big data. Coupled with ethical challenges [10], analysing big data confronts the researchers with many difficulties [4]. Charles and Gherman [9] underlined that in order to create value and competitive advantage, big data should be further considered in view of the dimensions of context, connectedness, and complexity. The growth of big data has created opportunities for pursuing new avenues of research in SOM. In particular, big data analytics may support improved policy- and decision-making and drive organisational performance.

It is vital to both comprehensively and quantitatively evaluate the development trend in research on big data for SOM, which can help not only practitioners and managers, but also academics interested in making informed decisions in their future research endeavours. Bibliometric analysis, which has been widely used across different fields, is a feasible means that can quantitatively and qualitatively assess trends in research fields over time. It can be used to systematically identify, organise, and analyse the main elements of a research topic [28, 38], as well as to clearly determine the development trends of a particular research field [2].

After decades of research developments in SOM, there is currently a growing interest in exploring the potential that big data pose for this field. This is supported not only by the increasing availability of data, but also by the methodological advances in a number of fields. Therefore, this research investigates the characteristics and trends of studies integrating big data and SOM through a bibliometric analysis so as to facilitate a comprehensive understanding of the state-of-the-art of current research directions and progress in the field.

2 Methods and Materials

Bibliometric analysis is a valuable research technique that can help in discovering the global research trends on a topic or field from multiple angles, providing an overview of future lines of research [24]. By bibliometric analysis, this paper reveals the characteristics and trends in studies integrating big data with SOM in view of publication outputs and major journals, subject categories, geographic and institutional distribution of publications, and keywords analysis.

The bibliometric analysis was conducted based on the Scopus database, which is the abstract and citation database of Elsevier. Although there are other databases available, Scopus was deemed to be one of the best choices in view of the fact that it is characterised by consistent citation metrics and precision in locating authors and institutions. Relevant data on big data for SOM research were downloaded on 13 March 2021 via the concurrent search for the keywords "big data", "operations management", and "service". These keywords were searched in the title, abstract, and keyword lists of each publication. Finally, only publications in English were selected for further analysis. The search criteria along with the Boolean expression was as follows: TITLE-ABS-KEY ("big data" "operations management" "service") AND (LIMIT-TO (PUBYEAR, 2020) OR LIMIT-TO (PUBYEAR, 2019) OR LIMIT-TO (PUBYEAR, 2018) OR LIMIT-TO (PUBYEAR, 2017) OR LIMIT-TO (PUBYEAR, 2016) OR LIMIT-TO (PUBYEAR, 2015) OR LIMIT-TO (PUBYEAR, 2014) OR LIMIT TO (PUBYEAR, 2013)) AND (LIMIT-TO (LANGUAGE, "English")).

3 Results

The concurrent search for the keywords "big data", "operations management", and "service" yielded 57 document results on 13 March 2021, all published between 2013–2020. In Sect. 3.1, we proceed to analyse these document results by means of summary statistics. Subsequently, in Sect. 3.2, we continue with a bibliometric analysis of the same material, while in Sect. 3.3, we focus on exploring the themes of the journal research articles only.

3.1 Descriptive Summary Statistics of Published Material

In this section, we provide an overall analysis of the 57 document results by means of various visualisations. Figure 1 shows the evolution of the number of publications in big data for SOM. Several observations are worth mentioning. First, all 57 documents were published after the year 2013, which means that at least in the Scopus database, there were no publications that integrated the three keywords together before this year. Second, we can observe an increased interest in the topic in recent years, with a peak in the year 2018 (16 publications).

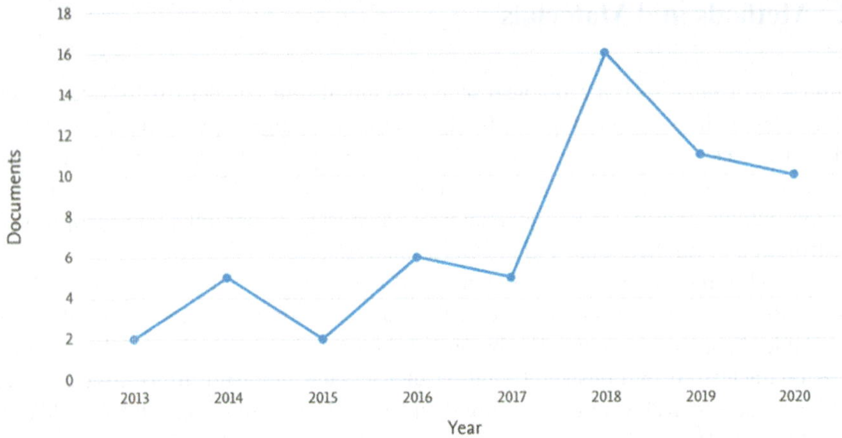

Fig. 1 Annual scientific production. (*Source* Scopus [33])

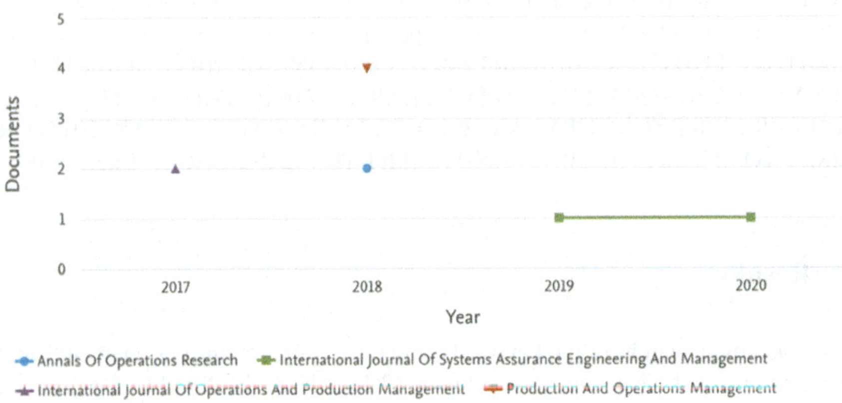

Fig. 2 Documents per year by source. (*Source* Scopus [33])

Figure 2 shows the number of documents per year by source, with a comparison of the document counts for the top four sources. The journals that have published most of the material on big data for SOM are Production and Operations Management (4 publications), Annals of Operations Research (2 publications), International Journal of Operations and Production Management (2 publications), and International Journal of Systems Assurance Engineering and Management (2 documents).

Figure 3 presents the documents by affiliation, comparing the document counts for the first 15 affiliations. Affiliation-wise, there are five institutions that lead the ranking with most document counts (2 publications each), namely Hong Kong Polytechnic University, University of Leeds, Pohang University of Science and Technology, Ulsan National Institute of Science and Technology, and Leeds University Business School.

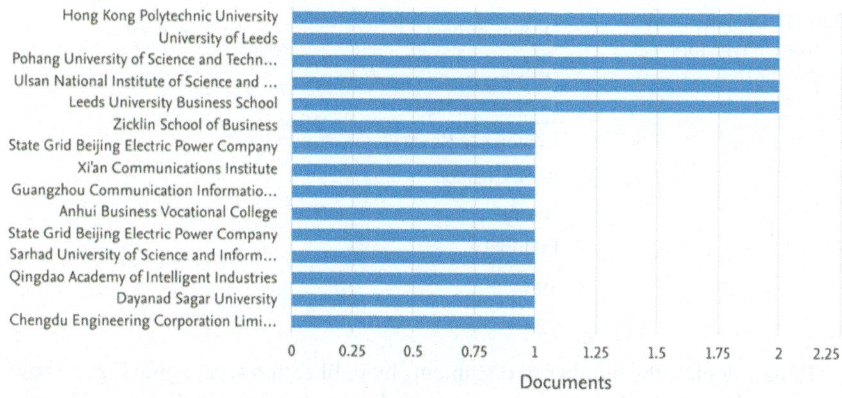

Fig. 3 Documents by affiliation. (*Source* Scopus [33])

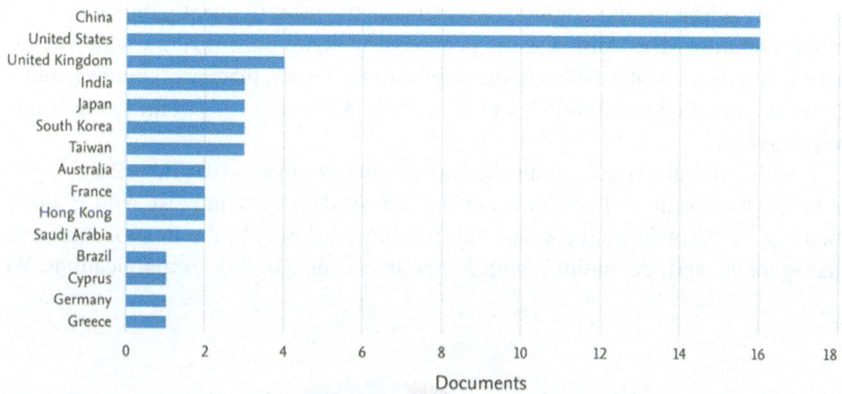

Fig. 4 Documents by country or territory. (*Source* Scopus [33])

These institutions represent a mixture of countries (China, United Kingdom, and South Korea). All remaining institutions each have one publication.

Figure 4 visually depicts the countries with the highest number of publications, comparing the document counts for 15 countries/territories. The countries of origin for the 57 documents were determined by considering the country of the corresponding author. It can be easily observed that China and the United States share the first place, with 16 publications each. As a matter of fact, a notable observation is that China and the United States together account for more than half (i.e., 56.14%) of the total number of publications. The United Kingdom occupies the third place with 4 publications, followed by India, Japan, South Korea, and Taiwan, each with 3 publications. All remaining countries each have one publication, while there are also 8 documents for which the corresponding author information is not available.

Table 1 Documents by publication type (*Source* Scopus [33])

Document type	No. of Documents
Article	22
Conference Paper	21
Conference Review	8
Book Chapter	3
Review	2
Editorial	1
Total	57

Table 1 depicts the number of documents by publication type, while Fig. 5 shows the same visually. In this sense, we can note that the literature is dominated by articles (22 documents, which constitute 38.6% of the publications), followed closely by conference papers (21 documents, representing 36.8% of the publications). Therefore, journal articles and conference papers are the most frequent publication types in the literature. The third place is occupied by conference reviews with 8 documents, which account for 14% of the publications. Lastly, there are 3 book chapters, 2 reviews, and 1 editorial, which represent 5.3%, 3.5%, and 1.8% of the publications, respectively.

Finally, an analysis of the published documents by subject area (Table 2 and Fig. 6) indicates that the area of "engineering" has received the most interest, with 29 documents or 23.2% of the publications. This is followed closely by the areas of "business, management, and accounting", with 25 documents or 20.0% of the publications. We

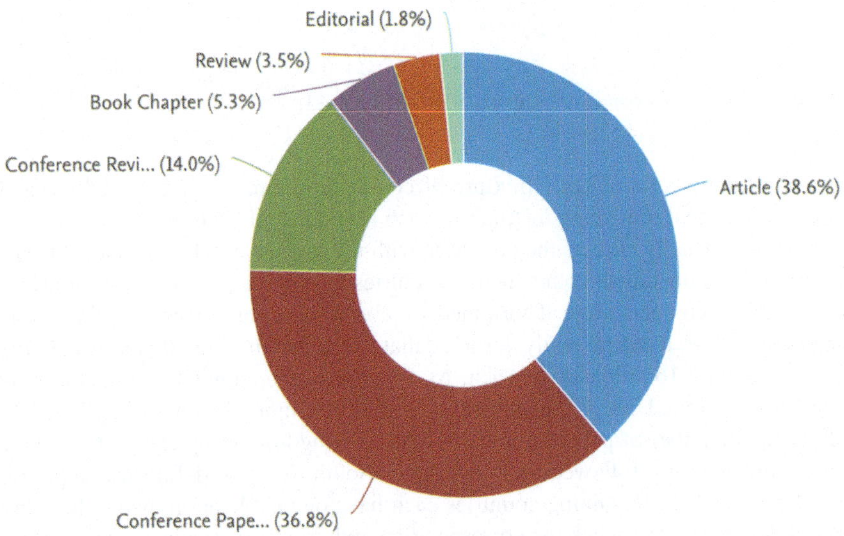

Fig. 5 Documents by publication type. (*Source* Scopus [33])

Table 2 Documents by subject area (*Source* Scopus [33])

Subject area	No. of Documents
Engineering	29
Business, Management, and Accounting	25
Computer Science	21
Decision Sciences	19
Social Sciences	8
Mathematics	6
Energy	5
Economics, Econometrics, and Finance	3
Earth and Planetary Sciences	2
Environmental Science	2
Physics and Astronomy	2
Biochemistry, Genetics, and Molecular Biology	1
Chemistry	1
Medicine	1

Note A publication can be classified under more than one subject area

then have "computer science" (with 21 documents or 16.8% of the publications) and "decision sciences" (with 19 documents or 14.2% of the publications).

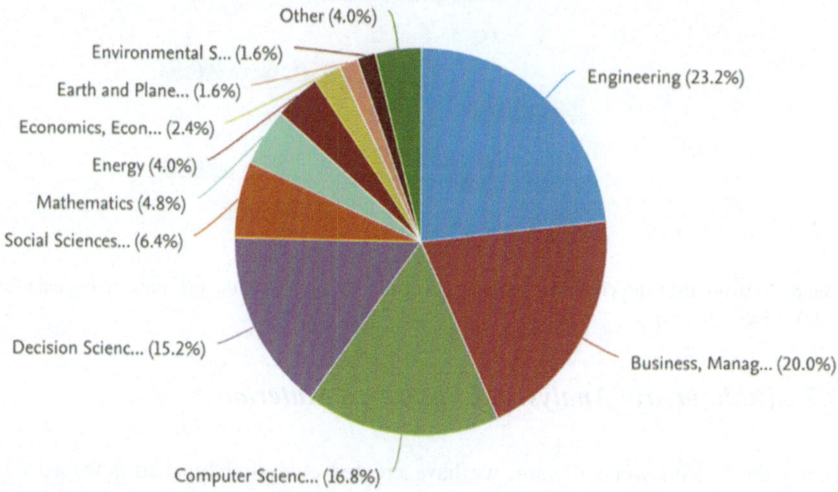

Fig. 6 Documents by subject area. (*Source* Scopus [33])

Fig. 7 Co-authorship network of countries

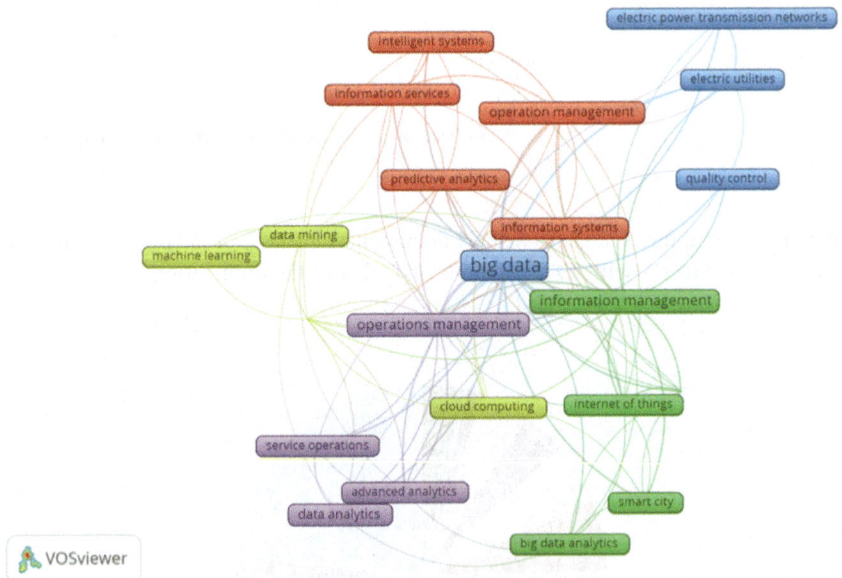

Fig. 8 Network map showing the relations between various topics in the literature on big data for SOM (57 documents)

3.2 Bibliometric Analysis of Published Material

Using the VOSviewer software, we have created co-authorship and keyword co-occurrence maps based on bibliographic data. The co-authorship analysis (Fig. 7) considered two as the minimum number of documents of a country; of the 25 countries identified, 11 met the threshold, although the largest set of connected items consisted of 8 countries. The keyword co-occurrence analysis (Fig. 8) has been performed using all the keywords as the unit of analysis, with minimum number of occurrences as three, and with full counting as the counting method.

Figure 7 reveals the network map of international cooperation among major countries (with the greatest total link strength) participating in research on big data for SOM. The colours indicate the clusters to which the countries are attributed according to the strength of their relationships, while the size of the circles is indicative of the number of publications held by each country. We can observe that there are 3 clusters in the figure. The first cluster (red colour) is dominated by the United States, and includes a mix of Eastern and Western countries, namely Taiwan and South Korea (in Asia), and France (in Europe). The second cluster (blue colour) is headed equally by Hong Kong and Japan. Lastly, the third cluster (green colour) is led by China, but also includes Australia.

Figure 8 shows that a total of 20 most common keywords have been identified. Keywords are labelled with coloured frames, whose size is positively correlated with the occurrence of the keyword in the document. Moreover, these keywords are grouped into five clusters that seem to assume a prominent role vis-à-vis "computational paradigms" (three items, yellow cluster), "big data for quality control and electric utilities" (four items, blue cluster), "information services and operations management" (five items, red cluster), "data analytics for SOM" (four items, purple cluster), and "big data analytics, internet of things, and smart city" (four items, green cluster).

3.3 Bibliometric Analysis of Journal Articles

In this section, we have proceeded to analyse only the journal articles on big data for SOM. Such decision was guided by both literature and practical considerations. First, conference papers generally do not provide enough information on the research conducted, as we encounter in full papers, being normally written with the aim of presenting preliminary results [29]. Book chapters, conference reviews, and reviews, also, do different work than journal articles, as do editorials; hence, these were also excluded from further analysis. This screening led to the consideration of 22 research articles for further processing, constituting 38.6% of the publications (Fig. 5).

Figure 9 shows the increasing number of journal articles in the field of big data for SOM in recent years, Fig. 10 positions the United States as the country with most of the journal articles publications, and Fig. 11 illustrates that the area of "business, management, and accounting" accounts for most of the journal article publications (27.7%), followed by "engineering" (21.3%), and "decision sciences" (19.1%).

A brief bibliometric analysis of the 22 journal articles composing the final sample of studies integrating big data with SOM identified a variety of keywords as the most common keywords (whose co-occurrence is at least two times) (Fig. 12). These keywords were further classified by the software into four clusters that seem to assume a prominent role vis-à-vis "big data analytics in supply chain management" (four items, red cluster), "information systems and SOM" (three items, blue cluster),

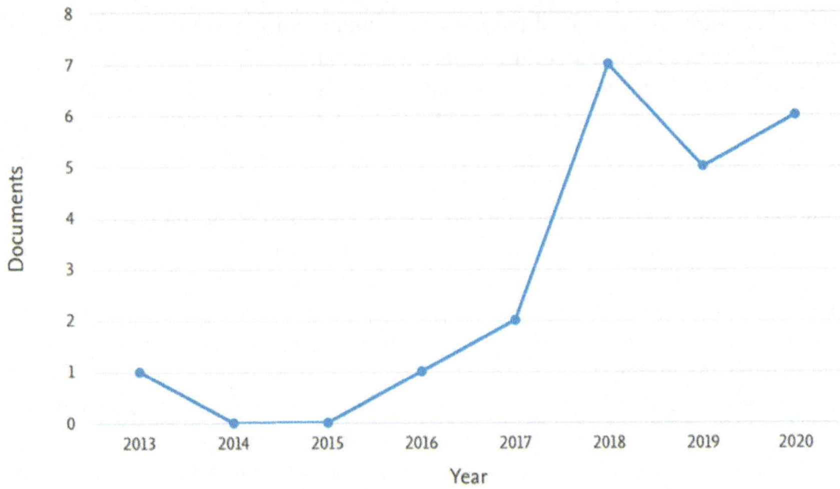

Fig. 9 Annual scientific production of studies on big data for SOM (*Source* Scopus [33])

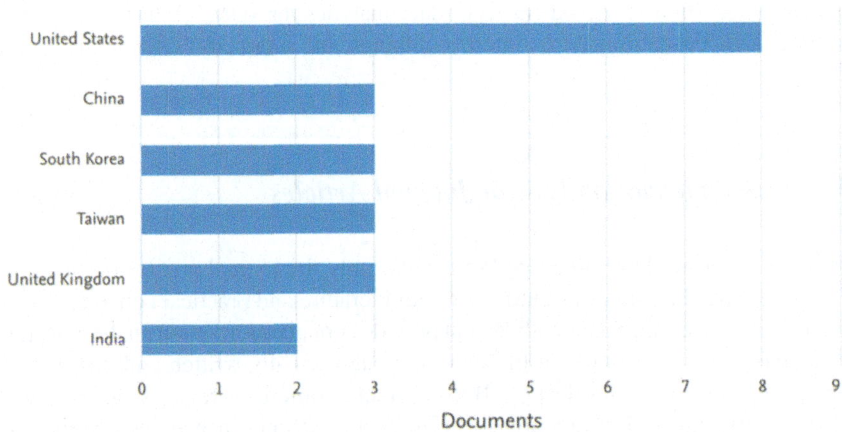

Fig. 10 Studies on big data for SOM by country or territory. (*Source* Scopus [33])

"internet of things and smart city" (four items, green cluster), and "computational paradigms" (two items, yellow cluster). Below, we present briefly the pool of 22 journal articles identified, which are arranged in chronological order, starting from the most recent one.

Ruan et al. [32] presented an IoT-based e-business model of intelligent vegetable greenhouses with details of the basic process and key nodes of the e-business model. The authors recognised key operation issues including big-data-driven pricing, planting structure and time optimisation, water and fertilizer integrated control, plant light supplement, and order-driven picking and packing. Kumar et al. [19] proposed a reliable, more accurate and efficient model based on the statistical analysis of the

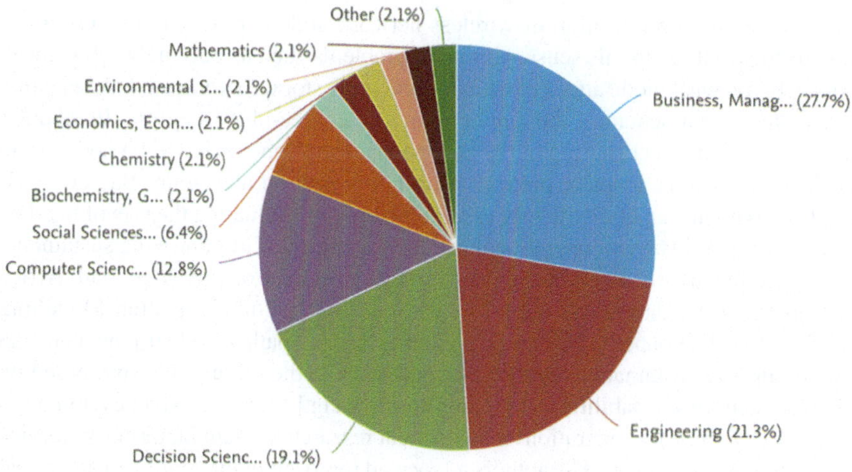

Fig. 11 Studies on big data for SOM by subject area. (*Source* Scopus [33])

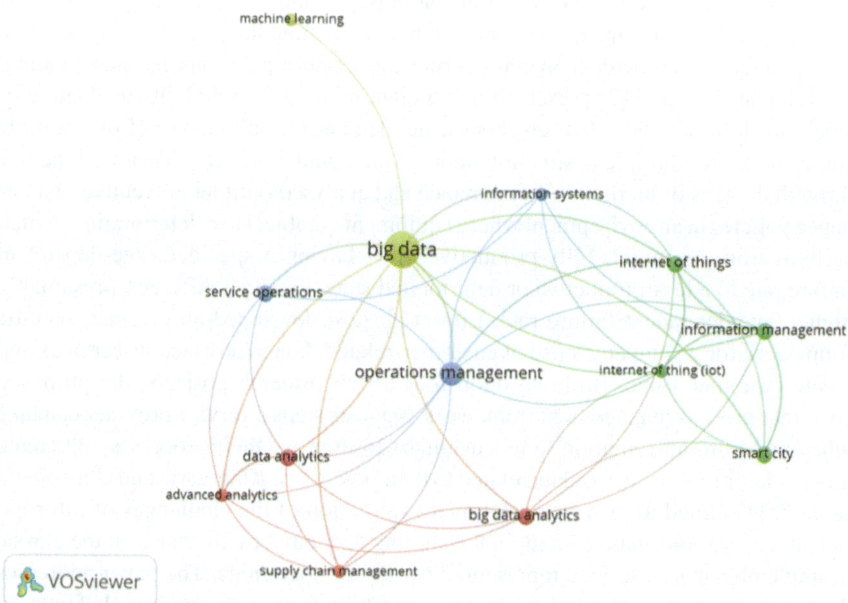

Fig. 12 Network map showing the relations between various topics in the journal article literature on big data for SOM (22 documents)

sensor-based data for occupancy detection. The paper also proposed one online and adaptive model-based online sequential extreme learning machine to perform occupancy detection on real-time data when complete data are not available, and learning is done with recent data points coming in the form of streams. Wang [37] focused on deployment and optimisation of wireless network node deployment and optimisation in smart cities. In this sense, aiming at problems such as poor network security connectivity, weak node attack resistance, and large storage overhead in the existing key management schemes, the author designed a three-phase key pre-distribution mechanism and direct sharing of the key management scheme based on node group deployment, via an adaptive particle swarm optimisation algorithm. Bag et al. [3] used the dynamic capability theory as a foundation for evaluating the role of big data analytics capability as an operational excellence approach in improving sustainable supply chain performance. The authors surveyed mining executives in South Africa and analysed the data using Partial Least Squares Structural Equation Modelling (PLS-SEM). The paper contributes to identifying two pathways that managers can use to improve sustainable supply chain outcomes in the mining industry, based on big data analytics capabilities. Roth and Rosenzweig [31] provided an examination of the rise of empirical operations management research in Manufacturing & Service Operations Management. The authors advocated for a tighter integration of analytical and empirical operations management knowledge in order to address the challenges and opportunities of the twenty-first century. Tamás and Koltai [36] aimed to review the relevant literature related to the development, improvement and application of learning curves in the age of big data, and to demonstrate the possible insight which its application can provide in manufacturing and service operations decision-making.

Datta and Goyak [12] presented an efficient method for reliability evaluation of stochastic flow networks that can pass various demands simultaneously from multiple source nodes to multiple destination nodes. March and Scudder [25] viewed the IoT through the lens of predictive maintenance and analysed optimal preventive maintenance policies in an environment where equipment is subject to a deterioration, which shifts it from its initial, fully-productive state, having a specified, age-dependent failure rate to a less-productive or deteriorated state, having a different, presumably higher, age-dependent failure rate. Lim et al. [23] developed an original, specific framework for a company's use of customer-related data to advance its services and create customer value. Building upon four action research projects, the proposed customer process management framework suggests steps a service provider can take when providing information to its customers to improve their processes and create more value-in-use by using data related to their processes. Albergaria and Chuaopetta Jabbour [1] aimed to provide an original exploration of the challenges of information and operations management in the sharing economy, by focusing on the classic example of a shared service represented by library operations. The paper addresses the organisational use of big data analytics capabilities, with the main goal of helping organisations make better business decisions, in terms of information and operations management issues. Carnerud and Bäckström [6] aimed to identify and depict the key areas around which research on quality has centred during the past 37 years and to explore longitudinal patterns in the identified key areas. The study identified seven central topics around which research on quality has centred during the time period analysed: Service Quality & Customer Satisfaction; Process design & Control; ISO Certification & Standards; TQM—Implementation, Performance &

Culture; QM—Practices & Performance; Reliability, Costs, Failure & Problems and Excellence—BEMs, Quality Awards & Excellence in Higher Education.

Focusing on the area of after-sales service, Boone et al. [5] developed a framework that seeks to define service parts performance goals for the purpose of outlining where scholars and practitioners can further examine where, how, and why big data applications can be employed to enhance service parts management performance. To objectively evaluate emergency physicians across facilities, Foster et al. [13] leveraged big data from an emergency physician management network and proposed data-driven metrics using a large-scale database. The proposed indices benchmark physicians from the perspectives of revenue potential, patient volume, patient complexity, and patient experience by controlling for exogenous factors at the facility level. As the authors acknowledge, the proposed framework can also be adapted to non-medical professional settings such as value chains, where employees often provide services in various profit- and cost-centres. Silva et al. [34] proposed a big data analytics-embedded experimental architecture for smart cities. The mentioned architecture facilitates the exploitation of urban big data in planning, designing, and maintaining smart cities, as well as it shows how big data analytics can be used to manage and process voluminous urban big data to enhance the quality of urban services. Cohen [11] discussed how the tremendous volume of available data collected by firms has been transforming the service industry, with particular focus on services in the sectors of finance/banking, transportation and hospitality, and online platforms (subscription services, online advertising, and online dating). Kuo et al. [20] applied big data mining and machine learning analysis techniques and used the Waikato Environment for Knowledge Analysis (WEKA) as a tool to discuss the convenience stores energy consumption performance in Taiwan. Shmueli and Yahv [35] introduced the use of Classification and Regression Trees for automated detection of potential Simpson's paradoxes in data with few or many potential confounding variables, and even with large samples (big data). The authors illustrate the approach via several real applications in e-governance and healthcare. Kim et al. [18] proposed an approach to analysing and utilizing vehicle operations management (VOM)-related data for designing VOM services. The feasibility and effectiveness of the proposed approach is demonstrated by means of a case study on the design of an eco-driving service.

By adopting an exploratory approach to the secondary research which examines vendors' offerings, Matthias et al. [26] focused on the application and exploitation of big data to create competitive advantage. To this aim, the authors presented a framework of application areas, and how these help the understanding of targeting and scoping specific areas for sustainable improvement. Mehmood et al. [27] aimed to advance knowledge of the transformative potential of big data on city-based transport models. In this sense, the authors developed a Markov-based model with several scenarios to explore a theoretical framework focused on matching the transport demands (of people and freight mobility) with city transport service provision using big data. Li et al. [22] developed and applied a framework to case examples that demonstrate how smart cities are redefining the characteristics of operations models around their scalability, analytical output, and connectivity. The paper contributes to our understanding of how smart cities can potentially transform operational models and sets out a research agenda for operations management in smart cities in the digital economy. Lastly, Huang and Rust [16] discussed the characteristics of information technology associated with consumer centricity.

4 Conclusions

In this chapter, we have aimed to review the literature on big data for SOM via an exploration of the Scopus database and by means of using the VOSviewer visualisation software for bibliometric analysis in order to highlight the research trends in the field. From a practical perspective, our analysis reveals patterns in scientific outputs and serves as a guide for global research trends in big data for SOM.

Overall, the findings reveal an increased interest in studies in the fields of urban planning and smart city decision management empowered by real-time data processing using IoT, big data analytics, and cloud computing technologies. Other research strands include big data for quality control, electric utilities, information services and information systems. Additionally, there is continued interest in exploring big data analytics to predict and mitigate the effect of supply chain risks and disruptions, which have been shown that can severely disrupt operations and supply chains. As Hazen et al. [15] stated, it is important for organisations to improve the quality of the analytical outputs of their decision-making processes by means of paying attention to the quality of the big data on which they base their decisions.

The review of the studies included in this research has further shown that there is an increased interest towards studying computational paradigms in the field of big data for SOM. Although great progress has been made so far, nonetheless, much more research is needed; in particular, more research is necessary to build big data-driven analytical models which are not only explainable and interpretable [14], but also deployable in the Cloud. Furthermore, there is a need for more interdisciplinary research that integrates traditional modes of enquiry in (service) operations management with, for example, behavioural approaches, with a blend of analytical and empirical methods.

Acknowledgements The authors are thankful to the reviewers for their valuable comments on the previous version of this work.

References

1. Albergaria, M., & Chuaopetta Jabbour, C. J. (2019). The role of big data analytics capabilities (BDAC) in understanding the challenges of service information and operations management in the sharing economy: Evidence of peer effects in libraries. *International Journal of Information Management, 51*, 102023.
2. Albort-Morant, G., Henseler, J., Leal- Millán, A., & Cepeda-Carrión, G. (2017). Mapping the field: A bibliometric analysis of green innovation. *Sustainability, 9*(6), 1011.

3. Bag, S., Wood, L. C., Xu, L., Dhamija, P., & Kayikci, Y. (2020). Big data analytics as an operational excellence approach to enhance sustainable supply chain performance. *Resources, Conservation and Recycling, 153*, 104559.
4. Bizer, C., Boncz, P., Brodie, M. L., & Erling, O. (2012). The meaningful use of big data: Four perspectives-four challenges. *ACM SIGMOD Record, 40*(4), 56–60.
5. Boone, C. A., Hazen, B. T., Skipper, J. B., & Overstreet, R. E. (2018). A framework for investigating optimization of service parts performance with big data. *Annals of Operations Research, 270*, 65–74.
6. Carnerud, D., & Bäckström, I. (2019). Four decades of research on quality: summarising, Trendspotting and looking ahead. *Total Quality Management & Business Excellence*, 1–23.
7. Charles, V., & Emrouznejad, A. (2018). Big Data for the Greater Good: An Introduction. In A. Emrouznejad & V. Charles (Eds.), *Big Data for the Greater Good* (pp. 1–18). Springer.
8. Charles, V., & Gherman, T. (2013). Achieving competitive advantage through big data. Strategic implications. *Middle-East Journal of Scientific Research, 16*(8), 1069–1074.
9. Charles, V., & Gherman, T. (2018). Big Data and Ethnography: Together for the Greater Good. In A. Emrouznejad & V. Charles (Eds.), *Big Data for the Greater Good* (pp. 19–34). Springer.
10. Charles, V., Tavana, M., & Gherman, T. (2015). The right to be forgotten–is privacy sold out in the big data age? *International Journal of Society Systems Science, 7*(4), 283–298.
11. Cohen, M. C. (2018). Big data and service operations. *Production and Operations Management, 27*(9), 1709–1723.
12. Datta, E., & Goyal, N. K. (2019). Evaluation of stochastic flow networks susceptible to demand requirements between multiple sources and multiple destinations. *International Journal of Systems Assurance Engineering and Management, 10*, 1302–1327.
13. Foster, K., Penninti, P., Shang, J., Kekre, S., Hegde, G. G., & Venkat, A. (2018). Leveraging big data to balance new key performance indicators in emergency physician management networks. *Production and Operations Management, 27*(10), 1795–1815.
14. Hansun, S., Charles, V., Gherman, T., Subanar, & Rini-Indrati, C. (2020). A tuned Holt-Winters white-box model for COVID-19 prediction. *International Journal of Management and Decision Making*, 1–22. In press.
15. Hazen, B. T., et al. (2014). Data quality for data science, predictive analytics, and big data in supply chain management: An introduction to the problem and suggestions for research and applications. *International Journal of Production Economics, 154*, 72–80.
16. Huang, M.-H., & Rust, R. T. (2013). IT-Related Service: A Multidisciplinary Perspective. *Journal of Service Research, 16*(3), 251–258.
17. Karmarkar, U. S., & Apte, U. M. (2007). Operations management in the information economy: Information products, processes, and chains. *Journal of Operations Management, 25*(2), 438–453.
18. Kim, M. J., Lim, C., & Kim, K. J. (2018). A data-driven approach to designing new services for vehicle operations management. *International Journal of Industrial Engineering: Theory Applications and Practice, 25*(5), 604–619.
19. Kumar, S., Singh, J., & Singh, O. (2020). Ensemble-based extreme learning machine model for occupancy detection with ambient attributes. *International Journal of Systems Assurance Engineering and Management, 11*, 173–183.
20. Kuo, C. F. J., Lin, C. H., & Lee, M. H. (2018). Analyze the energy consumption characteristics and affecting factors of Taiwan's convenience stores-using the big data mining approach. *Energy and Buildings, 168*, 120–136.
21. Laney, D. (2001). *3D Data Management: Controlling Data Volume, Velocity and Variety. Applications Delivery Strategies.* META Group (now Gartner) [online] http://blogs.gartner.com/doug-laney/files/2012/01/ad949-3D-Data-Management-Controlling-Data-Volume-Velocity-and-Variety.pdf.
22. Li, F., Nucciarelli, A., Roden, S., & Graham, G. (2016). How smart cities transform operations models: A new research agenda for operations management in the digital economy. *Production Planning and Control, 27*(6), 514–528.

23. Lim, C., Kim, M.-J., Kim, K.-H., Kim, K.-J., & Maglio, P. (2019). Customer process management: A framework for using customer-related data to create customer value. *Journal of Service Management, 30*(1), 105–131.
24. Liu, X. J., Zhang, L. A., & Hong, S. (2011). Global biodiversity research during 1900–2009: A bibliometric analysis. *Biodiversity and Conservation, 20*(4), 807–826.
25. March, S. T., & Scudder, G. D. (2019). Predictive maintenance: Strategic use of IT in manufacturing organizations. *Information Systems Frontiers, 21*, 327–341.
26. Matthias, O., Fouweather, I., Gregory, I., & Vernon, A. (2017). Making sense of Big Data–can it transform operations management? *International Journal of Operations and Production Management, 37*(1), 37–55.
27. Mehmood, R., Meriton, R., Graham, G., Hennelly, P., & Kumar, M. (2017). Exploring the influence of big data on city transport operations: A Markovian approach. *International Journal of Operations and Production Management, 37*(1), 75–104.
28. Mourao, P. R., & Vítor, D. M. (2020). Forest entrepreneurship: A bibliometric analysis and a discussion about the co-authorship networks of an emerging scientific field. *Journal of Cleaner Production, 256*, 120413.
29. Mubin, O., Arsalan, M., & Al Mahmud, A. (2018). Tracking the follow-up of work in progress papers. *Scientometrics, 114*, 1159–1174.
30. Provost, F., & Fawcett, T. (2013). *Data science for business: What you need to know about data mining and data-analytic thinking*. Sebastopol, California: O'Reilly.
31. Roth, A., & Rosenzweig, E. (2020). Advancing Empirical Science in Operations Management Research: A Clarion Call to Action. *Manufacturing and Service Operations Management, 22*(1), 179–190.
32. Ruan, J., Hu, X., Huo, X., et al. (2020). An IoT-based E-business model of intelligent vegetable greenhouses and its key operations management issues. *Neural Computing and Applications, 32*, 15341–15356.
33. Scopus (2021) Scopus database. Retrieved from https://www.elsevier.com/scopus on 13 March 2021.
34. Silva, B. N., Khan, M., Jung, C., Seo, J., Muhammad, D., Han, J., Yoon, Y., & Han, K. (2018). Urban planning and smart city decision management empowered by real-time data processing using big data analytics. *Sensors, 18*, 2994.
35. Shmueli, G., & Yahav, I. (2018). The forest or the trees? tackling Simpson's paradox with classification trees. *Production and Operations Management, 27*(4), 696–716.
36. Tamás, A., & Koltai, T. (2020). Application of learning curves in operations management decisions. *Periodica Polytechnica Social and Management Sciences, 28*(1), 81–90.
37. Wang, W. (2020). Deployment and optimization of wireless network node deployment and optimization in smart cities. *Computer Communications, 155*, 117–124.
38. Wong, S. L., Nyakuma, B. B., Wong, K. Y., Lee, C. T., & Lee, C. H. (2020). Microplastics and nanoplastics in global food webs: a bibliometric analysis (2009–2019). *Marine Pollution Bulletin, 158*, 111432.

Vincent Charles PhD, PDRF, FRSS, FPPBA, MIScT, is a Full Professor and the Director of the Center for Value Chain Innovation at CENTRUM Católica Graduate Business School, PUCP, Lima, Peru. Additionally, he holds multiple visiting professorship positions across the Globe. He is an experienced researcher in the fields of artificial intelligence, data science, and OR/MS. He has more than two decades of teaching, research, and consultancy experience in various countries, in the fields of applied analytics/optimisation and big data science, having been a Full Professor and Director of Research for more than a decade in triple-crown business schools. He holds Executive Certificates from the MIT, HBS, and IE Business School. He has published over 150 research outputs. He is a recipient of many international academic honours and awards. An AWS Certified Cloud Practitioner, AWS Accredited Educator, a certified Six Sigma Black Belt, and an Advance HE Certified External Examiner, UK. He has developed the 4^3 Architecture for Service Innovation. His area of focus includes productivity, quality, efficiency, effectiveness, competitiveness, innovation, and design thinking. He is a Co-Founder and the Chief Analytics Officer at Anthrolytics Ltd., UK.

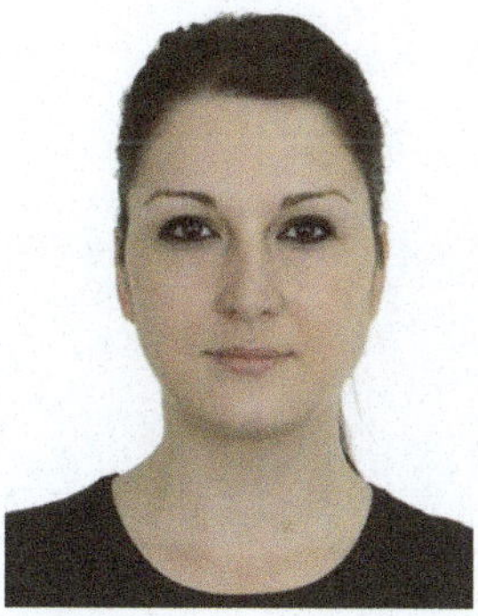

Tatiana Gherman MBA, PhD, is a Senior Lecturer and has more than a decade of teaching and research experience. She is currently researching how Machine Learning and Artificial Intelligence tools can support various business functions to help make business management more effective; with particular interest in how to design Artificial Intelligence techniques grounded in and informed by patterns of social interaction and communication. Working towards the creation of a new field of research grounded in ethno-data science. Her area of research interests further includes multi-attribute decision-making techniques and other advanced quantitative analytics at different levels. She has research publications in reputed journals and is one of the recipients of the 2019 GDN Informs/Springer Researcher Award. Her field of expertise includes Artificial Intelligence, Data Science, Big Data, Group Decision Support, Workplace Studies, Behavioural Studies, Quantitative Analytics (at different levels), Conversation Analysis, and Ethnomethodology.

Ali Emrouznejad PhD, is a Professor and Chair in Business Analytics at The University of Surrey, UK. His areas of research interest include performance measurement and management, efficiency and productivity analysis as well as big data and data mining. Dr. Emrouznejad is editor of Annals of Operations Research, associate/guest editor or member of editorial board in number of other journals including European journal of Operational Research, Journal of Operational Research Society, Socio-Economic Planning Sciences, IMA journal of Management Mathematics, OR Spectrum, RAIRO - Operations Research. He has published over 200 articles in top ranked journals; he is author of the book on "Applied Operational Research with SAS", editor of the books on "Performance Measurement with Fuzzy Data Envelopment Analysis" (Springer), "Managing Service Productivity" (Springer), "Big Data Optimization" (Springer), "Big Data for Greater Goods" (Springer) and "Handbook of Research on Strategic Performance Management and Measurement" (IGI Global). He is also co-founder of Performance Improvement Management Software (PIM-DEA), see http://www.Emrouznejad.com.

Strategy Formulation and Service Operations in the Big Data Age: The Essentialness of Technology, People, and Ethics

Vincent Charles, Ali Emrouznejad, and Tatiana Gherman

Abstract Studies have shown that the sensible operation of big data may yield powerful insights that can improve the organisations' strategic decision-making process and contribute to achieving an enhanced competitive advantage. In this manuscript, we explore the promise of big data in redefining strategy in service operations management (SOM) by means of investigating a rich range of bibliographic material. The SOM field has a plethora of research opportunities to capitalise on, which are enhanced by the presence of big data. SOM research in the big data age implies a shift in attention from being increasingly integrative across themes to being integrative across multiple disciplines, requiring the expertise of and tuning between different actors and expertise domains. Our aim is to stimulate debate in the field and set out a renewed research agenda by means of calling for additional considerations of strategic aspects, namely technology, people, and ethics, that can help guide and move the field forward.

Keywords Analytics · Big data · Service operations management · BD-SOM strategy · Competitive advantage

V. Charles (✉)
CENTRUM Católica Graduate Business School, Lima, Peru
e-mail: vcharles@pucp.pe

Pontifical Catholic University of Peru, Lima, Peru

A. Emrouznejad
Surrey Business School, The University of Surrey, Guildford, UK
e-mail: a.emrouznejad@surrey.ac.uk

T. Gherman
Faculty of Business and Law, University of Northampton, Northampton, UK
e-mail: tatiana.gherman@northampton.ac.uk

A. Emrouznejad and V. Charles (eds.), *Big Data and Blockchain for Service Operations Management*, Studies in Big Data 98,
https://doi.org/10.1007/978-3-030-87304-2_2

1 Introduction

This manuscript is concerned with the evolution of service operations management (SOM) in the age of big data and how big data can help to redefine the concept of *service operations strategy* to build an enhanced competitive advantage for organisations. Overall, SOM has had a rather unclear path in the literature, being mostly perceived as a branch of operations management (OM). Despite this, as Grönroos (1994) also acknowledged almost two decades ago, "*the business logic is different in service*" (p. 13); therefore, there is a constant need to refine our understanding of the principles and assumptions underpinning SOM, something that has been changing over the years, even more so during the big data age.

The relatively recent phenomenon posed by the exponential growth of big data has brought with it new challenges, one of the most intriguing of which deals with knowledge discovery and large-scale data-mining [26]. The presence of big data has been 'pushing' organisations to review their practices and identify opportunities that would allow them to base a substantial portion of their operational decisions on data, otherwise known as *data-driven decision-making* [9]. Big data are transforming the service industry, in areas ranging from finance/banking, to transportation and hospitality, education and government, among others. But at the same time, this transformation is only in its infancy and undoubtedly will require more cross- and inter-disciplinary approaches and collaborations, bringing together a variety of stakeholders, from researchers to data scientists, marketers, psychologists and behavioural analysts, regulators and policymakers, just to name a few. The potential held by big data to transform strategic decision-making in SOM is, indeed, substantial, but so seem to be the challenges. Today, big data analytics and high-speed computing are no longer the main concerns, but rather how "*to carefully exploit and unlock the power of big data while preserving fairness, trust, and consumers' happiness*" (Cohen [19], p. 1722). In this manuscript, we investigate a rich range of bibliographic material and identify strategic aspects that need to be considered in SOM in the context of big data, with a particular focus on technology, people, and ethics. We also postulate that if we want to understand the promise of big data in redefining SOM strategy, we must first understand how service management hypotheses and assumptions have changed as a result of innovations brought about by the big data age.

The remainder of the manuscript is organised as follows: Sect. 2 introduces and discusses the concept of SOM, while Sect. 3 explores traditional misconceptions in SOM and Sect. 4 details the current state-of-the-art in SOM. Subsequently, we introduce the concept of big data in Sect. 5, discuss the various types of analytics in Sect. 6, and explore the big data pipeline in Sect. 7. An extensive narrative of strategic considerations in SOM in the big data age is proffered in Sect. 8. Here, we further introduce the BD-SOM strategy. Lastly, Sect. 9 concludes with final thoughts and a discussion of future directions.

2 The Concept of Service Operations Management

Operations management has traditionally been associated with manufacturing production [5], having had little contact with customers and not always understanding their needs and desires. Activity scheduling charts and assembly lines, linear programming, queuing theory, and PERT are just some of the traditional areas that OM researchers have been generally concerned with, with Chopra et al. [18] noting that most of the early research "*focused on tactical issues such as line balancing, scheduling, production planning, inventory control, and lot sizing*" (p. 9).

In recent decades, however, the focus has shifted more strongly towards the services sector, which has experienced an accelerated growth. As Peinado et al. [71, p. 374] assessed, "*operations management is a discipline that originated to solve management problems in a factory environment, but since the mid-twentieth century researchers, lecturers and practitioners have begun to adapt the knowledge of the field to also support service operations.*" So, in time, the concept of OM has come to encompass the area of services; in the words of Manikas et al. [56], "*in the 1990s, business process reengineering, six-sigma, enterprise resource planning systems (ERP), and the internet became enablers of operations management. As the techniques described by operations have also been applied to service industries and other non-manufacturing areas of organisations, the scope of operations management increased to include service*" (p. 1442).

The globalisation of markets and the technological progress in information and communications technology have altered the management landscape not only for manufacturing enterprises, but also for the productive management of services. Services are increasingly important particularly in today's industrialised economies, where technological changes have translated into a greater number of people being employed in services, a sector that accounts for most of the countries' gross domestic product [30, 66]. As a matter of fact, Levitt's [53] observation about the importance and pervasiveness of services for business is just as true today as it was the day it was first uttered, almost half a century ago. The increasing importance of studying service management was echoed by researchers at different points in time: Miller et al. [61], Amoako-Gyampah and Meredith [2], Pannirselvam et al. [67], Nie and Kellogg [64], and Johnson [45]; these researchers concluded that there is still not enough research being done in the SOM field and called for more research on the topic, especially empirical research.

The history of SOM as a distinct topic in the academic literature has had its own trajectory. Generally, studies can be traced back to the 1970s [46], although there are some studies of service economy even prior to that time (e.g., Pearce [70], Penrose [72]). But for some reason, the evolution of SOM as a standalone discipline has remained unclear and the research community has continued to hold SOM as a part of OM. For example, in [60], Meredith et al. defined OM as consisting of 17 focus areas: (1) Aggregating planning, (2) Capacity Planning, (3) Distribution, (4) Facility Layout, (5) Facility location, (6) Forecasting, (7) Inventory control, (8) Maintenance, (9) Process design/technology, (10) Project management, (11) Purchasing,

(12) Quality, (13) Quality of working life, (14) Scheduling, (15) Services, (16) Strategy, and (17) Work measurement. As it can be noted, this study acknowledged services as a branch of OM. All in all, this perspective over SOM as a part of OM has meant that research on SOM has traditionally been scarce. At least, such was the situation in 2007, when Machuca et al. [55] performed the first in-depth study on the state of the then current SOM research.

But one may wonder how things would have changed since then. In a recent study, Manikas et al. [56] performed a review of the literature on OM using a data-driven approach; in this sense, the authors performed a historical analysis of major research topics and trends between 1961 and 2017 and found 18 topic areas that received most of the attention. One among these topics is the topic of *service operations*, which contains articles mostly focused on buyer–seller relationship, queuing theory, and service quality, with sub-themes such as multi-criteria evaluation of strategic issues in service quality, design of service, information technology in service, and globalisation of services. Furthermore, the authors concluded that service operations are a growing area of publications, with particular interest in lean management (waste reduction, improvement in resource usage, and so on). An interesting observation that the authors further made is that an evaluation of only a recent subset of the entire data that they used would show that emerging topics include healthcare efficiency and effectiveness, environmental and sustainability, security and infrastructure, reverse logistics, and big data methods. Despite the granular and useful analysis, we can see, however, that once more, the SOM's 'curse' of being perceived as a part of the OM literature has not changed till date.

In time, SOM has struggled to find its fit not only as a distinct discipline in the literature but also within the broader field of OM itself. This has been fueled by the constant concern in the minds of researchers regarding how to identify the distinctive contribution that SOM research makes compared to OM research. This SOM opacity is, of course, surprising, especially considering that service operations represent, as previously mentioned, the clear majority of economic activity (at least in developed countries) [80].

3 Traditional Misconceptions in Service Operations Management

A predominant view in the literature is that services are defined by four attributes: intangibility, heterogeneity, inseparability, and perishability (IHIP); these have commonly been accepted as the paradigm for services. There are studies, however, that have refuted such view and concluded that it is without merit. For example, Vargo and Lusch [82] argued that the IHIP characteristics "*(a) do not distinguish services from goods, (b) only have meaning from a manufacturing perspective, and (c) imply inappropriate normative strategies*" (p. 324). In a similar fashion, Edvardsson et al. [23] highlighted that "*the IHIP characteristics should not be used in the future as generic service characteristics*" (p. 115). Like these, there are many other myths and misconceptions about services that continue to exist. What this means for SOM is that not only has SOM research been characterised by opacity, but traditional misconceptions have also accompanied the field. In his work, Sampson [78, p. 183]) identified five main stereotypes around SOM. These are:

1. "*the operations function of firms is simply (or primarily) about managing physical products and product inventories*";
2. "*service operations are an unscientific type of operations, which is sometimes based on the observation that traditional manufacturing operations models do not often fit well in service contexts*";
3. "*there is a traditional assumption that there is some dichotomy between goods and services*" (e.g., Greenfield [34]),
4. "*some have asserted that service is inherently customer oriented and solution focused*" (e.g., Grönroos [36, p. 46], Vargo & Lusch [83, p. 138]),
5. "*service is adequately defined by what it is not*" [e.g., "nonmanufacturing", which is a residual definition (Morey [63]); or "nonownership" (Judd [47]).

All of the above have eventually led to confusion as to what a *service* is, as well as with regards to the very principles that lay as a foundation for SOM. This has also meant that SOM has had a rich history of key but not widely recognised contributions to both research and practice. Another interesting aspect that has contributed to the perpetuation of SOM stereotypes has been the fact that quite a lot of the research involving service management has come from the marketing discipline [54] (Rust, 2004). In this sense, marketing scholars have focused on addressing traditional OM topics (e.g., quality management, facility layout, process focus) in service contexts. Particular attention has been paid to quality management, which is by excellence an OM topic; nevertheless, the most widely used model of service quality and associated SERVQUAL instrument belong to the field of service marketing (Parasuraman et al. [68, 69]).

Sampson [78] proposed that the stereotypes surrounding SOM could be rectified by showing how service operations can be conceptualised, visualised, and analysed. To this aim, the author introduced a Process-Chain-Network (PCN) visual framework to clarify the fundamental concepts of SOM. As stated by the author, "*the framework is built upon PCN Diagrams that depict processes and interactions involving networks of entities. PCN Analysis includes identifying the value proposition of a given process network, assessing performance characteristics and value propositions of a process configuration, and identifying opportunities for process improvement and innovation*" (Sampson [78, p. 182]). Among others, his findings showed that "*the basic structural elements of PCN Diagrams can reveal commonality among seemingly disparate lines of business*", that "*every business has a mix of interactive processes and independent processing*", and that "*service interaction requires some degree of integration of processes across multiple entities and can, therefore, be more difficult to design and execute than independent processes*" (p. 194).

4 Current State-Of-The-Art in Service Operations Management

Previous review articles have documented the history and evolution of research on service operations. For example, Chase and Apte [16], Heineke and Davis [40] described the history of service operations, Machuca et al. [55], Smith et al. [80] provided a taxonomy of research in view of content topic, research methodology, journal, and author affiliation of service operations research, and Bretthauer [7], Chase [15], Hill et al. [42], Johnston [45], Roth and Menor [77] performed an analysis of the research gaps, limitations, and avenues for future research.

In their review paper of published work in the Journal of Service Management during 2010–2016, Victorino et al. [84] used the Delphi method to identify research themes in service operations that have great potential for exciting and innovative conceptual and empirical work. Their efforts resulted in the identification of the following key themes: service supply networks, evaluating and measuring service operations, performance, understanding customer and employee behaviour in service operations, managing servitisation, managing knowledge-based service contexts, managing participation roles and responsibilities in service operations, addressing society's challenges through service operations, and the operational implications of the sharing economy. The authors reviewed nearly 700 articles, which were classified based on the discipline focus into one of the following categories: "Operations," "Marketing," "Organisational Behavior," Management of Information Systems (MIS)," or "Interdisciplinary." Table 1 depicts the characteristics of the articles in each category.

Although limited to analysing only one journal, the findings reveal two important trends. First, the authors noted that over the years, the percentage of articles in each category has declined, with the exception of interdisciplinary research, which has seen a continuous upward trend. Second, another interesting observation that the authors make is that what most of the research articles in the literature seem to share is a call for more research that involves an interdisciplinary approach to understanding service operations, by drawing from fields such as marketing, human resources, and information systems, among others. This perspective has been further expanded in the approach termed as "service science," which focuses on combining knowledge from scientific management and engineering disciplines for service innovation (Spohrer & Maglio [81], as cited by Victorino et al. [84]).

Victorino et al.'s [84] study is accompanied by a follow-up paper by Field et al. [29], in which the authors take a broader view by means of including a wider literature to offer a comprehensive review of each research theme and posit future research questions for advancing the field of service operations. One of their main conclusions is that the field of service operations has many interesting research topics and questions that remain largely unexplored till date.

Table 1 Characteristics of SOM articles by discipline [84]

	Operations	Marketing	Organisational Behaviour	Management of Information Systems (MIS)	Interdisciplinary
Focus	Process or production-oriented	Customer-oriented	People-oriented	Information-technology oriented	Majority of articles found at the interface of operations and marketing
Topics	Productivity Capacity management Quality management	Customer behaviour Customer loyalty Customer segmentation	Service climate Leadership Employee satisfaction		Service quality Service design and innovation
% of Articles	20%	43%	11%	1%	26%

5 The Concept of Big Data

Before proceeding with an assessment of SOM in the context of big data, let us first briefly review the concept of 'big data'. Just like with the concept of SOM, ambiguity is also surrounding the concept of 'big data'. Initially coined in 1997, the term 'big data' (Cox & Ellsworth [21]) has evolved to become today the new normal. However, big data in themselves are not a new 'thing'.

In 1908, on the island of Crete, archaeologists discovered a disc of fired clay, which was dated to the middle or late Minoan Bronze Age, hence from around 2000 B.C. The disc, called the Phaistos Disc (Fig. 1) was dubbed the 'first Minoan CD-ROM'. The disk is round in shape and covered on both sides with a spiral of inscriptions, whose meaning remains a mystery till today. Nevertheless, the point is that this is an example of what data used to look like thousands of years before the advent of CDs. This is how society used to store and transmit data at some point in time in the distant past. A rudimentary attempt, by all accounts, considering the limited 'storage' available and the impossibility of rewriting it or making any operations with it.

So, (big) data have always been with us and, as a society, we have constantly aimed at finding ways to store them and passing them on. What has changed, however, is the fact that the increase in IT-related infrastructure over the past years, coupled with the emergence of complex analytics, has enabled us to store virtually any amount of data at a significantly reduced cost and analyse and interpret such data in ways that could not have been done before [24].

Today, there is no universally accepted definition of big data. Big data is a multi-dimensional concept. As cited by Charles and Emrouznejad [10] and Emrouznejad and Charles [25], Hammond [38] associated big data with evidence-based decision-making, Beyer and Laney [6] defined it as high volume, high velocity, and/or high

Fig. 1 The Phaistos Disk (or the 'first Minoan CD-ROM'). *Note. Source* https://araenil.files.wordpress.com/2011/06/phaistosdisklarge.jpg

variety information assets, and Ohlhorst [65] described big data as vast data sets which are difficult to analyse or visualise with conventional information technologies. Most commonly today, big data are defined in terms of data characteristics or dimensions, often with names starting with the letter 'V'. By some accounts, there are today as many as 10 Vs (Markus [58]). Laney [51] defined the initial four Vs, which have become central to understanding big data; these are as follows:

Volume: Big data are characterised by their extremely large volume. It is estimated that by the end of 2020, the volume of data would be around 40 zettabytes, or 300 times bigger than the volume of data in 2005 (Herschel & Mori [41]). To be noted that the focus is not only on human-generated data (which are mostly structured and represent a small fraction of the entire data being generated), but also data produced by devices, such as sensors and connected devices (which are mostly unstructured and account for most of the data out there).

Velocity: Big data arrive at un unprecedented speed. Velocity refers to a real-time or near-real time stream of data, which poses an issue for real-time processing. Real-time processing is essential for businesses looking to obtain a competitive advantage over their competitors; for example, IBM [44] stated that "*for time-sensitive processes such as catching fraud, big data must be used as it streams into your enterprise in order to maximize its value*". This means that velocity is a double-edged sword. On the one hand, the possibility to collect mountains of data at such high speed opens up new possibilities to derive more insights, but on the other hand, the time needed to translate the data into intelligent decisions remains a challenge [51].

Variety: The characteristic of 'variety' of big data refers to different types of data, such as structured, semi-structured, and unstructured, which arrive from a variety of sources [51] (IBM, 2016). As expected, the biggest challenge is posed by the semi-structured and unstructured data, which are hard to analyse due to not having an easily identifiable internal structure. Examples include photos, audio and video, web pages, wikis, and blogs, streaming data, emails, social media data, and so on.

Veracity: It refers to the trustworthiness of the data and the reason behind questioning the existence of inherent discrepancies in the data is rooted precisely in the unstructured feature of big data. Another reason is the presence of inaccuracies. Inaccuracies can be due to the data being intrinsically inaccurate or from the data becoming inaccurate through processing errors [51].

In their paper, Charles and Gherman [11] argued that the term big data is a misnomer, stating that while the term in itself refers to the large volume of data, Big Data is essentially about the phenomenon that we are trying to record and the hidden patterns and complexities of the data that we attempt to unpack. The authors advanced an expanded model of Big Data, wherein they included three additional dimensions, namely the 3 Cs: Context, Connectedness, and Complexity. The authors stated that understanding the *Context* is essential when dealing with Big Data, because "raw data could mean anything without a thorough understanding of the context that explains it" (p. 1072); *Connectedness* was defined as the ability to understand Big Data in its wider Context and in view of its ethical implications; and *Complexity* was defined from the perspective of having the skills to survive and thrive in the face of

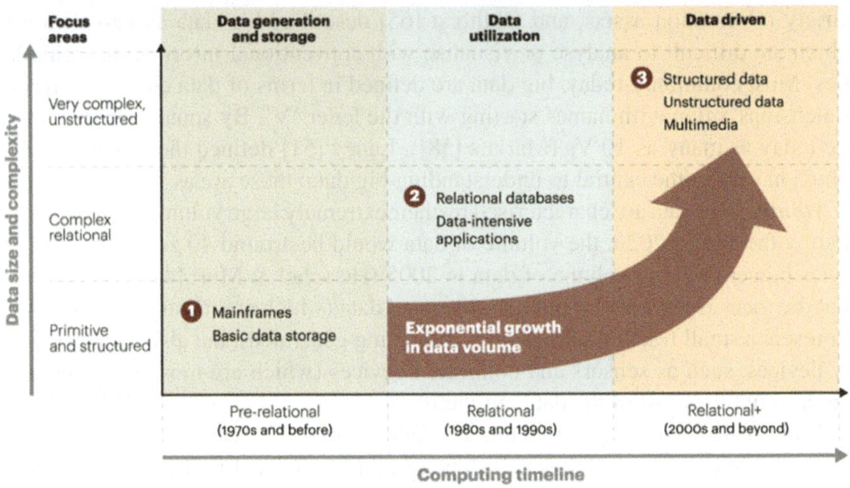

Fig. 2 The evolution of big data. *Note. Source* Epstein and Hagen [27]

complex data, by means of being able to identify the key data and differentiate the information that truly has an impact on the organisation.

Figure 2 depicts the evolution of big data from their initial primitive and structured forms housed locally, to the unstructured, highly complex forms housed in the Cloud.

Perhaps one more observation to make in relation to big data, at this point, has to do with the 'risk' accompanying them. We define risk in terms of its dual nature, materialised by understanding the concept both as a *danger* and as an *opportunity*. Big data can indeed translate into a big opportunity for organisations, helping decision-making processes become more effective. But one must not forget the other side of the coin, the challenges that come with big data and how big data's usefulness can be constrained by the ability of the researchers to ask the right questions and apply the right tools, all within an ethical framework.

For example, take the case of the Google Flu Trends Project, which has shown just how vulnerable the exploitation of big data can turn out to be, when not done properly. Google Flu Trends was a web service operated by Google, launched in 2008 and abandoned in 2015 (although problems started being apparent in 2009 itself). The project's goal was to develop a means of identifying the emergence of flu so that health resources could be mobilised to treat the illness immediately and prevent a possible outbreak. The web service used data readily available from Google's search engine, mainly data on the frequency of "flu" searches, it aggregated them, and attempted to predict future rates of flu across more than 25 countries. After apparent early success, the project had to be abandoned as the predictions were not accurate enough. It was found that the model was consistently over-inflating future occurrences and was less accurate than existing 'small data' strategies that utilised data on confirmed cases of flu. The Google Flu Trends' epic failure showed just how vital the elements of rigour, criticality, and the correct consideration of the wider context, among others, actually are. In the words of Lazer et al. [52, p. 1203], "the quantity of data does not mean that one can ignore foundational issues of measurement and construct validity and reliability and dependencies among data". In the absence of a specific criticality in the analysis performed and in the interpretation of the results obtained, big data cannot add value to any organisation.

The above considerations led Fung [31] to conclude that McKinsey's widely circulated definition of big data as "datasets whose size is beyond the ability of typical database software tools to capture, store, manage, and analyze" is not helpful in providing any answers to prevent future failures. Instead, he proposed a new approach to defining big data, called the OCCAM framework, which has the advantage of bringing to one's attention the assumptions underlying the concept. In view of the OCCAM framework, big data are [31]:

- "*Observational: much of the new data come from sensors or tracking devices that monitor continuously and indiscriminately without design, as opposed to questionnaires, interviews, or experiments with purposeful design.*"
- "*Lacking Controls: controls are typically unavailable, making valid comparisons and analysis more difficult.*"
- "*Seemingly Complete: the availability of data for most measurable units and the sheer volume of data generated is unprecedented, but more data creates more false leads and blind alleys, complicating the search for meaningful, predictable structure.*"
- "*Adapted: third parties collect the data, often for purposes unrelated to the data scientists', presenting challenges of interpretation.*"
- "*Merged: different datasets are combined, exacerbating the problems relating to lack of definition and misaligned objectives.*"

6 Types of Analytics

Data analytics is a multidimensional term, often interchangeably used with *data science, data mining*, and so on. The distinction between these terms is vague, but in essence, they all refer to the extraction of useful information from a preprocessed set of data. The techniques that can be used to this aim are varied and stem from across the disciplines, such as statistics (e.g., inferential statistics, various types of regression); machine learning (e.g., ensemble decision trees like random forest), in particular kernel methods (e.g., support vector machines); and biology (e.g., neural networks, genetic algorithms, and nature-inspired algorithms like ant colony algorithm and swan particle algorithm, among others).

There is a plethora of analytics that organisations can perform. Below, we discuss a more comprehensive framework (Fig. 3), comprising five types of analytics: descriptive, diagnostic, predictive, prescriptive, and cognitive. It should be noted at the outset, however, that there is no consensus over this taxonomy, with most frameworks considering only three types (descriptive, predictive, and prescriptive). Others consider four, namely descriptive, diagnostic, predictive, and prescriptive [Gartner

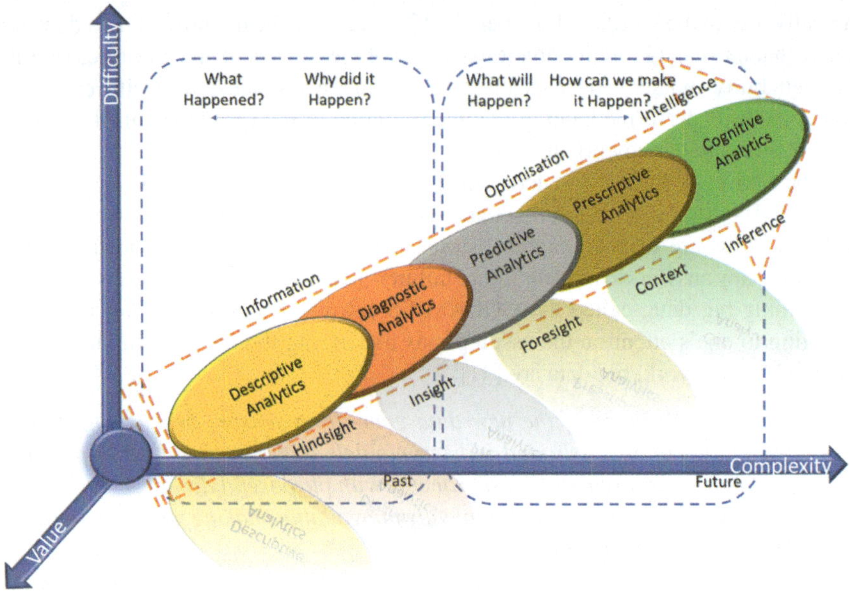

Fig. 3 Extended data analytics framework

Analytics Ascendancy Model (GAAM), Maoz [57]]; while others do not differentiate between descriptive and diagnostic analytics. Most recently, also, in the context of marketing strategy formulation, Eriksson et al. [28] proposed a new form of analytics, namely *creative analytics*, which is seen as a capability to address the potential of artistic creativity. Independent of these views, however, there are few observations worth making:

(a) each type of analytics provides distinct value propositions (hindsight, insight, foresight, context, or inference), serving distinct purposes (information, optimisation, or intelligence);
(b) the types of analytics are interconnected, but should not be viewed sequentially within a growth model framework; they can work in parallel and measure value differently.
(c) the more advanced the analytics get, the more complexity and difficulty are added, requiring greater resources, both computational and human resources; and normally, this should be accompanied by an increase in the value proposition made. But this does not always hold true and it depends on application. Sometimes there just is not enough information in the data to make higher levels of analytics (such as predictive, prescriptive, or cognitive) valuable.

The data analytics framework (Fig. 3) is an extension to the existing frameworks, by means of considering all five levels of analytics. We position this framework in a three-dimensional view along three axes: complexity, difficulty, and value. It is relevant to note that most literature has failed to differentiate between complexity and difficulty; here, we contribute by making a clear distinction between the two. We define *difficulty* as computational difficulty or the amount of resources needed to run computational algorithms (with particular focus on time and memory requirements); while *complexity* is seen as stemming from the challenges associated with extracting valuable knowledge in a meaningful manner. These two dimensions hold different ontological and epistemological stances; and, therefore, have different implications for a big data strategy. Next, let us explore the five types of analytics exhibited in Fig. 3.

Descriptive analytics: It is the most common and simple type of analytics. It seeks to understand what phenomenon took place over a certain period of time, answering the question "What happened?". Hence, the aim is to gain a view into historical data, an understanding of the significance and nature of past events—in one word, *hindsight*. It focuses on summarising past data, usually in the form of dashboards. Common approaches to perform descriptive analytics include visual analytics.

Diagnostic analytics: It seeks to understand why or what caused the phenomenon to happen in the first place; hence, the aim is to find the causes of past outcomes, answering the question "Why did it happen?". More specifically, it helps identify anomalies or outliers in the data; it drills into analytics, assisting in the identification of the data sources that help explain the anomalies; and it attempts to determine hidden causal relationships. Although relatively easy to be performed, this is a very important type of analytics, as it helps identify patterns in the data; so that, should the phenomenon happen again, the organisation can be prepared to act on time for a regrettable outcome to be avoided. Simply put, it provides *insight*. Common approaches to perform diagnostic analytics include clustering techniques, outlier detection, Naïve Bayes, and time-series data analytics.

Predictive analytics: It attempts to answer the question "What will happen"? or "What is likely to happen?". Hence, it seeks to understand what can happen in the future by utilising past data to make logical predictions about future outcomes. The goal is to achieve a state of self-awareness; namely, *foresight*. This type of analytics is also relatively easy to perform, but in practice, it can pose some serious challenges for organisations. This is because it requires added technology and a meaningful set of skills. Prediction and forecasting are, ultimately, only an estimate; and the accuracy of predictions and forecasts depends on both the quality of the data and the analyst's skills. Common approaches to perform predictive analytics include supervised machine learning techniques, such ensemble decision trees—random forest, support vector machines, and artificial neural networks, among others.

Prescriptive analytics: One of the most sought after type of analytics, prescriptive analytics aims to answer the question "How can we make it happen?". It combines the knowledge obtained from all the previous analyses to predict the likely outcome of various corrective measures. Otherwise stated, it advises organisations on all possible outcomes and the actions that are likely to optimise the desired outputs; it considers and builds upon the wider *context*. Common approaches to perform prescriptive analytics include simulation techniques, nature-inspired algorithms, and optimisation; it uses mathematical programming like linear, integer, and stochastic programming, Monte Carlo simulation, and game theory, among others. Prescriptive analytics can help improve decision-making in ways that previous analytics cannot do, creating a real competitive advantage for the organisation; but, the effort, organisational commitment, and resources needed to carry it out are substantial and not all organisations can afford them.

Cognitive analytics: This is the most advanced form of analytics, aiming to mirror human thinking, translating into pure intelligence. It is also known as "intelligent analytics". It mimics the human brain by drawing *inferences* from existing data and then inserting these inferences back into the knowledge base for future inferences—a self-learning feedback loop. Common approaches to perform cognitive analytics include a blend of artificial intelligence (AI), machine learning (ML) algorithms (more precisely, reinforcement learning), deep learning models, semantics, and game theory. Cognitive analytics blends traditional analytics techniques with AI and ML features for advanced analytics outcomes.

Perhaps one additional observation worth making at this point is that, in building analytical models, several requirements should be considered, depending on the application area [3]: business relevance, statistical performance (statistical significance and predictive power in terms of related performance metrics), interpretability, justifiability, operational efficiency, economic cost, and compliance with international regulation and legislation. To be noted that interpretability often needs to be balanced against statistical performance and this is an important trade-off to keep in mind. For example, neural networks are high performing (i.e., have higher accuracy), but offer no insight into the underlying patterns in the data, lacking clarity around inner workings. On the other hand, a linear regression model has limited modelling power, but is highly comprehensible and interpretable. Now, as Roda et al. [76] argued, sophisticated models, also known as black-box models, might not be more reliable than simpler models, also known as white-box models, this is especially the case when we deal with a phenomenon we know almost nothing about or do not yet fully understand (e.g., the recent COVID-19 pandemic), in which case it is more beneficial to have models that we can clearly explain how they behave, how they produce predictions, what the influencing variables are, and so on. In a nutshell, transforming data into meaningful knowledge is an art and not a blind application of analytical models. A comparison between white-box and black-box models is offered in Table 2.

7 The Big Data Pipeline

The value of data is unlocked only after they are transformed from their raw form into actionable knowledge, and when that value proposition is promptly delivered to relevant stakeholders. In the context of big data, an organisation may wish to rely on a data pipeline, which is a series of data processing steps, to successfully achieve this aim. A data pipeline is especially useful when the organisation deals with large amounts of data from multiple sources which are moreover generally stored in the Cloud, and when it requires real-time or near real-time complex data analysis. So, in

Table 2 A comparison between black-box models and white-box models

Black-box models	White-box models
Address highly non-linear structures	Address linear or stepwise linear or curve linear structures
Logically well-defined and mathematically complex	Logically and mathematically well defined, and simple
Large number of parameters; hence, models are high-dimensional	Small number of parameters
Large number of features	Relatively small number of features
Statistical hypothesis testing is irrelevant	Statistical hypothesis testing is relevant
High computational complexity	Low computational complexity
Lack clarity around inner workings	The input–output relationship is visible, and the process through which the output is produced is also visible
Large data set	Relatively small data set
Do not warrant any statistical distributional assumptions	Warrant statistical distributional assumptions
Modelling is usually a trial and error and iterative process	Modelling is less of a trial and error process and more of a systematic approach
Guided by rules of thumb	Guided by established criteria
Results depend on the hyperparameter tuning strategy	Results depend on the statistical estimation properties
Lower explainability or interpretability	Higher explainability or interpretability
Lower transparency and accountability	Higher transparency and accountability
Higher accuracy	Relatively lower accuracy

Source Hansun et al. [39]

considering a *big data pipeline*, big data principles need to be applied to the pipeline; and here, we are specifically referring to the consideration of the 4 Vs of big data: volume, variety, velocity, and veracity, all of which will impact the organisation's big data journey.

The full data pipeline for big data traditionally passes through several stages, as can be appreciated in Fig. 4. It consists of three main phases, namely data engineering, analytics, and delivery. Furthermore, the data engineering phase is split into three steps: data collection, ingestion, and preparation.

The data source layer is composed of the raw data arriving in the organisation. It includes all types of data: structured, semi-structured, and unstructured, which can come from static sources, as well as from real-time sources, such as IoT devices and sensors. The ingestion stage is about getting all the data needed, in a raw format, in a single repository called a *data lake*. The data storage layer is where the data are kept after being collected and ingested from the various sources. In view of the explosion of data volume available, sophisticated and accessible systems have been developed to help with this task. It should be mentioned that for simple, small data

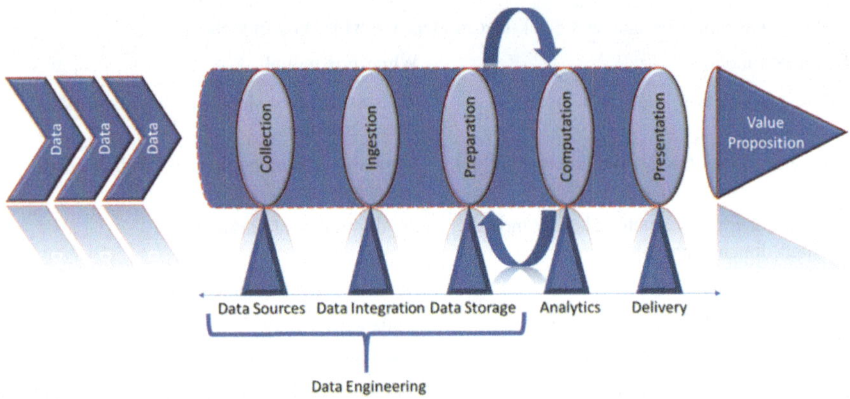

Fig. 4 The big data pipeline

pipelines, a simple workflow could be created, one that blurs the barriers between collection, ingestion, and preparation of data. Nevertheless, when dealing with big data, massive processing engines run in parallel, because the data pipeline views all data as streaming data, and it allows for flexible schemas. Therefore, the big data pipeline divides each data stream into smaller amounts that it processes in parallel, conferring extra computing power. It is, thus, recommended for these stages to be considered separately to avoid pressures on the system. Moreover, data ingestion is a critical and complex stage because of the dependencies to systems outside of the organisation's control; this means that special attention has to be paid to setting up a reliable and secure data ingestion process. The data storage layer is associated not only with storage, but also with data preparation, which is about getting the data ready for analysis. In this phase, efforts need to be put into cleaning, normalising, standardising, transforming, validating, and wrangling the data, with the aim of having a trustworthy dataset, with a well-defined schema. To be noted that, depending on application, data transformations can be done on load (a process known as extract-transform-load or ETL) or on read (extract-load-transform or ELT). In the case of big data, data lakes are flexible and can do ELT and schema on read.

The analytics layer is all about computation and turning data into insights. To be noted that in a big data context, we have a self-learning feedback loop, wherein inferences drawn from the existing data are inserted back into the knowledge base (data storage) for the generation of further inferences. Finally, the delivery layer has to do with how the outputs of the previous stage are passed on to the relevant stakeholders who can take action to benefit from them. Clear and concise communication is essential, as well as the collection of constant feedback. The output of this stage can take the form of reports, charts, figures, and key recommendations. Ultimately, what is needed here is to show how decisions and actions based on the analysis performed can lead to business improvement and enhanced competitive advantage.

8 Service Operations Management in the Big Data Age: Strategic Considerations

Having now reviewed the concept of 'big data', let us try to explore how big data has been transforming SOM in recent times. In order to do that, we must first recognise that organisations are increasingly complex, with manufacturing and service processes transpiring across economies, geographies and times zones, cultures and geopolitical conditions, and facing social, legal, and environmental challenges, just to name a few. The metamorphosis of organisations today is sharpened by the edge of continuous innovations and waves of globalisation, and is influenced by increasingly knowledge-based economies, all within a technologically-driven global context. With services becoming the most dynamic area of economic competition, it is not too bold to say that over the past two decades, many of the world's economies have turned into service economies.

The SOM research is somewhat eclectic, sitting on the border between production-oriented OM research and consumer-oriented marketing research (Sampson [78, p. 195]). As previously stated, this has been the source of the difficulty generally encountered by some SOM researchers when trying to apply traditional OM research paradigms to service contexts. Roth and Menor [77, p. 146] observed how "*many service management problems are fuzzy and unstructured, are multidimensional and complex, and are less conducive to normative, analytical modeling*". Similarly, Nie and Kellog [64, p. 352] asserted that "*service OM must be studied in different ways, using different theories, skills, competencies, and language*" than traditional manufacturing-oriented OM research. As expected, these issues are compounded by the exponential growth of big data.

Among others, Field et al. [29, p. 79] identified the following as being among the research questions of greater interest to be answered when it comes to evaluating and measuring service operations performance, in the context of big data:

- *What role should the customer play in the data collection process?*
- *What should be the operational processes around the collection of data?*
- *What are the best approaches for analysing big data to inform operational decision making?*
- *How does the proliferation of data sources and channels for dissemination (e.g. from non-customers and through social media, respectively) impact the evaluation of services and their operations?*
- *How can service providers use big data for service design and decision making?*

Big data has been changing how we think about services and service operations in fundamental ways. Let us explore how in view of the literature reviewed, while being conscious that the following is, nevertheless, an analysis that is limited by the frameworks explored. Since we have previously used the field of marketing as an example for various other considerations, let us start with a brief example in the same context. In his now seminal work, Grönroos [35, p. 7], a reference in the field of service and relationship marketing, identified five key facets of the service management perspective, namely overall management perspective, customer focus, holistic approach, quality focus, and internal development and reinforcement:

(1) "*It is an overall management perspective which should guide decisions in all areas of management (not only provide management principles for a separate function such as customer service)*";
(2) "*It is customer driven or market driven (not driven by internal efficiency criteria)*";
(3) "*It is a holistic perspective which emphasizes the importance of intraorganizational, cross-functional collaboration (not specialization and the division of labour)*";
(4) "*Managing quality is an integral part of service management (not a separate issue)*"; and
(5) "*Internal development of the personnel and reinforcement of its commitment to company goals and strategies are strategic prerequisites for success (not only administrative tasks)*".

Plumb and Zamfir [73], however, argued that in the context of a knowledge-based society, some of the key elements of the above traditional SOM perspective are no longer valid. Figure 5 below illustrates these old hypotheses in contrast with the new hypotheses. The authors based their assertions on the following arguments: (1) SOM is focused on knowledge rather that on the customer or the market, in view of the fact that knowledge is at the 'heart' of the society and of all activities that

Fig. 5 Old versus new hypotheses in SOM, in the marketing context. Based on Plumb and Zamfir [73]

people do, (2) SOM is a holistic perspective that is grounded in both intra- and inter-organisational collaboration, as knowledge is created and shared inside and outside of the organisation; and (3) SOM is based on the continuous development of people as an essential criterion for success, in the context in which knowledge and people are the most important strategic resources of any organisation.

Today, we can argue, however, that the arrival of the big data age has led to the removal of the need to 'pick a side' at the expense of the other. For example, an organisation does not need to choose between being customer-focused and being knowledge-focused. It can very well be both simultaneously (as it should be), a possibility which is enabled by the fact that big data has gone from a stock to a flow, from being rigid and unchangeable to being dynamic and fluid. This in turn, has translated into liquid information that can constantly inform and can feed into decision-making processes, in real-time, across the organisation. The fluidity brought about by big data means that the SOM strategy it feeds into must also become fluid and adaptable, and able to re-invent itself.

Figure 6 builds upon the studies by Barbosa and Azevedo [4], Charles and Gherman [11], Iansiti [43] and draws from literature on strategic management, to suggest an integrative framework that brings together big data and SOM (BD-SOM framework). This framework offers a starting point for understanding how organisations can leverage value and competitive advantage from their SOM strategy underpinned by big data.

In this framework, the SOM strategy is composed of two strategies: directed and emergent [43]. In a traditional way, the directed strategy is the one wherein the senior management considers the organisation of the internal resources, along with the external environment and the stakeholders' demands. It is about focusing on what the business wants and what the market needs, and it assumes some solid

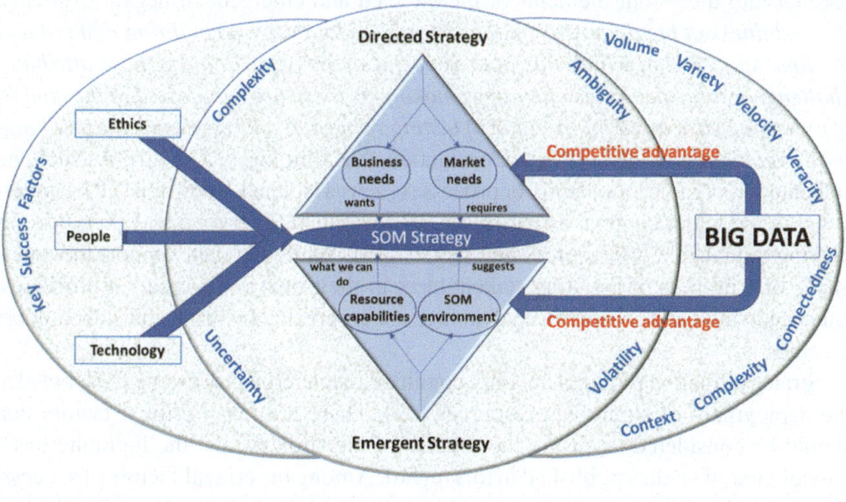

Fig. 6 The BD-SOM framework

planning. In other words, in setting up any kind of strategy, one must assess what the organisation has (resource capabilities and its SOM environment) and measure it against what the organisation needs. On the other hand, the emergent strategy is the one that develops, organises, and uses resources to attain operational excellence, competitive advantage, market share, and good performance. It is about a focus on what the organisation can do and what its SOM environment suggests it can do, and it assumes some flexibility and adaptability to changing conditions.

Strategy is seen as a continuous search for new market opportunity, while considering internal resources and limiting the negative impact to existing advantage [22]. Otherwise stated, it is essential for a strategy to match the organisation's internal skills and resources with external risks and opportunities in order to be effective [33]. The framework's overarching propositions should be considered in view of the VUCA dimensions that characterise today's business environment: volatility, uncertainty, complexity, and ambiguity, but also in view of the dimensions that characterise the big data age: volume, variety, velocity, veracity [51], connectedness, complexity, and context [11]. Under this framework, the organisation's competitive advantage will be born out of the interaction between the capabilities of big data and the elements of the directed and emergent strategies.

How have big data influenced strategy-making? The mechanisms by which big data are produced and disseminated introduce important changes in the ways information is generated and made relevant for organisations [20]. Big data blurs the demarcation between directed strategy and emergent strategy in SOM. As mentioned, big data have enabled the possibility of considering all the stakeholders' requirements simultaneously and eliminates the need to 'pick a side' at the expense of the other. Big data can help create competitive advantage by means of informing and building upon business needs, market needs, resource capabilities, and the SOM environment in the organisation. Big data have the potential to inform SOM strategy by means of feeding into these four elements of the directed and emergent strategy. "*Crucially, the usefulness of big data rests on their steady updatability, a condition that reduces the time span within which this data is useful or relevant. Jointly, these attributes challenge established rules of strategy making as these are manifested in the canons of procuring structured information of lasting value that addresses specific and long-term organizational objectives*" (Constantiou & Kallinikos [20, p. 44]. As such, no stakeholder, organisational unit, or process is seen as isolated. They must all be understood in the wider context, as part of the same context. Everyone and everything is playing a dual role of resources and targets [13]. While big data support the emergence of a process of assisting stakeholders in their own value-creation processes, enhanced value and competitive advantage are co-created by firms and stakeholders together.

Strategy creation requires not only cognitive managerial processing [37], but also the deployment of significant resources [62]. There are many critical factors that should be considered to ensure the success of the strategy, and the literature has a vast amount of studies published in this regard. Among the critical factors of success, we would like to draw particular attention to three of them: technology, people, and ethics, which we believe have an overarching role to play in the evolution of the SOM strategy in a big data world. We recognise that there is a plethora of studies looking into the implications of these three factors in the big data age. The brief analysis we provide here, however, differs from other studies by focusing on discussing these factors from a perspective aimed at instilling new insights.

8.1 Technology

Field et al. [29] argued that technology has an overarching role in the evolution of the service operations field. Technology plays a fundamental role in how processes are designed or redesigned (process design and, in the case of servitisation, product design). For example, while discussed under the servitisation theme, "*the use of technologies to monitor product usage and identify unusually high resource consumption also has implications for sustainability*" (p. 81). The opportunities for researchers to better understand the role of technology within SOM are vast.

> More generally, research questions have been posed about technology as a performance driver (e.g. IoT devices with enhanced analytical capabilities), a facilitator (e.g. apps that align customer behavior with desired actions), a substitute for customer and/or employee service efforts (e.g. SSTs), and a means to target increasingly smaller operating segments (e.g. virtual reality, artificial intelligence, and avatars applied to sharing economy operational processes). With rapidly changing technology being increasingly integral to the design and implementation of service operations, new technologies that inspire novel questions will continue to provide opportunities for leading-edge research (Field et al. [29, p. 81]).

We posit that technology should be researched by focusing not only on the "*what*" represented by big data technology (e.g., specific techniques or tools used for analysing big data), but also from the perspective of the "*how*" technology is used and dealt with by stakeholders involved in strategic decision-making (e.g., the group dynamics and management of relationships; how big data technologies in general feature in social action and interaction in strategy meetings) [32]. There is a lack of empirical studies focusing on stakeholders doing strategy work using big data technology in situ and how strategy is actually constructed as an interactional achievement.

On the other hand, the progress of new technologies, such as video, real-time messaging, chatbots and artificial intelligence, cryptocurrencies, and self-service, among others, has unlocked the potential for big changes in the day-to-day workings of service operations. Challenges are commensurate to opportunities. In the past, most research in SOM has relied on the analysis of structured data using traditional methods based on statistics and optimisation to improve services and businesses. Nevertheless, with the emergence of knowledge-based economies and the transition into a more digital platform, we have been experiencing an exponential increase in unstructured data. One area in SOM big data research that has received relatively little attention is precisely the use of unstructured data in SOM research. Nevertheless, as Manikas et al. [56] noted, studies have started to emerge in response to the new challenges, as evidenced by the calls for research and recent studies that use non-traditional methods, such as those based on data mining techniques [17]. Kuo and Kusiak [50] echoed the assertion by highlighting a shift from analytical models to data-driven models. For example, research points to the potential of text mining and the use of unstructured data for improved decision-making [56]. For examples of applications, interested readers can refer to the studies by Rendon et al. [75] for products and services acquisition, Vosbergen et al. [85] for healthcare operations, and Shi et al. [79] for transportation research. Without a doubt, technology will continue to play a vital role in the future of SOM research and more studies will emerge to leverage the latent power of data mining technologies, influencing SOM strategy-making.

8.2 People

Talking about services without placing people (both workers and customers) at the heart of strategic decision-making is meaningless. Historically and paradoxically, research in SOM has typically treated people as fixed, un-changing, or exogenous entities. However, this assumption has been increasingly challenged, particularly with the rise of the big data age.

Let us take the case of workers. Without a doubt, they are key to success; so, investing in training and upskilling them is a sensible decision to make. Exploring and taking advantage of big data requires thoughtful organisational change. A range of skills become necessary. For example, data engineers such as developers, administrators, and DevOps, who have a strong operational background in coding and software development, are needed for ingestion, enrichment, and transformation of data. But at the same time, data scientists such as business intelligence specialists and data analysts, who have a strong business understanding and are strong in visualisation techniques, are also essential for success, as they can gather insights, generate reports and dashboards, and communicate results in an insightful manner to a non-technical audience. Both types of stills are needed. Big data and big data analytics are not just about finding patterns in big data. It is not sufficient to discover that one phenomenon correlates with another. Context, meaning and interpretation become a necessity, and not a luxury [12]. This observation was also made by Maxwell [59, p. 182], who elegantly stated that:

> Analytics often happens in a black box, offering up a response without context or clear transparency around the algorithmic rules that computed a judgment, answer, or decision. Analytics software and hardware are being sold as a single-source, easy solution to make sense of today's digital complexity. The promise of these solutions is that seemingly anyone can be an analytics guru. There is real danger in this do-it-yourself approach to analytics, however. As with all scientific instruments and approaches, whether it be statistics, a microscope, or even a thermometer, without proper knowledge of the tool, expertise in the approach, and knowledge of the rules that govern the process, the results will be questionable.

Let us also shift briefly our attention to the case of customers, an equally important stakeholder. Charles [8] suggested that successful companies will be not only those that can discover current needs of the consumers, but also those that develop the capability to anticipate the future needs of the consumers. Charles [8] called this *Extraordinary Excellence Innovation.* Studying customer behaviour has never been an easy task and the presence of big data has brought both challenges and opportunities. While in the past, analytics would generally be performed by means of studying written artefacts which recorded past human behaviour, today, big data technology enables the recording of *current* human behaviour, *as this happens,* in situ (e.g., live feed data). Of course, analysing such data poses great computational challenges. The literature has noted that these could be tackled by a consideration of novel research approaches to studying big data, rooted in more qualitative grounds, such as ethnography [12]. Ethnography is a great complement to big data analytics, as such research supports the attempts at understanding the multiple realities of behaviour *in context* [74], hence, by definition, ethnography provides detailed snapshots of contextualised social realities, which can help make sense of the data. Ethnography is also a great tool to 'unpack' unstructured data. As technology keeps evolving, the necessity of 'collaboration' between big data analysts and ethnographers, and not only, will only become more obvious. They can help each other in the successful handcrafting and implementation of bigger projects for a bigger, greater good [12].

8.3 Ethics

Cohen [19] presented an overview of how big data have been shaping the service industry by enabling real-time service personalisation, with examples from finance/banking, transportation and hospitality, and online platforms. One of the concluding arguments of the analysis performed was that having access to large amounts of data can help firms to enhance the service quality by tailoring the offerings to the users' needs. Firms collect unprecedented amounts of granular data on consumers and on transactions and are developing quantitative-driven tools to improve operational decision-making, allowing them to provide real-time service personalisation at a large scale. Cohen [19] warns, however, of the dangers that such endeavours bring. In the author's words "*Firms that have access to this rich data can utilize it to price discriminate against customers. In addition, data-driven algorithms can include a machine bias that accentuates illicit discrimination. […] Furthermore, the availability of data on sensitive personal information attracts hackers. The number of breaches has increased and is now a major concern for most firms*" (p. 1721).

One of the major challenges for organisations in the big data age is to carefully exploit and unlock the power of big data and technology while preserving aspects of fairness and trust, among others. Here, we would further like to draw attention to the issue that, in the context of big data, one of the most sensitive and worrisome issues in SOM is the *privacy of personal information.* In 2015, Charles et al. [14] identified a gap in big data research, suggesting that while privacy had been extensively explored in different settings, it had not yet been sufficiently studied relative to the social and technological changes in the big data era.

> The social setting of our society has changed tremendously over the past few years and we are now witnessing the mutation of the very definitions and meanings of traditional concepts, such as privacy. We must admit that privacy is no longer what it used to mean. One of the effects of the big data age is a divorce of privacy from the person. Basically, this means that with more and more of our personal information being routinely collected and stored by companies or willingly disclosed via a computer and an internet connection, our privacy is now being traded as if it were to be a commodity (Charles et al. [14, p. 295]).

We believe that the issue of privacy of personal information will only continue to be exacerbated in a big data world, unless substantial efforts are made now to address it. In a big data context, there is a need more than ever to treat people as intrinsic values of the society and not just as mere parts of an impersonal SOM strategy. The SOM strategy needs to be designed to reflect that. Such a view evokes the deontological moral philosophy of Kant ([1785] [48]), which interested readers can explore in more detail.

9 The Future Ahead and Final Thoughts

The field of service operations continues to evolve on a constant basis. It is also rather clear that this transformation is only in its infancy. Only with an understanding of the past and current published SOM research, as well as of the opportunities and risks brought about by the big data age and today's business environment, can we identify what the future holds for the field of SOM.

A growing number of organisations nowadays engage in cross-disciplinary services, a situation where they try to exploit the interactions between different types of services. For example, Amazon operates in retail, cloud computing, media streaming, and food delivery services. Airbnb operates in the hotel industry and the dining reservation market; and IKEA operates in retail but has also acquired TaskRabbit, an American online and mobile marketplace that matches freelance labor with local demand, allowing consumers to find immediate help with everyday tasks, including cleaning, moving, delivery and handyman work. "*Having access to big data on different types of services can allow firms to exploit the multidimensionality of their users' interactions in order to reach a more comprehensive picture and to enhance the service quality, as well as the long-term profits*" (Cohen [19, p. 1722]). Without a doubt, there are significant opportunities for continued research on service operations, and in this manuscript, we have shown some of them.

Despite the range of opportunities that big data hold, it seems that big data applications have so far only delivered incremental changes and improvements. We are yet to see strategic, truly transformational changes. But the potential is there. Clearly, the SOM field has a plethora of research opportunities to capitalise on, which are enhanced by big data. This means that SOM research also implies a shift in attention from being increasingly integrative across themes to being integrative across multiple disciplines, requiring the expertise of and tuning between different stakeholders and expertise domains. This creates opportunities for ever more impactful research. This also means that we need to keep on questioning whether assumptions inherent to the SOM field will remain valid or not. We thus join and reiterate the call for SOM researchers to continue to create deep knowledge within the field, in collaboration with colleagues across disciplines, in an attempt to develop efforts that incorporate new insights and trends and build upon exciting and innovative research opportunities.

Our aim has been to stimulate debate in the field and set out a renewed research agenda by means of calling for additional considerations of strategic aspects that can help guide and move the field forward. For practitioners, the implication is that service organisations must get ready to respond to new challenges in a world impacted not only by the characteristics of today's business environment [i.e., volatility, uncertainty, complexity, and ambiguity—see VUCA], but also by the characteristics of the big data age [among which we mention context, complexity, and connectedness—see Charles and Gherman [11]]. We live in exponential times, with the world around us changing faster than ever before; a world in which speed is praised as the 'Holy Grail' of modern times. But what current developments are teaching us is that the cult of speed must be challenged, and that slowness does not necessarily have a negative connotation. As Kundera [49] wrote in his novel titled Slowness, "*when things happen too fast, nobody can be certain about anything, about anything at all, not even about himself*" (p. 135). Managers in charge of designing big data-informed SOM strategies need to learn to balance speed with thoughtfulness [1], two concepts elegantly reflected in the classical Latin adage *festina lente* (which translates as *hasten slowly*).

For researchers, the implication is that, as mentioned, there is a need for more cross- and inter-disciplinary empirically grounded research, more specifically, for new research approaches to study practice in truly insightful and impactful ways [for an example, please see Charles and Gherman [12]]. Changing times call for innovative measures. Such endeavours will require more than ever before the collaboration of a range of stakeholders, from researchers to data scientists, marketers, psychologists and behavioural analysts, regulators and policymakers, just to name a few. No one knows for sure what SOM research will look like in the future, but as researchers we have an opportunity to do things every day that make some futures more probable than others.

Acknowledgements The authors are thankful to the reviewers for their valuable comments on the previous version of this work.

References

1. Akkermans, H. A. (2018). *Service operations dynamics: Managing in an age of digitization, disruption and discontent*. Palgrave Pivot.
2. Amoako-Gyampah, K., & Meredith, J. R. (1989). The operations Management research agenda: An update. *Journal of Operations Management, 8*(3), 250–262.
3. Baesens, B. (2014). *Analytics in a big data world*. Wiley.
4. Barbosa, C., & Azevedo, A. (2015). Operations strategy frameworks in manufacturing, services and product-service systems. Paper presented at the 6th IESM Conference. Seville, Spain.
5. Bayraktar, E., Jothishankar, M., Tatoglu, E., & Wu, T. (2007). Evolution of operations management: Past, present and future. *Management Research News, 30*(11), 843–871.
6. Beyer, M. A., & Laney, D. (2012). *The Importance of 'Big Data': A Definition*. META Group (now Gartner). http://www.gartner.com/DisplayDocument?id=2057415&ref=clientFriendlyUrl. Accessed 1 Aug 2013.
7. Bretthauer, K. M. (2004). Service management. *Decision Sciences, 35*(3), 325–332.
8. Charles, V. (2016). 4^3 architecture for service innovation: Innovation doesn't come out of nowhere. *Strategia, 40*, 105–114.
9. Charles, V., Aparicio, J., & Zhu, J. (Eds.). (2020). *Data science and productivity analytics*. Springer.
10. Charles, V., & Emrouznejad, A. (2018). Big data for the greater good: An introduction. In A. Emrouznejad & V. Charles (eds), *Big data for the greater good* (pp. 1–18). Springer.
11. Charles, V., & Gherman, T. (2013). Achieving competitive advantage through big data. Strategic implications. *Middle-East Journal of Scientific Research, 16*(8), 1069–1074.
12. Charles, V., & Gherman, T. (2018). Big data and ethnography: Together for the greater good. In A. Emrouznejad & V. Charles (eds.), *Big data for the greater good* (pp. 19–34). Springer.
13. Charles, V., Gherman, T., & Paliza, J. C. (2019). Stakeholder involvement for public sector productivity enhancement: Strategic considerations. *ICPE Public Enterprise Half-Yearly Journal, 24*(1), 78–87.
14. Charles, V., Tavana, M., & Gherman, T. (2015). The right to be forgotten—Is privacy sold out in the big data age? *International Journal of Society Systems Science, 7*(4), 283–298.
15. Chase, R. B. (1996). The mall is my factory: Reflections of a service junkie. *Production and Operations Management, 5*(4), 298–308.
16. Chase, R. B., & Apte, U. M. (2007). A history of research in service operations: What's the big idea? *Journal of Operations Management, 25*(2), 375–386.
17. Choi, T. M., Wallace, S. W., & Wang, Y. (2017). Big data analytics in operations management. *Production and Operations Management, 27*(10), 1868–1883.
18. Chopra, S., Lovejoy, W., & Yano, C. (2004). Five decades of operations management and the prospect ahead. *Management Science, 50*(1), 8–14.
19. Cohen, M. C. (2018). Big data and service operations. *Production and Operations Management, 27*(9), 1709–1723.
20. Constantiou, I. D., & Kallinikos, J. (2015). New games, new rules: Big data and the changing context of strategy. *Journal of Information Technology, 30*(1), 44–57.
21. Cox, M., & Ellsworth, D. (1997). Application-controlled demand paging for out-of-core visualization. In *Proceedings of the 8th IEEE Conference on Visualization*. Los Alamitos, CA: IEEE Computer Society Press.
22. Day, G. S., & Wensley, R. (1988). Assessing advantage: A framework for diagnosing competitive superiority. *Journal of Marketing, 52*(2), 1.
23. Edvardsson, B., Gustafsson, A., & Roos, I. (2005). Service portraits in service research: A critical review. *International Journal of Service Industry Management, 16*(1), 107–121.
24. Emrouznejad, A. (Ed.). (2016). *Big data optimization: recent developments and challenges*. Cham.
25. Emrouznejad, A., & Charles, V. (Eds.). (2018). *Big data for the greater good*. Springer.

26. Emrouznejad, A., & Marra, M. (2016). Big data: Who, what and where? Social, cognitive and journals map of big data publications with focus on optimization. In A. Emrouznejad (ed.), *Big data optimization: Recent developments and challenges* (pp. 1–16). Springer.
27. Epstein, E., & Hagen, C. (n/a). *Big data and the creative destruction of today's business models*. Retrieved October 1, 2020, from https://at.kearney.com/web/guest/analytics/article/?/a/big-data-and-the-creative-destruction-of-today-s-business-models
28. Ericksson, T., Bigi, A., & Bonera, M. (2020). Think with me, or think for me? On the future role of artificial intelligence in marketing strategy formulation. *The TQM Journal, 32*(4), 795–814.
29. Field, J. M., et al. (2018). Service operations: What's next? *Journal of Service Management, 29*(1), 55–97.
30. Fitzsimmons, J. A., & Fitzsimmons, M. J. (2004). *Service management* (4th ed.). McGraw Hill Irwin.
31. Fung, K. (2014). Google flu trends' failure shows good data > Big Data. *Harvard Business Review*. https://hbr.org/2014/03/google-flu-trends-failure-shows-good-data-big-data
32. Gherman, T. (2018). *Spoken and embodied interaction in facilitated computer-supported workplace meetings* (Doctoral dissertation). Retrieved from Loughborough University Institutional Repository.
33. Grant, R. M. (1999). Transforming uncertainty into success: The strategic leadership forum 1999. *Strategy and Leadership, 27*(4/5), 31–36.
34. Greenfield, H. I. (2002). A note on the goods/services dichotomy. *The Service Industries Journal, 22*(4), 19.
35. Grönroos, C. (1994). From scientific management to management service. A management perspective for the age of service competition. *International Journal of Service Industry Management, 5*(1), 5–20.
36. Grönroos, C. (2000). *Service management and marketing: A customer relationship approach* (2nd ed.). Wiley.
37. Hambrick, D. C., & Mason, P. A. (1984). Upper echelons: The organization as a reflection of its top managers. *Academy of Manageent Review, 9*(2), 193–206.
38. Hammond, K. J. (2013). The value of big data isn't the data. *Harvard Business Review*. http://blogs.hbr.org/cs/2013/05/the_value_of_big_data_isnt_the.html. Accessed 13 July 2013.
39. Hansun, S., Charles, V., Gherman, T., Subanar, & Rini-Indrati, C. (2020). A tuned Holt-Winters white-box model for COVID-19 prediction. *International Journal of Management and Decision Making*, 1–22. In press.
40. Heineke, J., & Davis, M. M. (2007). The emergence of service operations management as an academic discipline. *Journal of Operations Management, 25*(2), 364–374.
41. Herschel, R., & Miori, V. M. (2017). Ethics and Big Data. *Technology in Society, 49*, 31–36.
42. Hill, A. V., Collier, D. A., Froehle, C. M., Goodale, J. C., Metters, R. D., & Verma, R. (2002). Research opportunities in service process design. *Journal of Operations Management, 20*(2), 189–202.
43. Iansiti, M. (2013). *Operations strategy*. Harvard Business Publishing.
44. IBM. (2013). *What is big data?* Retrieved July 20, 2012, from http://www-01.ibm.com/software/data/bigdata
45. Johnston, R. (1999). Service operations management: Return to roots. *International Journal of Operations and Production Management, 19*(2), 104–124.
46. Johnston, R. (2005). Service operations management: From the roots up. *International Journal of Operations & Production Management, 25*(12), 1298–1308.
47. Judd, R. C. (1964). The case for redefining services. *Journal of Marketing, 28*(1), 58.
48. Kant, I. ([1785] 1990). *Foundations of a metaphysics of morals*. McMillan.
49. Kundera, M. (1996). *Slowness*. Faber and Faber.
50. Kuo, Y., & Kusiak, A. (2018). From data to big data in production research: The past and future trends. *International Journal of Production Research, 57*(15–16), 4828–4853.
51. Laney, D. (2001). *3D Data Management: Controlling Data Volume, Velocity and Variety. Applications Delivery Strategies*. META Group (now Gartner). http://blogs.gartner.com/doug-laney/files/2012/01/ad949-3D-Data-Management-Controlling-Data-Volume-Velosity-and-Variety.pdf. Accessed 1 Aug 2013.

52. Lazer, D., Kennedy, R., King, G., & Vespignani, A. (2014). The parable of google flu: Traps in big data analysis. *Science 343 (6176)*, 1203–1205.
53. Levitt, T. (1972). Production line approach to service. *Harvard Business Review, 50*(5), 41–52.
54. Lovelock, C., & Gummesson, E. (2004). Whither services marketing? In search of a new paradigm and fresh perspectives. *Journal of Service Research, 7*(1), 20–41.
55. Machuca, J. A. D., Gonzalez-Zamora, M. D. M., & Aguilar-Escobar, V. G. (2007). Service operations management research. *Journal of Operations Management, 25*, 585–603.
56. Manikas, A., & Boyd., L., Guan, J., & Hoskins, K. . (2020). A review of operations management literature: A data-driven approach. *International Journal of Production Research, 58*(5), 1442–1461.
57. Maoz, M. (2013). How IT should deepen big data analysis to support customer-centricity. *Gartner G00248980.*
58. Markus, L. (2015). New games, new rules, new scoreboards: The potential consequences of big data. *Journal of Information Technology, 30*(1), 58–59.
59. Maxwell, C. R. (2013). Accelerated pattern recognition, ethnography, and the era of big data. In B. Jordan (ed.), *Advancing ethnography in corporate environments. Challenges and emerging opportunities* (pp. 175–192). Left Coast Press.
60. Meredith, J. R., Raturi, A., Amoako-Gyampah, K., & Kaplan, B. (1989). Alternative research paradigms in operations. *Journal of Operations Management, 8*(4), 297–326.
61. Miller, J. G., Graham, M. B. W., Freeland, J. R., Hottenstein, M., Maister, D. M., Meredith, J., & Schmenner, R. W. (1981). Production/operations management: Agenda for the 80s. *Decision Science, 12*(4), 547–571.
62. Mintzberg, F. (1978). Patterns in strategy formation. *Management Science, 24*(9), 934–948.
63. Morey, R. (1976). Operations management in selected nonmanufacturing organizations. *Academy of Management Journal, 19*(1), 120.
64. Nie, W., & Kellogg, D. L. (1999). How professors of operations management view service operations? *Production and Operations Management, 8*(3), 339–355.
65. Ohlhorst, F. (2013). *Big data analytics: Turning big data into big money*. Wiley.
66. Oliveira, P., Roth, A. V., Conley, M. V., & Voss, C. (2003). Knowledge assets in global services strategy. In D. V. Gibson, C. Stolp, P. Conceição, & M. V. Heitor (Eds.), *Systems and policies for the global learning economy* (pp. 511–536). Praeger Publishers.
67. Pannirselvam, G. P., Ferguson, L. A., Ash, R. C., & Siferd, S. P. (1999). Operations management research: An update for the 1990s. *Journal of Operations Management, 18*(1), 95–112.
68. Parasuraman, A., Zeithaml, V. A., & Berry, L. L. (1985). A Conceptual model of service quality and its implications for future research. *Journal of Marketing, 49*(4), 41–50.
69. Parasuraman, A., Zeithaml, V. A., & Berry, L. L. (1988). SERVQUAL: A multiple-item scale for measuring consumer perceptions of service quality. *Journal of Retailing, 64*(1), 12.
70. Pearce, E. (1957). *History of the standard industrial classification*. Executive office of the President, Bureau of the Budget, Office of Statistical Standards.
71. Peinado, J., Graeml, A. R., & Vianna, F. (2018). Operations management body of knowledge and its relevance to manufacturing and service organizations. *Revista De Gestão, 25*(4), 373–389.
72. Penrose, E. T. (1959). *The theory of the growth of the firm*. Wiley.
73. Plumb, I., & Zamfir, A. (2011). Noi ipoteze privind managementul serviciilor în lumea globală [*New hypotheses regarding service management in the global world*]. *Economie teoretică şi aplicată, XVIII*(1), 122–130.
74. Rallis, S. F., & Rossman, G. B. (2003). *Learning in the field: An introduction to qualitative research* (2nd ed.). Sage.
75. Rendon, R., Apte, A., & Apte, U. (2012). Services contracting in the DoD: A comparison of management practices in the Army, Navy, and Air Force. *Defense Acquisition Research Journal, 19*(1), 3–32.
76. Roda, W. C., Varughese, M. B., Han, D., & Li, M. Y. (2020). Why is it difficult to accurately predict the COVID-19 epidemic? *Infectious Disease Modelling, 5*, 271–281.

77. Roth, A. V., & Menor, L. J. (2003). Insights into service operations management: A research agenda. *Production and Operations Management, 12*(2), 145–164.
78. Sampson, S. E. (2012). Visualizing service operations. *Journal of Service Research, 15*(2), 182–198.
79. Shi, D., Guan, D., Zurada, J., & Manikas, A. (2017). A data-mining approach to identification of risk factors in safety management systems. *Journal of Management Information Systems, 34*(4), 1054–1081.
80. Smith, J. S., Karwan, K. R., & Markland, R. E. (2007). A note on the growth of research in service operations management. *Production and Operations Management, 16*(6), 780–790.
81. Spohrer, J., & Maglio, P. P. (2008). The emergence of service science: Toward systematic service innovations to accelerate co-creation of value. *Production and Operations Management, 17*(3), 238–246.
82. Vargo, S. L., & Lusch, R. F. (2004). The four service marketing myths: Remnants of a goods-based, manufacturing model. *Journal of Service Research, 6*(4), 324–335.
83. Vargo, S. L., & Lusch, R. F. (2010). Advancing service science with service-dominant logic. In P. P. Maglio, C. Kieliszewski, & J. C. Spohrer (Eds.), *Handbook of service science* (pp. 133–156). Springer.
84. Victorino, L., et al. (2018). Service operations: What have we learned? *Journal of Service Management, 29*(1), 39–54.
85. Vosbergen, S., Janzen, J., Stappers, P. J., van Zwieten, M., Lacroix, J., Idema, K., van den Brook, I., Kemps, H. M., Kraaijenhagen, R. A., & Peek, N. (2013). A qualitative participatory study to identify experiences of coronary heart disease patients to support the development of online self-management services. *International Journal of Medical Informatics, 82*(12), 1183–1194.

Vincent Charles PhD, PDRF, FRSS, FPPBA, MIScT, is a Full Professor and the Director of the Center for Value Chain Innovation at CENTRUM Católica Graduate Business School, PUCP, Lima, Peru. Additionally, he holds multiple visiting professorship positions across the Globe. He is an experienced researcher in the fields of artificial intelligence, data science, and OR/MS. He has more than two decades of teaching, research, and consultancy experience in various countries, in the fields of applied analytics/optimisation and big data science, having been a Full Professor and Director of Research for more than a decade in triple-crown business schools. He holds Executive Certificates from the MIT, HBS, and IE Business School. He has published over 150 research outputs. He is a recipient of many international academic honours and awards. An AWS Certified Cloud Practitioner, AWS Accredited Educator, a certified Six Sigma Black Belt, and an Advance HE Certified External Examiner, UK. He has developed the 4^3 Architecture for Service Innovation. His area of focus includes productivity, quality, efficiency, effectiveness, competitiveness, innovation, and design thinking. He is a Co-Founder and the Chief Analytics Officer at Anthrolytics Ltd., UK.

Ali Emrouznejad PhD, is a Professor and Chair in Business Analytics at The University of Surrey, UK. His areas of research interest include performance measurement and management, efficiency and productivity analysis as well as big data and data mining. Dr Emrouznejad is editor of Annals of Operations Research, associate/guest editor or member of editorial board in number of other journals including European journal of Operational Research, Journal of Operational Research Society, Socio-Economic Planning Sciences, IMA journal of Management Mathematics, OR Spectrum, RAIRO - Operations Research. He has published over 200 articles in top ranked journals; he is author of the book on "Applied Operational Research with SAS", editor of the books on "Performance Measurement with Fuzzy Data Envelopment Analysis" (Springer), "Managing Service Productivity" (Springer), "Big Data Optimization" (Springer), "Big Data for Greater Goods" (Springer) and "Handbook of Research on Strategic Performance Management and Measurement" (IGI Global). He is also co-founder of Performance Improvement Management Software (PIM-DEA), see http://www.Emrouznejad.com

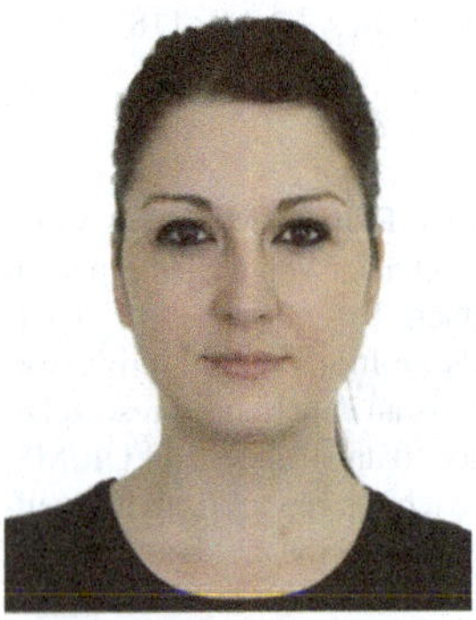

Tatiana Gherman MBA, PhD, is a Senior Lecturer and has more than a decade of teaching and research experience. She is currently researching how Machine Learning and Artificial Intelligence tools can support various business functions to help make business management more effective; with particular interest in how to design Artificial Intelligence techniques grounded in and informed by patterns of social interaction and communication. Working towards the creation of a new field of research grounded in ethno-data science. Her area of research interests further includes multi-attribute decision-making techniques and other advanced quantitative analytics at different levels. She has research publications in reputed journals and is one of the recipients of the 2019 GDN Informs/Springer Researcher Award. Her field of expertise includes Artificial Intelligence, Data Science, Big Data, Group Decision Support, Workplace Studies, Behavioural Studies, Quantitative Analytics (at different levels), Conversation Analysis, and Ethnomethodology.

Modeling Big Data Enablers for Service Operations Management

Mahdi Nasrollahi and Mohammad Reza Fathi

Abstract The purpose of this chapter is to identify big data enablers in Service Operations Management (SOM) and to analyze the interactions between them. First, we identify forty-four initial big data enablers in SOM from literature and consequent deliberations with experts. In the second step, thirteen main enablers were identified using factor analysis. Two different MCDM techniques, i.e. ISM and fuzzy-DEMATEL were used to identify driving enablers. Furthermore, common enablers from each technique, their hierarchies, and inter-relationships have been established. The enabler modeling using ISM and Fuzzy-DEMATEL shows that the most influential/driving enablers are top management, change management, organizational culture, structural factors, financial factors, feasibility study for big data adoption, and strategies and skills. On the other hand, velocity, veracity, data security and privacy, and variety were commonly identified as the most dependent big data enablers for SOM across two different modeling techniques. The findings allow decision makers to select the desired and drop undesired enablers in the implementation of big data initiatives to improve SOM performance. MCDM-based hierarchical models and causal diagrams for big data enablers depicting contextual inter-relationships have been proposed, which is a new effort to implement big data in SOM.

Keywords Big data · Service operations management (SOM) · Interpretive structural modeling (ISM) · Fuzzy-DEMATEL · Enablers

- SOM: Service Operations Management
- MCDM: Multi-Criteria Decision Making
- ISM: Interpretive Structural Modeling
- OM: Operations Management

M. Nasrollahi (✉)
Faculty of Social Sciences, Imam Khomeini International University, Qazvin, Iran
e-mail: m.nasrollahi@soc.ikiu.ac.ir

M. R. Fathi
College of Farabi, University of Tehran, Qom, Iran

A. Emrouznejad and V. Charles (eds.), *Big Data and Blockchain for Service Operations Management*, Studies in Big Data 98,
https://doi.org/10.1007/978-3-030-87304-2_3

- IOT: Internet of Things
- PCA: Principal Component Analysis
- SSIM: Structural Self-Interaction Matrix

1 Introduction

The transformation of industrialized economies from a manufacturing base to a service orientation is an ongoing phenomenon. Service operations management (SOM) is the term used to cover the activities, decisions and responsibilities of operations managers in service organizations [33]. SOM's history as a distinct topic in academic literature dates back to the 1970s [32]. Nevertheless, the SOM field has languished in relative obscurity, at least in relation to traditional operations management (OM) that focuses on manufacturing management [65]. Service operations have great appeal, and they're all around us. Nevertheless, there are some unfortunate stereotypes about the management of service operations. First, there is a stereotype that companies' operations function is primarily about managing physical products and inventories of products. The second stereotype to dispel is that service operations are an unscientific type of operations, sometimes based on the observation that traditional manufacturing operation models do not often fit well in service contexts. Third, there is a traditional assumption that there is a certain dichotomy between goods and services. Fourth, some have claimed that service is inherently customer-oriented and solution-focused. Finally, based on the Fifth Stereotype, we are trying to overcome the confusing idea that service is adequately defined by what it is not [57]. So, despite a wealth of important history in this field, there is a need for widely recognized contributions to research and practice in SOM. Good service operations management will provide the right services, experiences and outcomes for the customer and will also mean a better experience for the staff. It will also deliver numerous organizational and economic benefits at micro and macro levels respectively [33].

With the continued dissemination of instrumentation technology in daily life, the amount of data available is increasing and the task of extracting valuable information is becoming increasingly sophisticated [10]. On the one hand, IOT,[1] the industrial IOT, cloud computing, wireless sensor networks, cyber-physical systems and social media are commonly used terminologies that change the operational management landscape [13, 63], and On the other hand, the intense market competition has made organizations intuitive to a more accurate and informed decision-making process backed by big data analytics to stay ahead in completion [78].

Big data allows organizations to analyze an immense volume and variety of both structured and unstructured data across a wide range of networks to support evidence-based decision-making and real-time action [73]. Today, most firms collect and store large amounts of data on their customers and keep records of all their past transactions [14]. According to Manyika et al. [41], in 2010 companies collected and stored

[1] Internet of Things.

around seven data exabytes and buyers generated thirteen Exabyte. If organizations can control data collected internally or externally, their operational capabilities could be optimized [20]. So it is clear that big data and its related technologies will form advances in every area.

Nowadays, organizations have accepted the usefulness of big data in decision-making processes and as a result, many operational managers are trying to use big data in their activities in real-time. Also, Today's businesses need to use new information technologies in their service system construction to provide convenient services to stakeholders. Big data unleashed the data-driven paradigm and paved the way to challenge new classes of problems that were not solvable in the past [66]. Using the capabilities and hidden potential of big data to make better decisions can lead to an organization's improved profitability [16]. Proactivity and anticipating needs, mitigating risk and fraud, delivering relevant products, personalization and service, optimizing and improving the customer experience are some of the benefits that can be realized by using big data in SOM. Researchers in a study highlighted that using big data for business decision-making has many benefits and reported 26% improvement in company performance over the past three years and expected 41% improvement in performance over the next three years [18]. In another research, McAfee and Brynjolfsson [44] found that "companies in the top third of their industries in the use of data-driven decision making were, on average, 5% more productive and 6% more profitable than their competitors".

Since the implementation and integration of big data initiatives to improve SOM processes is a fairly new area, not much work has appeared in the literature highlighting the enablers of big data success in SOM. Keeping this as a motivation, the present study attempts to identify and model the enablers for the success of big data in SOM. A systematic model showing hierarchy and inter-relationships among enablers and classifying the enablers into cause and effect groups will be particularly valuable for decision-makers to use their resources strategically to successfully embed big data in SOM. In particular, the study attempts to achieve the objectives listed below:

- Identifying the enablers of big data initiatives in SOM;
- Reduction and integrating of enablers using factor analysis and developing the hierarchical structure;
- Establishing hierarchy and inter-relationship among enablers using ISM and recognizing their causal relationships using fuzzy-DEMATEL.
- Using MICMAC[2] [26] analysis for analyzing the driving power and dependence of enablers.
- Discussing theoretical and managerial implication of this research and suggesting future research directions.

The rest of the chapter is structured as follows. Section 2 describes the design and methodology of the research. Section 3 highlights the key enablers for big data

[2] Matriced 'Impacts croises-multipication applique' and classment (cross-impact matrix multiplication applied to classification).

success in SOM. While Sect. 4 provides an overview of ISM and highlights the detailed procedure for modeling the enablers using ISM and MICMAC; describing the fuzzy-DEMATEL and using it for modeling enablers is discussed in Sect. 5. Finally, Sect. 6 presents the discussion on the findings of this work's analysis, and Sect. 7 presents conclusions, implications, and future scope.

2 Research Methodology

The main purpose of this study is to model big data enablers in SOM. To fulfill this objective, we applied field data with a focus on service organizations and experts with experience or expertise in this field. We need to perform a preliminary factor analysis and item reduction of big data enablers in SOM to achieve this purpose. Completing the factor analysis and reduction of items is important for several reasons. First, the number of identified enablers is 44 items and is too long and some enablers have repetitive meaning. Second, several enablers required consistent clarification.

For factor analysis, a principal component analysis (PCA) was selected to consolidate enablers. We applied an internet-based questionnaire to gather data from a sample of experts around the world. The questionnaires were sent to respondents through an email process and messages in LinkedIn and Researchgate. From 8 to 16 February 2019, 3023 questionnaires were sent in full. A total of 317 questionnaires were received over a period of 40 days, 65 were eliminated as not completed and finally, 252 experts completed the survey. Thus, data from 252 experts were included in the dataset yielding a valid response rate of 8.3%. All data were analyzed with version 25.0 of IBM SPSS. Table 1 shows the demographics of the sample.

Table 2 shows the type and location of the industry of the respondent. The respondents were with many differences from various industries.

Figure 1 shows the steps of study.

3 Identification of Enablers for Big Data in SOM

3.1 Identifying the Initial Enablers

Based on the deliberations with academic and industry domain experts and conducting a comprehensive review of the relevant literature, 44 initial enablers for the successful implementation of big data in SOM have been identified. Since the role of academic and industry domain experts and their answers is very important for the identification of big data enablers and selection of experts is important for conducting high-quality deliberations, these experts were selected from different countries to give their informed opinions on the state of big data and its enablers in SOM. 23 experts from 8 different countries responded to structured email questions.

Table 1 Sample demographics

Item		N	%
Age	Under 24	9	3.6
	25–34	84	33.3
	35–44	102	40.5
	45–54	29	11.5
	Over 55	28	11.1
Work experience	1–5	65	25.8
	6–10	74	29.4
	11–15	37	14.7
	16–20	20	7.9
	Over 21	56	22.2
Job title	Managing Director	30	11.9
	Vice president	16	6.3
	Consultant	28	11.1
	University Professor/Lecturer	37	14.7
	Marketing and sales manager	11	4.4
	Purchasing manager	8	3.2
	Service Operations manager	58	23.0
	Information system manager	12	4.8
	Researcher	43	17.1
	Other	9	3.6
Education	Bachelor's degree (e.g. B.A., B.S.)	0	0
	Master's degree (e.g. M.A., M.S., M.Ed.)	90	35.7
	Professional degree (e.g. M.D., D.D.S., D.V.M.)	18	7.1
	Doctorate (e.g. Ph.D., Ed.D.)	117	46.4
	Postdoctoral	27	10.7
	Total	252	100

Also, a workshop with 4 leading academic experts was arranged in January 2019 and experts were interviewed in person. The comments have been analyzed and a synthesis of the views in combination with literature presented in this section. Table 3 shows the demographics of the experts.

- *Restructuring of data management processes*

The restructuring of data management processes should be in line with the priorities of the organization [46]. Because implementing and processing big data is a critical and crucial issue for organizations to succeed, data management processes need to be designed to create competitive advantage in receiving, processing and analyzing data.

Table 2 Sample industry

Description		N	%
Transportation and warehousing		10	4.0
College, university, and adult education		68	27.0
Wholesale and retail		12	4.8
Health care and social assistance		19	7.5
Hotel and food services		16	6.3
Legal services		4	1.6
Real estate, rental and leasing		2	0.8
Information services and data processing		49	19.4
Other information industry		9	3.6
Software		8	3.2
Scientific or technical services		4	1.6
Publishing		1	0.4
Telecommunications		3	1.2
Broadcasting		3	1.2
Finance and insurance		6	2.4
Arts, entertainment, and recreation		2	0.8
Government and public administration		8	3.2
Other service industry		28	11.1
Owner	Multinational corporation (MNC)	18	7.1
	Private sector (e.g. most businesses and individuals)	149	59.1
	Public sector (e.g. government and semi-government)	66	26.2
	Not-for-profit sector	19	7.5
Location	Asia	51	20.1
	Europe	75	29.8
	Africa	4	1.6
	Oceania (Australia, New Zealand, etc.)	11	4.4
	America	84	33.3
	Middle East	27	10.7
Total		252	100

- *Reengineering of the processes*

To embrace the new technology, like big data, would need to train employees to use the new technology, rewrite the service operations manuals and, definitely, reengineer the processes within the organization [46]. Changing to a smart, modern and agile enterprise in today's competitive market means stakeholders are setting up a global mind-change: sharing information is an essential ingredient in destroying obstructions that exist in a traditional approach to information management. Reengineering information management processes is therefore required for data processing.

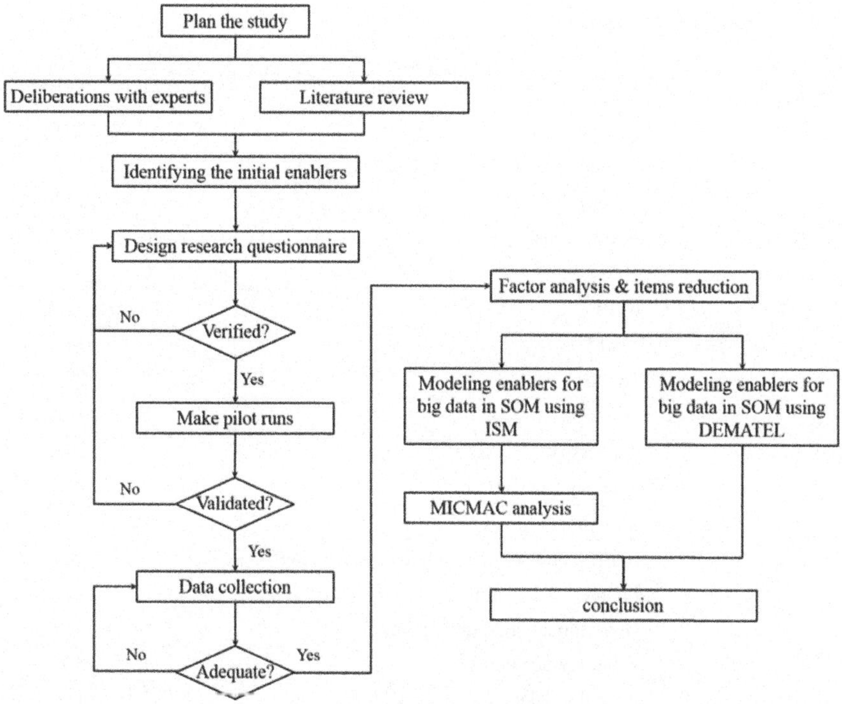

Fig. 1 Research process

- *Digitalization of data*

Data digitization means that an organization has become business-critical in understanding what data it creates and processes in its various activities [15]. Decisions on what data/information to store and what should not be based on sound business rationale as opposed to technical considerations. Digitalization of data is associated with three main issues: first, a plethora of dirty data, the quality of which may be impossible for other investigators to evaluate,second, weak or non-existent long-term persistence guarantees for many data sources; and finally, inconsistent metadata practices that may render reuse of data impossible, despite their intention to do the opposite [19].

- *Business (Artificial) Intelligence*

The expression big data has become synonymous with Business Intelligence (BI) among some suppliers and for others, BD was an incorporation of the traditional BI but with the addition of new elements such as predictive analyses, data mining, operating tools/approaches as well as research and science [21]. Big Data has a complex nature requiring powerful technologies and advanced algorithms. The traditional static Business Intelligence tools can no longer be efficient for Big Data applications

Table 3 Expert's demographics

Item		N	%
Location	Asia	6	22.2
	Europe	8	29.6
	Oceania (Australia, New Zealand, etc.)	3	11.1
	America	4	14.8
	Middle East	6	22.2
Work experience	1–5	0	0
	6–10	10	37.0
	11–15	4	14.8
	16–20	8	29.6
	Over 21	5	18.5
Industry	University	8	29.6
	Telecommunications	3	11.1
	Software	4	14.8
	Information services and data processing	7	25.9
	Finance and insurance	3	11.1
	Government and public administration	1	3.7
	Other service industry	1	3.7
Education	Bachelor's degree (e.g. B.A., B.S.)	3	11.1
	Master's degree (e.g. M.A., M.S., M.Ed.)	5	18.5
	Professional degree (e.g. M.D., D.D.S., D.V.M.)	3	11.1
	Doctorate (e.g. Ph.D., Ed.D.)	12	44.4
	Postdoctoral	4	14.8
	Total	27	100.0

[49]. By adopting BI, organizations become more able to load all data into a consistent data warehouse and then apply different analytical methods to exploit the hidden insight into the data [43].

- *Mobile technologies*

As the internet and mobile technologies evolve, all features of our lives are transformed [42]. Organizations can use advances in mobile technologies to explore and explore the opportunities offered by social media, business intelligence (BI), and service orientation to build an agile ecosystem in and around the business. Very immense data sets were generated quickly in various formats with internet and mobile communication. Mobile technologies are thus increasingly being used as a data source to enable big data analytics [25].

- *Desire to reduce costs*

The desire to reduce costs has driven the use of big data in service organizations [56]. Organizations have learned that using big data can reduce overall costs and respond to changes in demand [3]. The significant increase in cost reduction demand has encouraged evaluation of the use of new technologies and tools in service organizations, such as big data. Information assets require cost-effective, innovative forms of information processing for enhanced insight and decision-making.

- *Commitment to resilience*

Commitment to resilience is about the ability to meet unforeseen circumstances and surprises [6]. Being resilient in the information age requires a lot of work and training, and is very important to succeed. Resilience commitment is a healthy focus on avoiding surprises and the ability to bounce back from failures [48]. Commitment to resilience is likely to lead to an active awareness in organizations that are seen as opening up to new information and using new tools to process it.

- *Preoccupation with failure*

Preoccupation with failure means an unyielding focus of a firm on trying to learn from failure or near-failures so it can transform learning to avoid and/or minimize future failures [48]. Equally important as the focus on failures is the preoccupation of early signals of failure. This focus on early signals and warnings is important to risk, as signals and warnings are closely linked to the uncertainties and the knowledge dimensions, which are essential elements of the new risk perspectives. In a probability-based risk perspective highlighting historical failure data, the risk description is not sensitive to changes in the same way.

- *Sensitivity to operations*

Operational sensitivity is awareness of the interrelationships of value-creating processes and their optimal levels of operation [48]. The key here is to be sensitive to what happens during the operation. Information is continuously collected during the service operations and the risk description is updated and appropriate risk management calls for action. The risk is monitored during the operation and is sensitive to everything that happens, depending on how we understand the risk. Thus, operational sensitivity can enhance the usefulness of big data and encourage organizations to apply it.

- *Reluctance to simplify*

A collective mindset of avoiding blind spots by consistently seeking different points of view and promoting skepticism towards simplistic solutions to problems [48]. Following this characteristic, we will not allow judgments of data-driven risks to be based solely on the result of quantitative risk expression. Members of all organizations handle complex tasks by simplifying how the current situation is interpreted. These simplifications essentially allow members to ignore data and keep going [74].

Reluctance to simplify requires encouraging and supporting the diversity of experience, perspective, and opinion, and emboldening the team to express differing opinions,questioning care plans and sharing suboptimal events [6]. All of these features require massive data generation and processing.

- *Deference to expertise*

A deemphasizing of hierarchical decision making so that people with the greatest knowledge and expertise assume primary decision making responsibilities in the event of a failure [48]. This characteristic could be manifested by pointing the problem to those experts who have the necessary knowledge to be able to take proper action. Deference to expertise requires pushing decision-making down and around to the people with the most related knowledge and expertise [30] that confirm the obligation to use big data to promote knowledge and expertise.

- *Continuous creation of data*

Continuous data creation has created new challenges and opportunities for organizations. With continuous data creation, big data processing and management have become increasingly crucial in most data-driven fields [53]. Continuous data creation from both structured and unstructured sources requires the use of cutting-edge techniques to fully exploit the potential embedded in data to provide effective tools for assessing and extracting actionable information.

- *Scale of data*

The scope and scale of data elements in the big data world are hard to conceive of Bizer et al. [8]. New information technologies have raised the data scale to an unprecedented level. The need for data scale may become a critical driver of consolidation [41], which may be an opportunity for operations management.

- *Data integration*

The effective large-scale analysis usually expects different data from multiple sources to be collected. Data integration systems require considerable schema mapping effort before the system can be used. Different stakeholders may collect data in different formats, so the company needs to create a common data warehouse to store homogeneous information. Integration of data is a process used to combine data from disparate sources into meaningful and valuable insights. It allows users, organizations, and applications to merge different data types for use as personal or business processes and/or functions [60]. These characteristics can intensify the success of big data in organizations.

- *Data variety*

As market competition is becoming stiff, decision-makers are becoming increasingly dependent on big data to gain visibility in organizational spending, identify performance trends and optimize performance [31]. Big Data consists of structured data, semi-structured data and unstructured data generated from various distributed sources and in multiple formats [49]. During any service operation, there is a surge in the creation of data in all possible formats. These data are service operational data, service time-based data, social networking data, customer service data, marketing data, etc. These diverse data will enrich the databases of the organization and can lead to a successful data process.

- *Interoperability*

Interoperability is a system specialty that acts as an interface to communicate, exchange meaningful and accurate information with other systems without making any special efforts on the part of end-users [4]. Systems interoperability facilitates the exchange of data and knowledge with other organizations [43]. It also allows service providers to exchange huge amounts of data between subsystems and develop solutions to improve service processes. Ensuring interoperability is particularly important because it allows the combination of different datasets [35].

- *Speed of data generation*

The increased number of digital devices in service operations caused a constant increase in data generation speed. Thus, it is heading us to handling large amounts of data. This speed of data generation will continue and it is expected to increase at an exponential level over the next decade [45] so that 90% of the data in the world today has been created in the last two years alone [1]. This speed in data generation ultimately leads to an increase in data and data variety scale and helps opportunities for big data success.

- *Data transmission time*

The differences in data transmission times will affect the service system due to real-time spent on data transmission. Reducing data transmission time can lead to improved performance [28]. Where real-time data transmission is required, it is necessary to use protocols that require less time. In general, system performance depends on the performance of network data transmission time.

- *Data processing time*

The rapid development of new technologies has significantly reduced data processing time and makes it possible to deal with large data that would not have been feasible a few years ago [7]. Furthermore, with increasing data volumes, the time to transfer a data unit can exceed its actual data processing time, so we need to reduce data processing time to achieve higher computing resource utility and big data analysis capacity. Distribution strategies for workloads can significantly save data processing time for big data.

- *Reliability*

Service operation management needs a system that, in addition to its advanced functionality, guarantees high reliability of generated data. In other words, the amount of data does not mean that one can ignore data reliability [42]. Besides, companies assessed BD vendors mostly on their security, reliability, and uptime capabilities [7].

Thus, the reliability of data can have a significant impact on the implementation of big data in SOM.

- *Accessibility*

Accessibility of big data is the one conflicting criterion. Accessibility in service operations management means the ability to reach all relevant information. With easy accessibility over standard internet devices, the convenience of using electronic services has made a big step forward, making big data attractive not only for companies but also for private customers [10]. Exclusive control over data sources and restricting accessibility can become an abuse of big data and restrict competition by posing unfair barriers to the organization.

- *Quality of information*

As businesses and organizations become more technologically dependent, information quality becomes an increasing concern because information quality is critical to effective resource utilization and has a decisive impact on the service capabilities of the organization. Progressive service companies are enhancing information quality to ensure that top management has the information they need to make quality decisions [43]. They also use information quality measurements to predict organizational outcomes [47]. Better information quality will improve the dissemination of information and lead to a better response to market demands.

- *Data circulation rules*

To date, fierce market competition and the rapid growth of big data processing among multiple organizations have resulted in data circulation for decision-makers. Big data circulation among diffident parties becomes a common form of big data utilization [46]. Efficient data circulation is critical to the implementation of Big Data. Rules to enable data circulation needs to give particular attention to licensing, access policies for documents and public records, and privacy. Among these, the presence of rewarding data opening policies has a crucial value for the transparent and efficient management of services.

- *Proper leadership*

The proper leadership feature is a contingency issue in the big data implementation process's leadership-related aspects [43]. There is nothing more pragmatic than ensuring proper leadership attitudes before ever touching software, algorithms, and data. This provides a low-risk/high-impact approach to identifying and prioritizing valid big data implementation opportunities in SOM. Thus, the necessary skills and proper leadership will lead to optimized performance, profits and costs.

- *Knowledge Management Systems*

Knowledge management systems enable companies to have direct contact with customers and other stakeholders, to have a deep knowledge of their requirements and needs, and to collect the required data. Different authors emphasize that knowledge management systems are essential for feeding and managing the process of big data analysis [43]. Knowledge management systems are required to support the dissemination, collection, and application of big data and knowledge, as KMSs enable greater freedom and accuracy in decision-making [71].

- *Adequate System Capabilities*

To ensure the effective use of big data, it is necessary to increase the system capabilities to meet the characteristics of big data that exceed the capabilities of a normal system. Data scalability, changing data nature, and performance considerations are gradually challenging system capabilities. Some organizations also created a system to meet the dynamic user needs [60]. The high speed of data generation should be added to the complexity of decision making and may have some advantages for big data.

- *Established Culture of Collaboration*

A good culture of collaboration is reflected in a group of people with different functional expertise who come together to complement their skills and work towards a common goal [60]. Collaborative culture is changing business models and the nature of the job. It also prepares organizations to be stunned by dramatic results never before seen in the big data field.

- *Good Working Attitude*

The tendency to respond negatively or positively to a certain situation is called working attitude. Committed, focused, hard-working, good morality, efficient, ethical, and trustworthy are some indicators of working attitudes that can enable big data projects [60].

- *Internet of things*

The extension of the internet of physical things in such a way that it is possible to share, process and use data at the same time as it is possible to interact with human beings and the virtual world, with value creation being the end product [62]. The Internet of Things creates a world of interconnected sensory devices containing sensors that can collect and store data from their respective real-world environment. It has enabled devices around the world to acquire and store data to be able to use it at a later stage [7]. Data analysis is one important component of the Internet of Things that can influence the analysis of a large amount of the generated data.

- *IT infrastructure*

IT infrastructure sets data production rules [29]. IT infrastructure has been identified as a key opportunity for the organization as each organization has increasingly used modern IT technology to support its data collection and processing programs [10].

- *Big data quality management*

Because of the extensive data generation with high speed in today's organizations, ensuring the best data quality has become highly imperative. The expertise of the organization in maintaining the quality of corporate data can have a positive impact on the success of big data analytics. Poor quality data has a significant impact on the operational performance of an organization [54], customer dissatisfaction, and ineffective decision-making [38]. Thus, data quality can have a significant impact on the implementation of big data in SOM.

- *Data capturing and storage*

The increased capacity of contemporary computers allows the gathering, storage and analysis of large amounts of data [22]. Capturing big data from various sources and its secure storage is an enormous task that organizations have to deal with. Stored data represent observations and findings over time about operations [38]. Service companies are capturing data from both internal and external sources. These data include already existing data from different departments in IT systems, data captured in operating processes, data provided by business partners along the value chain, data from the environment and open data. All data will be stored in databases, so storing huge data can be a challenging task. Some individual companies have stored data far higher than their sector average, potentially giving them more potential to capture value from big data [41]. Thus an adequate mechanism for meaningful data capture and storage and retrieval in SOM can play a defining role for big data.

- *Data security and privacy*

Big data security is another significant challenge for organizations. Customer information, business data, budgets, and government records are an irresistible target for terrorists, cyber-thieves, and hackers. Databases usually contain complex structured, semi-structured, or unstructured data on which traditional methods of database protection render ineffective. Big databases can thus pose a potential threat to personal and firm security [77]. Data privacy needs to be maintained at various stages such as big data collection, storage, and processing to ensure success with big data initiatives in SOM.

- *Top management commitment and willingness*

Top management commitment is needed to create a shared vision in the organization [38] and inculcating a data-driven mindset across the entire SOM should be a key thrust for top management [75]. Also, top management commitment is the main driver of big data implementation programs. Big data activities should be given equal importance similar to other organizational goals by integrating all SOM members. The top management should provide ongoing support in the strategic and action plans for the successful execution of big data initiatives.

- *Financial support for big data initiatives*

Various activities and processes associated with big data such as capturing and storage of data, mining, cleansing, processing, interpretation [52], acquiring data scientists with a high level of knowledge and expertise, and purchase of analytics technologies requires considerable investment,so the big data initiatives are costly affair. It is a challenge for small businesses to divert huge funds into such nascent innovations.

Thus dedicated financial support from top management may be highly imperative for the success of big data in SOM.

- *Big data/data science skills*

Big data applications allow real-time optimization for certain areas of operations management, but due to limited analytical capabilities and time consumed in carrying out big data analysis, organizations often fail to leverage their benefits. The need for qualified data scientists is therefore the key to success. Big data analytics initiatives may not bear meaningful returns in the absence of the necessary technical expertise [72]. The lack of analytical prowess is a major shortcoming in realizing the full potential of big data analytics in SOM [38].

- *Appropriate selection of big data technologies and techniques*

Many big data tools and technologies are not only expensive but may not even be suitable in the context of a particular organization. Also, various analytical techniques such as predictive analytics, data mining, business intelligence, machine learning techniques, etc. Can be used in different contexts Therefore, it is important to first understand the organizational needs and expectations before actually embedding big data within the organization [38].

- *Feasibility study for big data adoption*

Because big data is seen as a fundamental tool for improving decision-making, there is a tendency among organizations to use it without the proper understanding of the big data landscape. Some organizations face the data collection challenge and their database may be too small to be statistically significant to generate any meaningful data analysis results [5]. Thus, organizations must first understand the feasibility of big data in their organization before they hurry to use and procure high-cost big data tools [64].

- *Aligning big data strategies*

Big data applications require cross-functional integration and human-technology cross-company interface. Consequently, it becomes a huge challenge for management to lead not only the entire organization but also the SOM to embed and use big data technologies in routine operations [38]. This requires aligning strategies on big data, and the leadership must be able to view the data captured strategically.

- *Organizational structure*

Organizational structure may play a significant role in the success of big data initiatives across SOM. The structure design of the organization can have a direct impact on the implementation of service innovation [29]. New concepts of organizations such as horizontal organizations are perceived to be more conducive to increased teamwork, reduced bureaucracy, increased communications, opportunities for professional development and greater job satisfaction [51]. Therefore, organizational structure can be expected to play an important role in affecting the success of big data initiatives in SOM.

- *Employee involvement*

Employees are responsible for actually working and creating services, hence incorrect allocation of employees to different tasks, improper feedback mechanisms, insufficient job enrichment, and training, and boring workplaces are significant contributors to poor performance in service operations [58]. Employee involvement can in many ways create a strengthening process in which the implementation of big data in various operations produces significant results. Organizations must understand that employees not only seek to be part of a system, but also help them implement it in such a way that a sense of meaning is derived from the day-to-day work they do [11].

- *Change Management Program*

To make big data initiatives successful, the decision-makers must work with the right people, the right data and the right techniques for problem-solving. These capabilities are created by appropriate changes programs. Change management focuses on engaging the various stakeholders, evaluating the readiness of the organization in the event of changes in the business environment, facilitating employee training and learning needs, and controlling the change process. Change management is critical to the success of big data because it showed that around 60% of big data projects will not go beyond piloting and experimentation and will eventually be neglected due to the lack of proper change management programs [38].

- *Information technology integration*

Big data involves huge volumes of data capture and accurate analysis. In it, information technology plays a vital role. It is therefore important to closely integrate information technology with all the organizational processes [38]. The flow of information and services in both short-term and long-term operations is facilitated, enabled and supported by various information technologies. To make use of big data from different sources for service operations management system, it is necessary to integrate information technologies that are associated with various entities such as cloud, network, and data.

- *Interpretability of the analysis*

Big data analytics capabilities of an organization are of limited value if the decision-makers cannot understand the analysis [37]. Eventually, if the results are to be implemented in the SOM, a decision maker must interpret the results of the analysis. Usually, it requires scrutinizing all the underlying assumptions and retracing the analysis. Also, many common sources of errors such as software bugs, model assumptions, and erroneous data can lead to incorrect analysis. It will, therefore, be helpful if additional information can be provided explaining how each result was derived and based precisely on what inputs.

Table 4 Reliability statistics

Cronbach's alpha	Cronbach's alpha based on standardized items	N of items
0.907	0.910	44

3.2 Validity, Reliability and Normality Tests

Convenience samples of 22 experts were interviewed after completing the questionnaire's 44-item draft version. These experts were asked about the wording of each questionnaire item and the relevance of the items. The experts were asked how strongly they felt that each item should be retained in the questionnaire. No questions were deleted, but the wording of each item was reviewed based on their responses, and some questions were modified accordingly. Cronbach's alpha was applied to verify the internal consistency of the constructs and to measure the measurement reliability. The results in Table 4 show that the alpha value is above 0.7 and constructs are consistent accordingly. Data normality must be evaluated before running the statistical tests. The normality of the data was tested at a meaning level of 5% using the K-S test. The level of significance for all constructs was greater than 5% so the data had a normal distribution.

3.3 Sampling Adequacy Test

The adequacy of the dataset was determined by use of Bartlett's test for sphericity and the Kaiser–Meyer–Olkin (KMO) measure of sampling adequacy. To determine whether the sample was adequate for further analysis, Bartlett's test of sphericity should be significant and the KMO index should be at least 0.60. As specified in Table 5, The Bartlett's test for sphericity was significant ($P < 0.5$) and the KMO index was 0.902; therefore, the sample was determined to be adequate for further analysis.

Table 5 KMO and Bartlett's test

Kaiser–Meyer–Olkin measure of sampling adequacy		0.902
Bartlett's test for sphericity	Approx. Chi-Square	1598.721
	df	903
	Sig	0.000

3.4 *Factor Analysis*

Principal component analysis (PCA) is a method of consolidating mutually correlated variables to a smaller number of variables, known as principal components, by linear combination with minimal loss of information contained in the observed data [36]. The main applications of factor analytic techniques are: (1) to reduce the number of variables and (2) to detect structure in the relationships between variables, that is to classify variables. One challenge was to take care of all the missing data in the survey. An average value of 3 was allocated for all the missing blanks in the survey, which would not influence the results in any way. After the PCA is run, Eigenvalue is analyzed to aid in selecting the number of factors that have to be extracted as part of the next step. Table 6 aids in deciding the number of factors to be extracted. From Table 6, it is seen that variable 13 has a variance of 1.035 and variance of 1.0 and above shows that there is considerate variability in the factor. Thus, it leads to conducting a 13 factor extraction.

Factor rotation maximizes the loadings of items based on their corresponding factor while minimizing the loadings on all other factors. Rotation optimizes the factor structure and improves the interpretability of the factor solution. The Varimax rotation was used on these 13 domains. The research team grouped the items from the matrix of pattern correlation in an attempt to propose the underlying theme or enabler within each domain. Items for which there was significant factor loading (above 0.4) [67] and apparent redundancy were combined to reduce the total number of items. The unrotated and Varimax rotation output were analyzed and the variables were categorized under enablers based on their maximum absolute value [24]. The result was a decrease from 44 enablers in the initial study to 13 enablers (Table 7).

As outlined in Table 7, by using factor analysis, 44 initial enablers were summarized into 13 enablers. A name was selected for each enabler according to the type and structure of the enablers. Figure 2 showed the hierarchical structure of big data enablers in SOM, and Table 8 also shows the reliability of these 13 enablers.

4 Modeling Enablers for Big Data in SOM Using ISM

Interpretive structural modeling (ISM) is a well-established method for identifying relationships among specific items and dealing with complex issues which were proposed by J. Warfield in 1974. It is an interactive process that helps to impose order and direction in a comprehensive model on the complexity of the relationships between the variables of a system. For problems, such as understanding the factors that enable big data in service operation management, several of them may affect each other at different levels. However, the direct and indirect relationships between the enablers affecting big data are far more accurate than the individual enablers considered in seclusion.

Table 6 Total variance explained

Component	Initial eigenvalues			Extraction sums of squared loadings			Rotation sums of squared loadings		
	Total	% of variance	Cumulative %	Total	% of variance	Cumulative %	Total	% of variance	Cumulative %
1	19.741	36.558	36.558	19.741	36.558	36.558	6.196	11.474	11.474
2	4.906	9.086	45.644	4.906	9.086	45.644	5.568	10.311	21.786
3	4.196	7.771	53.414	4.196	7.771	53.414	5.197	9.624	31.410
4	3.883	7.191	60.606	3.883	7.191	60.606	4.708	8.718	40.127
5	3.303	6.117	66.723	3.303	6.117	66.723	4.561	8.446	48.573
6	2.716	5.030	71.753	2.716	5.030	71.753	4.514	8.360	56.933
7	2.331	4.317	76.070	2.331	4.317	76.070	3.351	6.205	63.137
8	2.206	4.085	80.155	2.206	4.085	80.155	3.281	6.075	69.213
9	1.927	3.569	83.724	1.927	3.569	83.724	2.804	5.193	74.405
10	1.499	2.775	86.500	1.499	2.775	86.500	2.785	5.157	79.562
11	1.428	2.645	89.144	1.428	2.645	89.144	2.776	5.141	84.703
12	1.187	2.198	91.343	1.187	2.198	91.343	2.454	4.544	89.247
13	1.035	1.916	93.259	1.035	1.916	93.259	2.166	4.012	93.259
14	0.979	1.813	95.072						
15	0.820	1.519	96.591						
16	0.519	0.962	97.553						
17	0.442	0.819	98.371						
18	0.355	0.657	99.028						
19	0.309	0.572	99.600						
20	0.216	0.400	100.000						

(continued)

Table 6 (continued)

Component	Initial eigenvalues			Extraction sums of squared loadings			Rotation sums of squared loadings		
	Total	% of variance	Cumulative %	Total	% of variance	Cumulative %	Total	% of variance	Cumulative %
21	0.000	0.000	100.000						
	. . .	. . .	. . .						
44	0.000	0.000	100.000						

Table 7 Rotated component matrix

Enabler	Component												
	1	2	3	4	5	6	7	8	9	10	11	12	13
Best rules to enable data circulation	<u>0.712</u>	0.340	0.006	0.028	0.046	0.329	0.237	−0.092	−0.083	0.105	0.218	0.385	0.000
Big data/data science skills	<u>0.659</u>	0.089	0.162	−0.137	0.002	0.125	0.071	0.214	0.115	0.018	0.169	0.074	0.132
Information technology integration	<u>0.837</u>	0.146	0.207	0.007	−0.015	0.230	−0.044	0.161	−0.054	0.124	0.269	0.121	0.035
Aligning big data strategies	<u>0.642</u>	0.184	0.106	0.273	0.029	−0.048	0.126	0.098	−0.036	−0.119	−0.056	0.043	−0.075
Feasibility study for big data adoption	0.142	<u>0.856</u>	0.235	0.152	0.193	0.035	0.213	0.165	−0.186	−0.122	0.286	0.121	0.153
Deference to expertise	0.101	0.014	<u>0.559</u>	0.114	−0.205	0.091	0.310	0.213	−0.217	−0.020	0.038	0.351	0.008
Sensitivity to operations	0.125	0.179	<u>0.460</u>	0.321	0.321	0.197	0.125	0.115	0.260	0.132	−0.074	0.112	−0.112
Preoccupation with failure	0.067	0.108	<u>0.780</u>	0.255	0.291	0.169	0.224	−0.197	0.325	0.090	0.374	0.244	0.019
Top management commitment and willingness	0.167	0.196	<u>0.791</u>	0.295	0.114	0.097	0.082	0.272	−0.244	−0.034	0.121	0.230	0.015
Proper leadership	0.002	0.057	<u>0.696</u>	0.273	0.162	<u>0.489</u>	0.035	0.213	0.081	0.069	0.160	−0.153	−0.022
Change management program	−0.003	0.239	−0.190	<u>0.664</u>	0.164	0.057	0.068	−0.026	−0.050	0.082	0.099	0.076	0.294

(continued)

Table 7 (continued)

Enabler	Component												
	1	2	3	4	5	6	7	8	9	10	11	12	13
Restructuring data management processes	0.001	0.256	−0.154	<u>0.704</u>	0.238	0.292	0.145	0.109	0.137	0.218	0.164	−0.197	0.145
Reengineering of the processes	0.029	0.106	0.273	<u>0.642</u>	−0.048	0.126	0.098	−0.036	−0.119	−0.056	0.043	−0.075	0.184
Knowledge management systems	0.155	0.215	−0.066	−0.123	<u>0.582</u>	0.078	0.074	0.184	0.035	−0.004	0.304	0.049	0.051
Organizational structure	0.038	0.237	0.049	−0.004	<u>0.925</u>	0.144	0.123	0.118	−0.043	−0.023	−0.047	0.050	0.100
Employee involvement	−0.021	0.070	0.016	0.173	<u>0.642</u>	0.192	−0.051	0.090	−0.197	0.103	0.121	0.186	−0.032
Digitalization of data	−0.030	0.010	0.020	0.089	−0.119	<u>0.770</u>	0.186	−0.018	−0.018	−0.079	−0.080	0.036	0.060
Adequate system capabilities	0.313	0.097	0.027	0.094	0.328	<u>0.538</u>	0.194	0.278	0.033	0.067	−0.109	−0.019	0.144
Appropriate selection of technologies and techniques	0.223	0.296	0.153	0.029	0.140	<u>0.796</u>	0.136	0.088	0.137	−0.022	−0.065	−0.126	0.256
IT infrastructure	0.177	0.094	0.259	0.085	0.094	<u>0.742</u>	0.084	0.106	−0.335	−0.078	0.228	0.285	0.123
Business (artificial) intelligence	0.251	0.112	−0.140	0.025	0.298	<u>0.617</u>	0.080	−0.039	0.036	0.331	0.174	0.255	0.261
Internet of things	0.192	0.185	0.299	−0.114	0.030	<u>0.502</u>	−0.203	0.084	−0.167	0.165	−0.084	0.080	0.064

(continued)

Table 7 (continued)

Enabler	Component												
	1	2	3	4	5	6	7	8	9	10	11	12	13
Mobile technologies	0.277	0.018	0.138	0.020	−0.017	<u>0.651</u>	0.057	−0.255	0.043	0.212	0.023	0.183	0.064
Data capturing and storage	0.180	0.360	0.082	0.261	0.047	0.102	<u>0.763</u>	0.065	−0.082	−0.171	0.074	0.115	0.109
Continuous creation of data	0.023	0.149	0.103	0.001	0.072	−0.030	<u>0.537</u>	0.315	0.304	0.321	0.011	−0.096	−0.195
Scale of data	0.359	0.109	−0.017	0.251	0.319	0.105	<u>0.694</u>	−0.076	−0.065	0.266	0.018	0.080	0.054
Data integration	0.168	0.107	0.232	−0.210	0.255	0.127	<u>0.604</u>	0.041	0.100	0.147	−0.092	−0.090	−0.069
Big data quality management	0.194	−0.148	0.192	0.123	0.128	0.102	0.034	<u>0.576</u>	0.028	0.002	−0.245	0.227	−0.050
Reliability	−0.015	0.260	0.047	−0.077	0.232	0.215	0.052	<u>0.515</u>	0.179	0.045	0.252	0.207	−0.252
Accessibility	0.107	0.199	0.097	0.187	0.283	0.248	0.142	<u>0.671</u>	0.100	0.155	0.030	0.055	0.039
Interpretability of analysis	0.041	0.139	0.072	0.012	0.119	−0.070	0.131	<u>0.857</u>	−0.132	0.163	0.254	0.108	0.149
Quality of information	0.181	<u>0.411</u>	0.151	0.294	0.126	0.282	−0.150	<u>0.739</u>	0.200	0.028	0.258	0.165	−0.257
Speed of data generation	0.219	0.178	−0.017	0.107	0.134	0.232	0.027	−0.033	<u>0.718</u>	0.209	0.053	0.001	0.260
Data transmission time	0.106	0.257	0.031	0.026	0.269	0.222	−0.299	0.183	<u>0.488</u>	−0.039	0.022	0.125	0.103
Data processing time	0.003	0.128	0.158	0.367	0.122	0.166	−0.349	0.059	<u>0.628</u>	0.040	−0.087	−0.124	0.109
Data variety	−0.079	0.170	−0.081	0.150	0.209	0.153	0.393	0.270	0.012	<u>0.640</u>	−0.338	−0.043	0.234

(continued)

Table 7 (continued)

Enabler	Component												
	1	2	3	4	5	6	7	8	9	10	11	12	13
Interoperability	0.072	0.245	−0.028	0.321	−0.286	0.088	0.159	0.194	0.121	<u>0.784</u>	0.003	0.054	0.044
Financial support for big data initiatives	0.122	0.206	0.029	0.034	−0.006	0.145	0.032	0.241	−0.065	0.214	0.854	−0.008	0.057
Desire to reduce costs	0.156	−0.107	0.350	0.027	−0.087	0.006	−0.272	0.014	0.171	0.060	<u>0.788</u>	0.093	0.074
Data security and privacy	−0.086	−0.188	0.132	−0.080	0.038	0.040	−0.023	0.086	0.070	−0.009	0.162	<u>0.897</u>	−0.059
Established culture of collaboration	0.174	−0.070	−0.025	0.280	0.161	0.073	−0.005	0.150	0.328	−0.044	0.241	0.192	<u>0.554</u>
Good working attitude	0.294	0.168	0.283	0.195	0.259	−0.046	0.158	0.177	−0.118	0.255	0.055	0.063	<u>0.693</u>
Reluctance to simplify	0.055	0.087	−0.029	0.391	0.124	−0.022	0.241	0.243	−0.093	0.023	−0.005	0.259	<u>0.696</u>
Commitment to resilience	0.298	0.115	0.109	0.369	0.149	0.104	−0.034	0.043	0.108	0.123	0.178	0.067	<u>0.639</u>

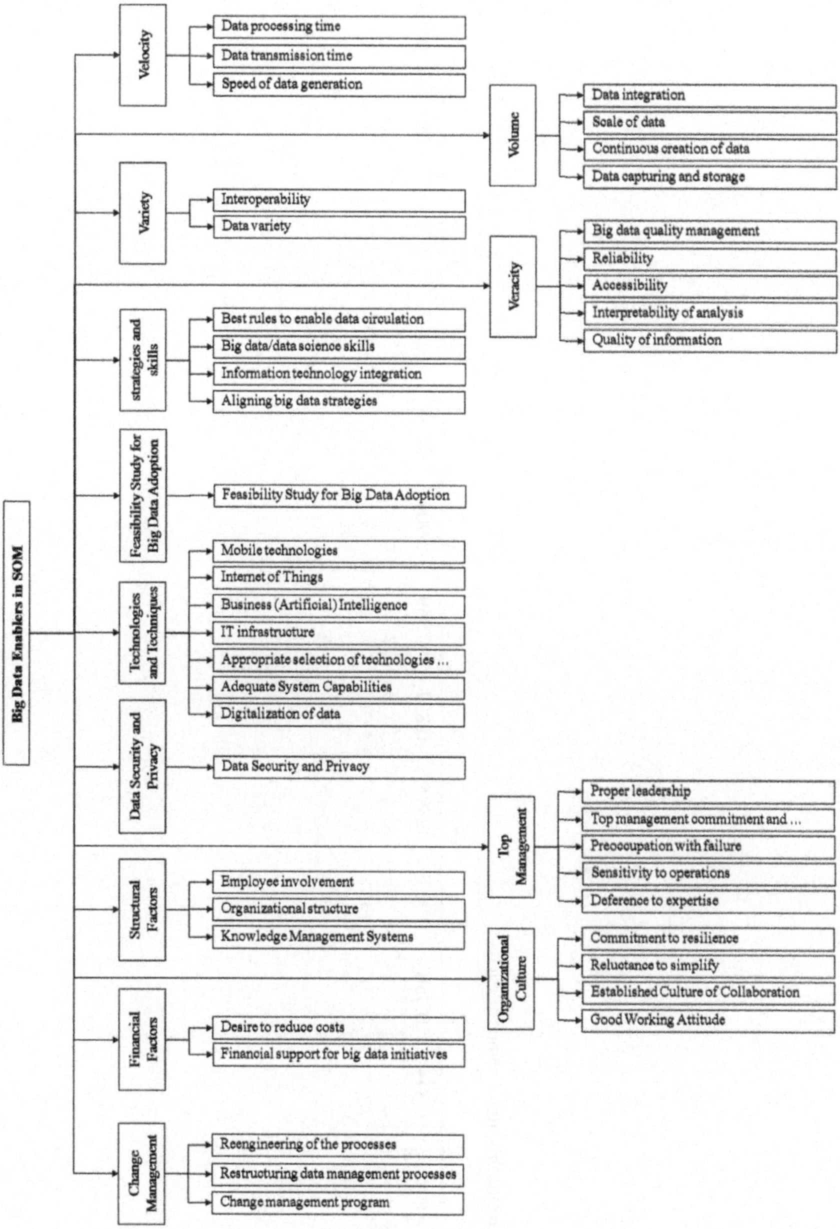

Fig. 2 Hierarchical structure of enablers for big data in SOM

Table 8 Cronbach's alpha value for each

Enabler	1	2	3	4	5	6	7	8	9	10	11	12	13
Cronbach's alpha	0.844	0.956	0.923	0.895	0.877	0.939	0.904	0.798	0.947	0.916	0.928	0.964	0.938
N of Items	4	1	5	3	3	7	4	5	3	2	2	1	4

The ISM methodology has been extensively utilized in various areas of SOM by researchers. Talib et al. [69], proposed an interpretive structural modeling approach for modeling the practices of total quality management in the service sector. In implementing cloud-based medical systems in hospitals, Chen and Chen [12] used interpretive structural modeling to identify the relationship between the resistance factors encountered during the organization innovation process. In another research, a Total interpretive structural modeling of strategic performance management for Indian telecom service providers was proposed [76]. Evaluating the Indian retail service quality enablers using interpretive structural modeling was the purpose of a research [26]. The purpose of a research was to develop relationship amongst the identified clusters of critical success factors of Six Sigma for non-formal service sectors using interpretive structural modeling [68]. Various steps leading to the development of the big data ISM model in SOM are discussed below.

4.1 Identification of Enablers for Big Data in SOM

Thirteen enablers for the success of big data initiatives in SOM have been identified using factor analysis after literature review and discussions with domain experts. These are: (1) strategies and skills, (2) feasibility study for big data adoption, (3) top management, (4) change management, (5) structural factors, (6) technologies and techniques, (7) volume, (8) veracity, (9) velocity, (10) variety, (11) financial factors, (12) data security and privacy and, (13) organizational culture. The detailed descriptions of these enablers have been presented in previous section.

4.2 Structural Self-Interaction Matrix (SSIM)

Once the enablers were identified, it was necessary to determine contextual relationships of 'leads to' or 'influences' between the enablers to develop the SSIM. This means that one enabler leads to or influences another. In total twelve experts, including three professionals with diverse service industry experience and five highly proficient academics with mixed experience of teaching, researching, and advising firms on service operations related matters, were chosen to provide their expert views. While developing the SSIM (Table 9), the following notations have been used:

V—Enabler i help achieves or influences enabler j
A—Enabler j helps achieve or influences enabler i
X—Enablers i and j helps achieve or influence each other
O—Enabler i and j are not related

For instance, based on expert opinion, strategies and skill (1) and feasibility study for big data adoption (2) of analysis influence each other, hence 'X' has been put in the cell (1, 2). Furthermore, top management (3) will influence strategies and skill

Table 9 Structural self-interaction matrix for big data enablers

E. no	.Enabler	13	12	11	10	9	8	7	6	5	4	3	2
1	Strategies and skills	O	V	A	V	O	V	V	V	O	A	A	X
2	Feasibility study for big data adoption	A	O	A	O	O	O	V	V	A	A	A	
3	Top management	V	O	V	O	O	V	V	O	V	V		
4	Change management	X	V	O	O	O	V	O	V	X			
5	Structural factors	X	V	O	O	V	O	V	V				
6	Technologies and techniques	O	V	A	V	V	V	V					
7	Volume	O	V	A	X	X	V						
8	Veracity	O	V	A	A	O							
9	Velocity	O	V	A	X								
10	Variety	O	O	O									
11	Financial factors	O	V										
12	Data security and privacy	A											

(1), thus 'A' has seemed in the cell (1, 3); while 'V' in the cell (1, 6) implies that strategies and skill (1) will affect technologies and techniques (6). Ultimately, the 'O' in the cell (1, 5) indicates that the enablers strategies and skill (1) and structural factors (10) are unrelated.

4.3 *Reachability Matrix from the SSIM*

The reachability matrix is derived from the SSIM. It contains the relationship among the factors in binary form. The SSIM is transformed into the reachability matrix using the following rules;

If the (i, j) entry in the SSIM is V, then the (i, j) entry in the reachability matrix is assigned 1 and the (j, i) entry is assigned 0.
If the (i, j) entry in the SSIM is A, then the (i, j) entry in the reachability matrix is assigned 0 and the (j, i) entry is assigned 1.
If the (i, j) entry in the SSIM is X, then both (i, j) and (j, i) entry in the reachability matrix are assigned 1.
If the (i, j) entry in the SSIM is O, then the (i, j) entry in the reachability matrix is assigned 0 and the (j, i) entry is assigned 0.

An initial reachability matrix (Table 10) is obtained by applying the above rules to the SSIM. From this matrix, a final reachability matrix (Table 11) is constructed taking into account the transitivity rule, which states that if a variable 'A' is related to 'B' and 'B' is related to 'C', then 'A' is necessarily related to 'C'.

Table 10 Initial reachability matrix

E. no.	Enabler	1	2	3	4	5	6	7	8	9	10	11	12	13
1	Strategies and skills	1	1	0	0	0	1	1	1	0	1	0	1	0
2	Feasibility study for big data adoption	1	1	0	0	0	1	1	0	0	0	0	0	0
3	Top management	1	1	1	1	1	0	1	1	0	0	1	0	1
4	Change management	1	1	0	1	1	1	0	1	0	0	0	1	1
5	Structural factors	0	1	0	1	1	1	1	0	1	0	0	1	1
6	Technologies and techniques	0	0	0	0	0	1	1	1	1	1	0	1	0
7	Volume	0	0	0	0	0	0	1	1	1	1	0	1	0
8	Veracity	0	0	0	0	0	0	0	1	0	0	0	1	0
9	Velocity	0	0	0	0	0	0	1	0	1	1	0	1	0
10	Variety	0	0	0	0	0	0	1	1	1	1	0	0	0
11	Financial factors	1	1	0	0	0	1	1	1	1	0	1	1	0
12	Data security and privacy	0	0	0	0	0	0	0	0	0	0	0	1	0
13	Organizational culture	0	1	0	1	1	0	0	0	0	0	0	1	1

4.4 Partitioning the Reachability Matrix in Different Levels

The reachability matrix is divided into different levels through successive iterations. For this, each enabler's reachability set and antecedent set are found. An enabler's reachability set consists of itself and all the other enablers that are influenced by it, whereas an enabler's antecedent set consists of itself and all the enablers that influence it. For all enablers, the intersection of these sets is derived. The enablers, for whom the reachability set and intersection set are the same, are assigned the highest level in the hierarchy. In Table 12, Data security and privacy (12) has the same reachability and intersection sets. Hence it will occupy the topmost level in the ISM hierarchy. Once an individual's hierarchy or group of enablers is set, they are not considered for analysis in subsequent iterations. Using this procedure, the hierarchy of each enabler has been set. The successive iterations (2–7) are presented in Table 13.

4.5 Developing ISM Model

The ISM model presented in Fig. 3 is constructed using the final reachability matrix (Table 11) and the enabler's hierarchical level shown in Tables 12 and 13. The model shows that top management (3) is undoubtedly the most driving enabler for the success of big data in SOM.

Table 11 Final reachability matrix

E. No	Enabler	1	2	3	4	5	6	7	8	9	10	11	12	13	Driving power
1	Strategies and skills	1	1	0	0	0	1	1	1	1[a]	1	0	1	0	9
2	Feasibility study for big data adoption	1	1	0	0	0	1	1	1[a]	0	1[a]	0	1[a]	0	7
3	Top management	1	1	1	1	1	1[a]	1	1	1[a]	1[a]	1	1[a]	1	13
4	Change management	1	1	0	1	1	1	1[a]	1	1[a]	1[a]	0	1	1	11
5	Structural factors	1[a]	1	0	1	1	1	1	1[a]	1	1[a]	0	1	1	11
6	Technologies and techniques	0	0	0	0	0	1	1	1	1	1	0	1	0	6
7	Volume	0	0	0	0	0	0	1	1	1	1	0	1	0	5
8	Veracity	0	0	0	0	0	0	0	1	0	0	0	1	0	2
9	Velocity	0	0	0	0	0	0	1	1[a]	1	1	0	1	0	5
10	Variety	0	0	0	0	0	0	1	1	1	1	0	1[a]	0	5
11	Financial factors	1	1	0	0	0	1	1	1	1	1[a]	1	1	0	9
12	Data security and privacy	0	0	0	0	0	0	0	0	0	0	0	1	0	1
13	Organizational culture	1[a]	1	0	1	1	1[a]	1[a]	1[a]	1[a]	0	0	1	1	10
	Dependence	7	7	1	5	4	8	11	12	10	10	2	13	4	94

[a] Values which are changed due to transitivity

Table 12 Iteration 1

Enabler	Reachability set	Antecedent set	Intersection set	Level
1	1, 2, 6, 7, 8, 9, 10, 12	1, 2, 3, 4, 5, 11, 13	1, 2	
2	1, 2, 6, 7, 8, 10, 12	1, 2, 3, 4, 5, 11, 13	1, 2	
3	1, 2, 3, 4, 5, 6, 7, 8, 9, 10, 11, 12, 13	3	3	
4	1, 2, 4, 5, 6, 7, 8, 9, 10, 12, 13	3, 4, 5, 13	4, 5, 13	
5	1, 2, 4, 5, 6, 7, 8, 9, 10, 12, 13	3, 4, 5, 13	4, 5, 13	
6	6, 7, 8, 9, 10, 12	1, 2, 3, 4, 5, 6, 11, 13	6	
7	7, 8, 9, 10, 12	1, 2, 3, 4, 5, 6, 7, 9, 10, 11, 13	7, 9, 10	
8	8, 12	1, 2, 3, 4, 5, 6, 7, 9, 10, 11, 13	8	
9	7, 8, 9, 10, 12	1, 3, 4, 5, 6, 7, 9, 10, 11, 13	7, 9, 10	
10	7, 8, 9, 10, 12	1, 2, 3, 4, 5, 6, 7, 9, 10, 11	7, 9, 10	
11	1, 2, 6, 7, 8, 9, 10, 11, 12	3, 11	11	
12	12	1, 2, 3, 4, 5, 6, 7, 8, 9, 10, 11, 12, 13	12	I
13	1, 2, 4, 5, 6, 7, 8, 9, 12, 13	3, 4, 5, 13	4, 5, 13	

Table 13 Iteration 2

Enabler	Reachability set	Antecedent set	Intersection set	Level
1	1, 2	1, 2, 3, 4, 5, 11	1, 2	V
2	1, 2	1, 2, 3, 4, 5, 11, 13	1, 2	V
3	3	3	3	VII
4	4, 5, 13	3, 4, 5, 13	4, 5, 13	VI
5	4, 5, 13	3, 4, 5, 13	4, 5, 13	VI
6	6	1, 2, 3, 4, 5, 6, 11, 13	6	IV
7	7, 9, 10	1, 2, 3, 4, 5, 6, 7, 9, 10, 11, 13	7, 9, 10	III
8	8	1, 2, 3, 4, 5, 6, 7, 9, 10, 11, 13	8	II
9	7, 9, 10	1, 3, 4, 5, 6, 7, 9, 10, 11, 13	7, 9, 10	III
10	7, 9, 10	1, 2, 3, 4, 5, 6, 7, 9, 10, 11	7, 9, 10	III
11	11	3, 11	11	VI
13	4, 5, 13	3, 4, 5, 13	4, 5, 13	VI

4.6 MICMAC Analysis

MICMAC analysis is used to classify enablers according to their driving power and dependence. The enablers are classified into four clusters, namely autonomous, dependent, linked and independent, as shown in Fig. 4. The grouping of the variable in this cluster is as below.

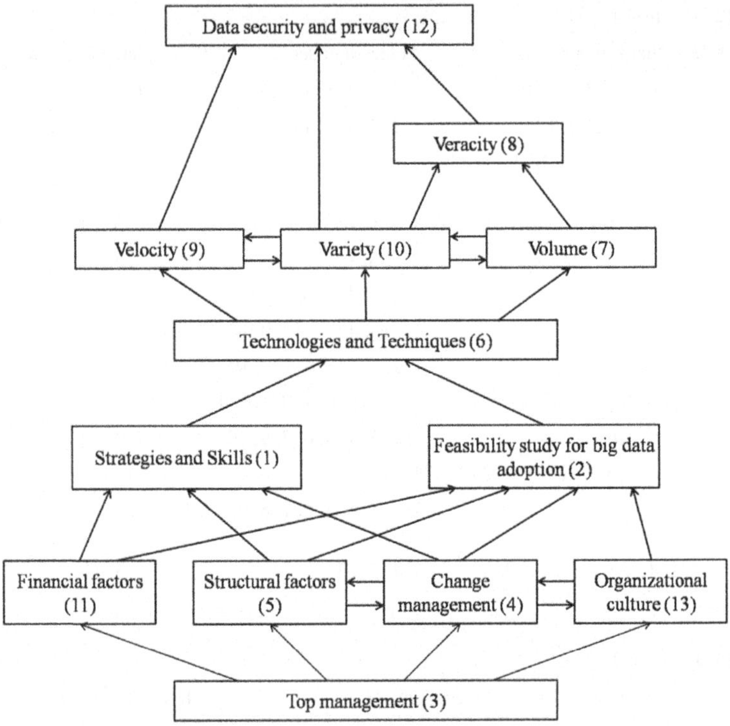

Fig. 3 ISM model of big data enablers in SOM

I. Autonomous variables: These variables have weak dependence as well as weak driving power. They are relatively isolated from the system. There are no variables in this cluster.
II. Dependent variables: The variables in this cluster have weak driving power but have strong dependence on other factors to drive them. Enablers 1, 2, 3, 4, 8, 13, 14 form this cluster.
III. Linkage variables: The variables in this cluster have high dependence as well as high driving power. There are no variables in this cluster.
IV. Independent variable: The variables in this cluster have strong drive power and weak dependence. Enablers 5, 6, 7, 9, 10, 11, 12 fall in this cluster.

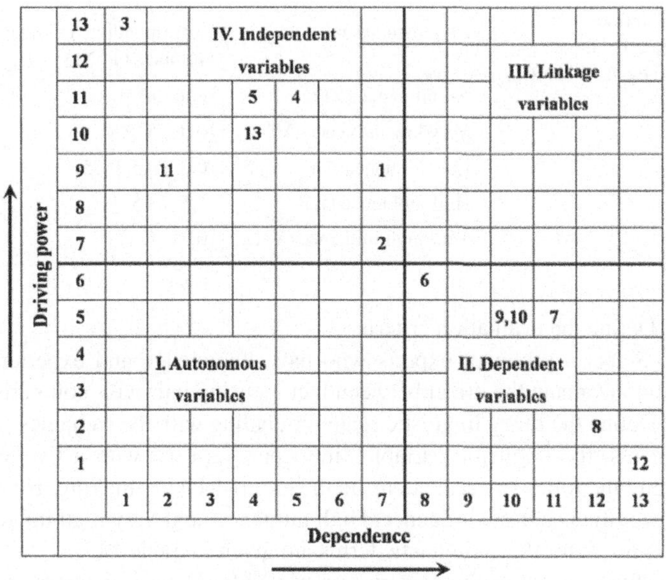

Fig. 4 MICMAC diagram of big data enablers in SOM

5 Modeling Enablers for Big Data in SOM Using Fuzzy-DEMATEL

5.1 Fuzzy-DEMATEL Method

The DEMATEL method was introduced between 1972 and 1976 by Battelle Memorial Institute of Geneva. DEMATEL is recognized as one of the best practical tools and a comprehensive method for building and analyzing a structural model consisting of causal relationships between complex factors [38]. It can discern the enabler's cause-effect relationship when modeling a problem [9]. The technique is based on the theory of graphs that facilitates the analysis by visualization of complex problems [39]. Many evaluation criteria are certainly imperfect and probably uncertain factors, so it can be helpful to apply fuzzy theory to the DEMATEL method for problem-solving.

The fuzzy DEMATEL technique has been extensively applied in many researches, including sustainable project management in construction [34], evaluating the enablers in solar power developments [40], analyzing supplier development program enablers [40], assessment of critical enablers for flexible supply chain performance measurement system [70], a framework of enablers of strategic alignment in supply chain [59], and analyzing environment sustainability enablers [27] among others. The fuzzy DEMATEL method can be applied using the following basic steps [2, 34, 61].

Table 14 The fuzzy linguistic scale for the respondents' evaluations [61]

Linguistic terms	Corresponding Triangular Fuzzy Numbers (TFNs)
No influence (NO)	(0, 0, 0.25)
Very low influence (VL)	(0, 0.25, 0.5)
Low influence (L)	(0.25, 0.5, 0.75)
High influence (H)	(0.5, 0.75, 1)
Very high influence (VH)	(0.75, 1, 1)

Step 1. Define the evaluation criteria.
Step 2. Select a group of experts who have knowledge and experience about problem to evaluate the effect between factors using pair-wise comparison.
Step 3. Define the fuzzy linguistic scale for dealing with the vagueness of human assessments, the linguistic variable "influence" is used with a five-level scale containing the following scale items in the group decision-making: No influence, Very low influence, Low influence, High influence, and Very high influence. The fuzzy numbers for these linguistic terms are given in Table 14.
Step 4. Obtain an initial direct relation matrix with pair wise comparison. Develop the initial fuzzy direct-relation matrix Z^k by having evaluators introduce the fuzzy pair-wise influence relationships between the components in a $n \times n$ matrix where k is the number of experts. Accordingly, the direct-relation matrix is established as $Z^k = \left[Z_{ij}^k\right]$ where Z is a $n \times n$ non-negative matrix; Z_{ij} represents the direct impact of factor i on factor j; and, when $i = j$, the diagonal elements $Z_{ij} = 0$.

For simplicity, denote Z^k as

$$Z_{ij}^k = \left(l_{ij}, m_{ij}, u_{ij}\right)$$

$$Z^k = \begin{matrix} C_1 \\ C_2 \\ \vdots \\ C_n \end{matrix} \begin{bmatrix} [0,0] & \otimes z_{12}^k & \ldots & \otimes z_{1n}^k \\ \otimes z_{21}^k & [0,0] & \ldots & \otimes_{2n}^k \\ \vdots & \vdots & \ddots & \vdots \\ \otimes z_{n1}^k & \otimes z_{n2}^k & \ldots & [0,0] \end{bmatrix} \tag{1}$$

Step 5. Obtain the normalized fuzzy direct-relation matrix "D" using Expression (2) related to the overall fuzzy direct-relation matrix Z.

$$D = \frac{Z^k}{\max\limits_{1 \le i \le n} \sum_{j=1}^{n} z_{ij}}, i, j = 1, 2, \ldots, n \tag{2}$$

Step 6. Compute the total-relation matrix T using Expression (3), where $n \times n$ identity matrix is represented with I. Upper and lower values are calculated separately.

$$T = D(I - D)^{-1} \tag{3}$$

where $T = D + D^2 + \cdots + = \sum_{j=1}^{\infty} D^i$.

Step 7. Determine row (r_i) and column (c_j) sums for each row i and column j from the T matrix, respectively, with following equations.

$$T = \left[t_{ij}\right]_{n \times n} \quad i, j = 1, 2, \ldots, n \tag{4}$$

$$r_i = \sum_{1 \leq j \leq n}^{n} t_{ij} \quad \forall i \tag{5}$$

$$c_j = \sum_{1 \leq i \leq n}^{n} t_{ij} \quad \forall j \tag{6}$$

Step 8. The causal diagram is built with the horizontal axis $(r_i + c_j)$ and the vertical axis $(r_i - c_j)$. The horizontal axis "Prominence" refers the importance degree of the factor, whereas the vertical axis "Relation" shows the extent of the influence. If the $(r_i - c_j)$ axis is positive, the factor is in the cause group. Otherwise, if the $(r_i - c_j)$ axis is negative; the factor is in the group of effects. Causal diagrams can convert complex relationships of factors into an easy-to-understand structural model, providing awareness for problem solving.

5.2 Modeling Enablers

The section uses fuzzy DEMATEL to classify enablers into cause and effect groups. This research first designed a questionnaire for fuzzy DEMATEL composed of two parts in accordance with the procedural steps described in Sect. 5.1. The first part outlines the definition of each enabler for easy understanding and response. The second part is a pair-wise comparison to evaluate each enabler's influence. Then, respondents were asked to compare the importance of each criterion using the following linguistics variables: "no importance", "very low importance", "low importance", "high importance", and "very high importance" (Table 15). The average linguistic scores of the expert's opinions were shown in Table 16. Based on Table 16, the normalized fuzzy direct-relation matrix "D" (Table 17) was computed using Expression 2. Then, Total-relation fuzzy matrix T (Table 18) was constructed from the normalized fuzzy direct-relation matrix using Expression 3. Next to this, the structural model was established. After having built matrix T, $r_i + c_j$ and $r_i - c_j$ were determined. In Expressions (4)–(6), r_i and c_j are obtained with the sum of the rows and columns of matrix T. While $r_i + c_j$ donates the importance of enabler i, $r_i - c_j$ indicates the net effect of enabler i. Results were presented in Table 19.

Table 15 Linguistic assessment of the evaluators' opinion (average)

E. no.	1	2	3	4	5	6	7	8	9	10	11	12	13
1	NO	VL	L	L	H	VH	L	L	H	L	L	VH	VL
2	L	NO	NO	NO	NO	H	L	H	L	L	NO	H	NO
3	VH	H	NO	H	H	H	VL	H	H	L	VH	L	VH
4	VH	H	NO	NO	H	H	L	L	H	H	VL	H	H
5	VH	H	L	H	NO	VH	L	VL	L	L	NO	VH	L
6	NO	NO	NO	VL	H	NO	VH	VH	VH	VH	NO	VH	NO
7	H	NO	NO	VL	H	NO	NO	H	H	VH	NO	VH	NO
8	NO	NO	NO	NO	L	L	NO	NO	H	L	NO	H	NO
9	NO	NO	NO	NO	NO	L	L	NO	NO	L	NO	VH	NO
10	NO	NO	NO	L	NO	L	H	H	L	NO	NO	H	NO
11	VH	VH	H	H	L	VH	L	H	H	L	NO	VH	H
12	H	NO	NO	L	VL	H	H	VH	H	VH	VL	NO	NO
13	H	VL	VH	VH	H	VL	NO	VL	VL	VL	H	NO	NO

Table 16 Average linguistic scores of the expert's opinions

E. no.	1	2	3		12	13
1	(0 0 0.25)	(0 0.25 0.5)	(0.25 0.5 0.75)	...	(0.75 1 1)	(0 0.25 0.5)
2	(0.25 0.5 0.75)	(0 0 0.25)	(0 0 0.25)	...	(0.5 0.75 1)	(0 0 0.25)
3	(0.75 1 1)	(0.5 0.75 1)	(0 0 0.25)	...	(0.25 0.5 0.75)	(0.75 1 1)
. . .	. . .	. . .	. . .	. . .	. . .	. . .
12	(0.5 0.75 1)	(0 0 0.25)	(0 0 0.25)	...	(0 0 0.25)	(0 0 0.25)
13	(0.5 0.75 1)	(0 0.25 0.5)	(0.75 1 1)	...	(0 0 0.25)	(0 0 0.25)

Table 17 Normalized initial direct-relation fuzzy matrix

E. no.	1	2	3		12	13
1	(0 0 0.02)	(0 0.02 0.04)	(0.02 0.04 0.07)	...	(0.07 0.09 0.09)	(0 0.02 0.04)
2	(0.02 0.04 0.07)	(0 0 0.02)	(0 0 0.02)	...	(0.04 0.07 0.09)	(0 0 0.02)
3	(0.07 0.09 0.09)	(0.04 0.07 0.09)	(0 0 0.02)	...	(0.02 0.04 0.07)	(0.07 0.09 0.09)
. . .	. . .	. . .	. . .	. . .	. . .	. . .
12	(0.04 0.07 0.09)	(0 0 0.02)	(0 0 0.02)	...	(0 0 0.02)	(0 0 0.02)
13	(0.04 0.07 0.09)	(0 0.02 0.04)	(0.07 0.09 0.09)	...	(0 0 0.02)	(0 0 0.02)

Table 18 Total-relation fuzzy matrix

E. no.	1	2	3		12	13
1	(0.015 0.049 0.207)	(0.006 0.046 0.180)	(0.025 0.058 0.180)	…	(0.089 0.164 0.331)	(0.005 0.042 0.170)
2	(0.026 0.061 0.195)	(0.001 0.006 0.115)	(0.001 0.005 0.104)	…	(0.057 0.109 0.267)	(0.000 0.005 0.109)
3	(0.086 0.147 0.298)	(0.054 0.099 0.244)	(0.011 0.029 0.160)	…	(0.056 0.142 0.347)	(0.073 0.113 0.232)
.	.	.	.	.	.	.
.	.	.	.	.	.	.
.	.	.	.	.	.	.
12	(0.050 0.094 0.241)	(0.002 0.015 0.137)	(0.002 0.010 0.122)	…	(0.023 0.069 0.241)	(0.002 0.013 0.130)
13	(0.062 0.114 0.255)	(0.012 0.055 0.174)	(0.071 0.105 0.195)	…	(0.020 0.071 0.244)	(0.011 0.030 0.144)

Table 19 Fuzzy values of r_i, c_j, $r_i + c_j$ and $r_i - c_j$

E. no.	r_i	c_j	$r_i + c_j$	$r_i - c_j$
1	(0.511 1.202 3.229)	(0.567 1.066 2.869)	(1.078 2.267 6.098)	(−0.056 0.136 0.360)
2	(0.301 0.668 2.379)	(0.256 0.557 2.116)	(0.557 1.226 4.495)	(0.045 0.111 0.263)
3	(0.733 1.514 3.720)	(0.200 0.408 1.820)	(0.933 1.923 5.540)	(0.533 1.106 1.899)
4	(0.583 1.246 3.398)	(0.359 0.855 2.670)	(0.942 2.102 6.067)	(0.224 0.391 0.728)
5	(0.572 1.238 3.222)	(0.428 0.955 2.927)	(1.000 2.194 6.149)	(0.144 0.283 0.295)
6	(0.503 0.937 2.551)	(0.630 1.314 3.379)	(1.133 2.251 5.929)	(−0.127 −0.377 −0.828)
7	(0.423 0.856 2.592)	(0.442 1.056 3.001)	(0.865 1.912 5.593)	(−0.019 −0.199 −0.409)
8	(0.212 0.481 2.010)	(0.586 1.263 3.296)	(0.797 1.744 5.306)	(−0.374 −0.782 −1.286)
9	(0.188 0.415 1.800)	(0.647 1.396 3.663)	(0.835 1.811 5.463)	(−0.458 −0.981 −1.862)
10	(0.387 0.813 2.545)	(0.597 1.341 3.369)	(0.984 2.154 5.914)	(−0.211 −0.528 −0.824)
11	(0.782 1.566 3.785)	(0.270 0.610 2.100)	(1.051 2.177 5.885)	(0.512 0.956 1.685)
12	(0.448 0.972 2.825)	(0.864 1.662 3.777)	(1.312 2.634 6.602)	(−0.416 −0.690 −0.953)
13	(0.428 1.052 2.898)	(0.224 0.478 1.966)	(0.651 1.530 4.864)	(0.204 0.573 0.932)

Table 20 Crisp values for The direct and indirect influences (r_i, c_j, $r_i + c_j$ and $r_i - c_j$)

E. No	Enabler	r_i	c_j	$r_i + c_j$	$r_i - c_j$
1	Strategies and skills	3.651	3.500	7.15	0.15
2	Feasibility study for big data adoption	2.703	2.430	5.13	0.27
3	Top management	4.146	2.043	6.19	2.10
4	Change management	3.747	3.051	6.80	0.70
5	Structural factors	3.740	3.253	6.99	0.49
6	Technologies and techniques	3.277	3.851	7.13	−0.57
7	Volume	3.103	3.402	6.50	−0.30
8	Veracity	2.249	3.762	6.01	−1.51
9	Velocity	2.054	3.938	5.99	−1.88
10	Variety	3.008	3.866	6.87	−0.86
11	Financial factors	4.230	2.530	6.76	1.70
12	Data security and privacy	3.300	4.385	7.68	−1.09
13	Organizational culture	3.400	2.232	5.63	1.17

Finally, Using Converting Fuzzy data into Crisp Scores (CFCS) defuzzification technique [50], as a popular defuzzification technique, $r_i + c_j$ and $r_i - c_j$ were defuzzified and obtained Best Non-fuzzy Performance (BNP) values. In the CFCS process, the defuzzified factor is represented by the geometric center of the area limited by the curve that represents its membership function. Crisp values of r_i, c_j, $r_i + c_j$ and $r_i - c_j$ are shown in Table 20.

The cause-effect diagram that defines the complicated causal relationships among various enablers was constructed based upon the values of prominence ($r_i + c_j$) on x axis and relation ($r_i - c_j$) on y axis While refer to the strength of influence among criteria and the influence relation among criteria respectively and is shown in Fig. 5.

6 Discussion

The corporate environment is moving rapidly towards data-driven service operations and big data can greatly serve this purpose. Therefore, the identification of enablers for the integration of big data in SOM assumes significant proportions. This can greatly assist decision-makers in prioritizing key enablers. The chapter responds to this call to undertake more research work to facilitate the integration of big data in SOM. The study investigates the prospects of developing big data integration frameworks with SOM based on well-established methodologies. The primary focus of this chapter is to establish a hierarchy of enablers along with classifying them according to their driving power and dependence to provide decision-makers with a clear framework for focusing their efforts on a few key enablers rather than all enablers. The study is unique in its approach as it enlists the most prominent enablers

Fig. 5 The causal diagram of enablers for big data in SOM

of big data in SOM and categorizes them into four categories viz., autonomous, driving, linkage and dependent variables based on their interrelationships. Successful integration of big data with SOM requires focusing on many key areas related to management systems, technical expertise, technological resources, organizational culture, etc. The hierarchical leveling and inter-relationships between these enablers have been established using ISM. Furthermore, the Fuzzy-DEMATEL approach is used to classify the enablers into groups of causes and effects. The MCDM models have helped to prioritize the big data enablers according to their relative importance to the system.

In this study, forty-four initial enablers are used for the present work based on literature review and expert discussions with industry practitioners. The SPSS software was used to facilitate analysis, to reduce items, and to examine the interrelations between variables. Finally, 44 initial enablers are classified into 13 enablers and used for further analysis and modeling.

The ISM model (Fig. 1) and MICMAC diagram (Fig. 2) classifies various enablers of big data in SOM according to their hierarchical levels and driving power/dependency. From the MICMAC analysis, it can be seen that the autonomous cluster has no variables, which means that no variable can be assumed to be isolated from the entire structure and the management has to pay attention to all the identified big data enablers. The next cluster contains dependent variables which have high dependence and weak driving power. The variables in this cluster are data security and privacy (12), veracity (8), volume (7), velocity (9), variety (10) and, technologies and techniques (6). These variables though are critical to big data success but their high dependence means that they need other enablers to drive them. The next cluster represents linkage variables. There are no variables for this cluster in the developed model. The last cluster contains independent variables namely, top management (3), structural factors (5), change management (4), organizational culture (13), financial factors (11), strategies and skills (1) and, feasibility study for big data adoption (2). These enablers have high driving power but very weak dependence. These enablers play a prominent role in ensuring that big data is successfully adopted across the entire SOM and its true value is realized.

Based on the ISM digraph model and MICMAC analysis, it can be seen that the factor top management (3) has highest driving power and lowest dependence plays a vital role in driving the entire big data enablers in SOM. It can be further stated that the absence of this factor may not drive other big data enablers in SOM. The enabler's priority can be decided based on their driving power and dependence. This will help to push the organizational efforts in SOM for the success of big data.

According to Table 20 the enabler data security and privacy (12) has highest $(r_i + c_j)$ value of 7.685 followed by enablers 1 > 6 > 5 > 10 > 4 > 11 > 7 > 3 > 8 > 9 > 13 > 2. Considering the values of their respective $(r_i - c_j)$ scores the enablers namely top management (3), financial factors (11), organizational culture (13), change management (4), structural factors (5), feasibility study for big data adoption (2) and, strategies and skills (1) are categorized in cause group while the enablers volume (7), technologies and techniques (6), variety (10), data security and privacy (12), veracity (8) and, velocity (9) occupy the effects group.

The factors in the cause group hold great importance as they can significantly influence the whole system. Thus criteria in the cause group should be given more attention. Among all the enablers in the cause group, top management has the highest $(r_i - c_j)$ score of 2.101, which signifies its high importance in the context of successful integration of big data in SOM. The enabler financial factors (11) has second-highest $(r_i - c_j)$ score 1.701 while being on the middle side in terms of $(r_i + c_j)$ score due to its high influence on other enablers and being less influenced by them. Also, its influence given index (r_i) is amongst the top three enablers. Organizational culture (13), structural factors (11), feasibility study for big data adoption (2), and strategies and skills (1) with $(r_i - c_j)$ scores of 1.169, 0.490, 0.275, and 0.150 respectively are the other two enablers in the cause group. The low $(r_i - c_j)$ scores of these enablers can be justified from the fact that each of these four enablers have almost identical influence given and influence received scores.

Enablers belonging to the effect group tend to be easily influenced by other enablers. Among all the enablers in the effect group, velocity (9) has the least $(r_i - c_j)$ score of -1.882. However, its influence received index (c_j) of 3.938 are the second-highest amongst all enablers. It signifies that this enabler is most influenced by other enablers but is still highly important to the success of big data in SOM. The enabler veracity (8) is second in the priority list in the effect group with $(r_i - c_j)$ score of -1.502. It has the fourth least $(r_i + c_j)$ score of 6.022 which can be attributed to its influence given (r_i) score of 2.260 which is second least among all enablers. The remaining enablers in the order of their priority in the effect group are data security and privacy (12), variety (10), technologies and techniques (6), and volume (7).

The two different techniques show that basically seven factors viz. top management (3), structural factors (5), change management (4), organizational culture (13), financial factors (11), strategies and skills (1) and, feasibility study for big data adoption (2) are commonly identified as the most influential/driving enablers. The findings of the seven most influential big data enablers can be further analyzed from the positioning of these in MICMAC and causal relationship as shown in the Fig. 4 and Fig. 5 respectively. Moreover, it is a well-known fact that without the top management (3) contributions big data for SOM cannot be successfully implemented. In addition to this important enabler, an organization's thinking for big data initiatives has to be restructured, therefore, structural factors (5) has emerged as another driving enablers, and this can only happen once the entire management including all stakeholders agree to accept the big data based change initiatives which brings out change management (4) as another influential enabler for big data in SOM. Undoubtedly, these enablers alone cannot work successfully without the proper cultural and financial support. This has been revealed very clearly from the analysis that financial factors (11) and big organizational culture (13) must be considered parallel besides management support and structure of the organization.

Similarly, on the other hand, four enablers viz. velocity (9), veracity (8), data security and privacy (12), and variety (10) have been commonly identified across two different modeling techniques as the most dependent big data enablers for SOM. Therefore, the common driving big data enablers should be considered and common dependent enablers should be dropped from initial consideration for big data implementation in SOM. For efficient implementation of big data in SOM, top management should focus primarily on driving enablers that will reinforce big data initiatives as well as drive other enablers.

7 Conclusions and Future Direction

Big data use has steadily gained prominence in various SOM domains due to fierce market competition. The use of big data in SOM has therefore propagated from theory to industrial practice. The study findings can, therefore, serve as a decision-making tool for top management and SOM practitioners. Two MCDM models were developed in this study to analyze the interactions between the big data enablers. The MICMAC matrices provide some valuable insights into the relative importance and interdependencies among the big data enablers. The theoretical frameworks developed in this study can greatly assist decision-makers in visualizing big data enablers through a system approach, shortening a complex system of big data enablers into a structured format. Furthermore, SOM practitioners can prioritize common driving enablers based on the need for implementation based on their driving powers and dependence. This study will also assist managers in identifying highly sensitive variables with high driving power as well as dependence that require ongoing supervision to effectively and thoroughly implement big data in SOM.

The MCDM models proposed in this work to identify big data enablers can provide a more realistic assessment of potential enablers for the successful adoption of big data in SOM. This chapter contributes significantly to the establishment of inter-relationships among various enablers of big data success. The usefulness of the proposed models in imposing hierarchy and direction on the inter-relationships between different enablers of big data can be tremendously valuable to decision-makers. Top management must emphasize enablers with strong driving power for the efficient implementation of big data in SOM. Practitioners working in the area of big data may also drive useful understanding from the chapter's presented hierarchical model of big data enablers. The models presented serve as a platform for establishing different inter-relationships using other qualitative approaches. The work reported here can also be mapped in a similar way to establish big data inhibitors in SOM. In the future, big data initiatives can also be linked further to the SOM's specific domain such as demand management, risk management, customer relationship management, and so on.

The authors' search through the literature wasn't systematic, therefore, these 44 enablers aren't a comprehensive list and there can be other enablers that the authors have not considered. Also, the present study employs ISM and Fuzzy-DEMATEL methods to model the enablers in SOM for the success of big data. The models developed were not statistically validated. To alleviate these limitations, a structured equation modeling (SEM) analysis can be performed based on a sufficient number of responses collected using a structured questionnaire. More qualitative techniques like analytical hierarchy process [23], analytical network process [17], and best–worst method [55] can be used to determine and rank the importance (weight) of enablers to focus on the critical few enablers.

References

1. Akerkar, R. (2014). Analytics on big aviation data: Turning data into insights. *International Journal of Computer Science and Applications, 11*(3), 116–127.
2. Akyuz, E., & Celik, E. (2015). A fuzzy DEMATEL method to evaluate critical operational hazards during gas freeing process in crude oil tankers. *Journal of Loss Prevention in the Process Industries, 38*, 243–253.
3. Almodovar, M. C. (2015). Information management and hospital enterprise information systems. In P. R. Barach, J. P. Jacobs, S. E. Lipshultz, & P. C. Laussen (Eds.), *Pediatric and congenital cardiac care* (p. 458). Springer.
4. Anand, A., & Bansal, G. (2017). Interpretive structural modelling for attributes of software quality. *Journal of Advances in Management Research, 14*(3), 256–269.
5. Auschitzky, E., Hammer, M., & Rajagopaul, A. (2014). *How big data can improve manufacturing*. Retrieved from http://digitalstrategy.nl/wp-content/uploads/2014.01-A-How-big-data-can-improve-manufacturing-_-McKinsey-Company.pdf
6. Aven, T., & Krohn, B. S. (2014). A new perspective on how to understand, assess and manage risk and the unforeseen. *Reliability Engineering and System Safety, 121*, 1–10.
7. Bi, Z., & Cochran, D. (2014). Big data analytics with applications. *Journal of Management Analytics, 1*(4), 249–265.

8. Bizer, C., Boncz, P., Brodie, M. L., & Erling, O. (2012). The meaningful use of big data: Four perspectives—Four challenges. *ACM Sigmod Record, 40*(4), 56–60.
9. Büyüközkan, G., & Çifçi, G. (2012). A novel hybrid MCDM approach based on fuzzy DEMATEL, fuzzy ANP and fuzzy TOPSIS to evaluate green suppliers. *Expert Systems with Applications, 39*(3), 3000–3011.
10. Cardoso, J., Fromm, H., Nickel, S., Satzger, G., Studer, R., & Weinhardt, C. (Eds.). (2015). *Fundamentals of service systems*. https://doi.org/10.1007/978-3-319-23195-2
11. Carmeli, A., Brammer, S., Gomes, E., & Tarba, S. Y. (2017). An organizational ethic of care and employee involvement in sustainability-related behaviors: A social identity perspective. *Journal of Organizational Behavior, 38*(9), 1380–1395. https://doi.org/10.1002/job.2185
12. Chen, P. T., & Chen, J. H. (2015). Implementing cloud-based medical systems in hospitals and strategic implications. *Technology Analysis and Strategic Management, 27*(2), 198–218.
13. Choi, T. M., Wallace, S. W., & Wang, Y. (2018). Big data analytics in operations management. *Production and Operations Management, 27*(10), 1868–1883.
14. Cohen, M. C. (2018). Big data and service operations. *Production and Operations Management, 27*(9), 1709–1723. https://doi.org/10.1111/poms.12832
15. Dahlberg, T., & Nokkala, T. (2015). A framework for the corporate governance of data—Theoretical background and empirical evidence. *Business, Management and Education, 13*(1), 25–45.
16. Demirkan, H., & Delen, D. (2013). Leveraging the capabilities of service-oriented decision support systems: Putting analytics and big data in cloud. *Decision Support Systems, 55*(1), 412–421.
17. Dev, N. K., Shankar, R., Gupta, R., & Dong, J. (2019). Multi-criteria evaluation of real-time key performance indicators of supply chain with consideration of big data architecture. *Computers and Industrial Engineering, 128*, 1076–1087.
18. Economist Intelligence Unit. (2012). *The deciding factor: Big data & decision making*. Retrieved from http://www.capgemini.com/sites/default/files/resource/pdf/The_Deciding_Factor__Big_Data___Decision_Making.pdf
19. Edwards, R. (2015). Knowledge infrastructures and the inscrutability of openness in education. *Learning, Media and Technology, 40*(3), 251–264.
20. Emrouznejad, A. (Ed.). (2016). *Big data optimization : Recent developments and challenges* (Vol. 18). T4—Recent developments and challenges U6. https://doi.org/10.1007/978-3-319-30265-2
21. Emrouznejad, A., & Charles, V. (Eds.). (2019). *Big data for the greater good*. Springer International Publishing.
22. Emrouznejad, A., & Marra, M. (2016). Big data: Who, what and where? Social, cognitive and journals map of big data publications with focus on optimization. In *Big data optimization: Recent developments and challenges* (pp. 1–16).
23. Emrouznejad, A., & Marra, M. (2017). The state of the art development of AHP (1979–2017): A literature review with a social network analysis. *International Journal of Production Research, 55*(22), 6653–6675.
24. Floyd, F. J., & Widaman, K. F. (1995). Factor analysis in the development and refinement of clinical assessment instruments. *Psychological Assessment, 7*(3), 286–299.
25. Fosso Wamba, S., Gunasekaran, A., Papadopoulos, T., & Ngai, E. (2018). Big data analytics in logistics and supply chain management. *International Journal of Logistics Management, 29*(2), 478–484.
26. Gopalan, R., Sreekumar, S., & Satpathy, B. (2016). Evaluating the Indian retail service quality enablers using interpretive structural modelling. *International Journal of Business Performance Management, 17*(4), 365–393.
27. Goyal, S., Routroy, S., & Singhal, A. (2019). Analyzing environment sustainability enablers using fuzzy DEMATEL for an Indian steel manufacturing company. *Journal of Engineering, Design and Technology, 17*(2), 300–329.
28. Gu, T., Zuo, C., Liao, Q., Yang, Y., & Li, T. (2013). Improving MapReduce performance by data prefetching in heterogeneous or shared environments. *International Journal of Grid and Distributed Computing, 6*(5), 71–82.

29. Günther, W. A., Rezazade Mehrizi, M. H., Huysman, M., & Feldberg, F. (2017). Debating big data: A literature review on realizing value from big data. *Journal of Strategic Information Systems, 26*, 191–209.
30. Gupta, S., Altay, N., & Luo, Z. (2017). Big data in humanitarian supply chain management: A review and further research directions. *Annals of Operations Research,* 1–21.
31. Hazen, B. T., Boone, C. A., Ezell, J. D., & Jones-Farmer, L. A. (2014). Data quality for data science, predictive analytics, and big data in supply chain management: An introduction to the problem and suggestions for research and applications. *International Journal of Production Economics, 154*, 72–80.
32. Johnston, R. (1999). Service operations management: Return to roots. *International Journal of Operations & Production Management, 19*(2), 104–124.
33. Johnston, R., Clark, G., & Shulver, M. (2012). *Service operations management: Improving service delivery* (4th ed.).
34. Kiani Mavi, R., & Standing, C. (2018). Critical success factors of sustainable project management in construction: A fuzzy DEMATEL-ANP approach. *Journal of Cleaner Production, 194*, 751–765.
35. Kitchin, R. (2014). *The data revolution: Big data, open data, data infrastructures and their consequences.*
36. Konishi, S. (2014). Introduction to multivariate analysis: Linear and nonlinear modeling. In *Introduction to multivariate analysis: Linear and nonlinear modeling.*
37. Labrinidis, A., & Jagadish, H. V. (2012). Challenges and opportunities with big data. *Proceedings of the VLDB Endowment, 5*(12), 2032–2033.
38. Lamba, K., & Singh, S. P. (2018). Modeling big data enablers for operations and supply chain management. *International Journal of Logistics Management, 29*(2), 629–658.
39. Lin, R. J. (2013). Using fuzzy DEMATEL to evaluate the green supply chain management practices. *Journal of Cleaner Production, 40*, 32–39.
40. Luthra, S., Govindan, K., Kharb, R. K., & Mangla, S. K. (2016). Evaluating the enablers in solar power developments in the current scenario using fuzzy DEMATEL: An Indian perspective. *Renewable and Sustainable Energy Reviews, 63*, 379–397.
41. Manyika, J., Chui, M., Brown, B., Bughin, J., Dobbs, R., Roxburgh, C., & Hung Byers, A. (2011). *Big data: The next frontier for innovation, competition and productivity.*
42. Matthias, O., Fouweather, I., Gregory, I., & Vernon, A. (2017). Making sense of Big Data—Can it transform operations management? *International Journal of Operations & Production Management, 37*(1), 37–55.
43. Mazza, E. M. (2016). *The role of digital technology in the innovation process of companies: a systematic literature review in the innovation management field*. The Polytechnic University of Milan. Retrieved from https://www.politesi.polimi.it/handle/10589/133780
44. McAfee, A., & Brynjolfsson, E. (2012). Big data : The management revolution. *Harvard Business Review, 90*(10), 60–68. Retrieved from http://tarjomefa.com/wp-content/uploads/2017/04/6539-English-TarjomeFa-1.pdf
45. Ming, Z., Luo, C., Gao, W., Han, R., Yang, Q., Wang, L., & Zhan, J. (2013). BDGS: A scalable big data generator suite in big data benchmarking. In *Advancing big data benchmarks* (pp. 138–154).
46. Molinari, A., Maltese, V., Vaccari, L., Almi, A., & Bassi, E. (2014). Big data and open data for a smart city. *IEEE-TN Smart Cities White Papers.* https://doi.org/10.13140/RG.2.2.33170.56001
47. Niemi, E. (2011). Designing a data governance framework. In *Proceedings of the IRIS Conference*, Oslo, Norway.
48. Ojha, D., Gianiodis, P. T., & Manuj, I. (2013). Impact of logistical business continuity planning on operational capabilities and financial performance. *International Journal of Logistics Management, 24*(2), 180–209.
49. Oussous, A., Benjelloun, F. Z., Ait Lahcen, A., & Belfkih, S. (2018). Big Data technologies: A survey. *Journal of King Saud University—Computer and Information Sciences, 30*, 431–448.

50. Perçin, S. (2018). Evaluating airline service quality using a combined fuzzy decision-making approach. *Journal of Air Transport Management, 68*, 48–60.
51. Powell, L. (2002). Shedding a tier: Flattening organisational structures and employee empowerment. *International Journal of Educational Management, 16*(1), 54–59.
52. Raghavendra, R., Ranganathan, P., Talwar, V., Wang, Z., & Zhu, X. (2008). No "power" struggles: Coordinated multi-level power management for the data center. *ACM SIGOPS Operating Systems Review, 42*(2), 48–59.
53. Ray, J., Johnny, O., Trovati, M., Sotiriadis, S., & Bessis, N. (2018). The rise of big data science: A survey of techniques, methods and approaches in the field of natural language processing and network theory. *Big Data and Cognitive Computing, 2*(3), 22.
54. Redman, T. C. (1998). The impact of poor data quality on the typical enterprise. *Communications of the ACM, 41*(2), 79–82.
55. Rezaei, J. (2015). Best-worst multi-criteria decision-making method. *Omega (United Kingdom), 53*, 49–57.
56. Rodríguez, L., Giannakis, M., Da Cunha, C., Rodríguez, L., Giannakis, M., Da, C., & Investigating, C. (2018). Investigating the enablers of big data analytics on sustainable supply chain. *IPSERA*. Retrieved from https://hal.archives-ouvertes.fr/hal-01982533
57. Sampson, S. E. (2012). Visualizing service operations. *Journal of Service Research, 15*(2), 182–198.
58. Sanders, A., Elangeswaran, C., & Wulfsberg, J. (2016). Industry 4.0 implies lean manufacturing: Research activities in industry 4.0 function as enablers for lean manufacturing. *Journal of Industrial Engineering and Management*, *9*(3), 811–833.
59. Sangari, M. S., & Abbasi, M. (2017, April). An integrated framework of enablers of strategic alignment in supply chain: A fuzzy DEMATEL approach. In *Second international conference on industrial management* (pp. 1–10).
60. Sejahtera, F., Wang, W., Indulska, M., & Sadiq, S. (2018, June). Enablers and inhibitors of effective use of big data: Insights from a case study. In *Proceedings of the 22nd Pacific Asia Conference on Information Systems* (pp. 27–32). Retrieved from https://aisel.aisnet.org/pacis2018/27
61. Seker, S., & Zavadskas, E. K. (2017). Application of fuzzy DEMATEL method for analyzing occupational risks on construction sites. *Sustainability, 9*(11), 1–19.
62. Sharma, S. K., & Wang, X. (2017). Live data analytics with collaborative edge and cloud processing in wireless IoT networks. *IEEE Access, 5*, 4621–4635.
63. Singhal, K., Feng, Q., Ganeshan, R., Sanders, N. R., & Shanthikumar, J. G. (2018). Introduction to the special issue on perspectives on big data. *Production and Operations Management, 27*(9), 1639–1641.
64. Sivarajah, U., Kamal, M. M., Irani, Z., & Weerakkody, V. (2017). Critical analysis of Big Data challenges and analytical methods. *Journal of Business Research, 70*, 263–286.
65. Smith, J. S., Karwan, K. R., & Markland, R. E. (2007). A note on the growth of research in service operations management. *Production and Operations Management, 16*(6), 780–790.
66. Song, I. Y., & Zhu, Y. (2016). Big data and data science: What should we teach? *Expert Systems, 33*(4), 364–373.
67. Tabachnick, B. G., & Fidell, L. S. (2018). *Using multivariate statistics* (7th ed.). Pearson.
68. Talankar, A., Verma, P., & Seth, N. (2015). Modelling the clusters of critical success factors of Six Sigma for non-formal service sectors using interpretive structural modelling. *International Journal of Six Sigma and Competitive Advantage, 9*(2–4), 222–240.
69. Talib, F., Rahman, Z., & Qureshi, M. N. (2011). An interpretive structural modelling approach for modelling the practices of total quality management in service sector. *International Journal of Modelling in Operations Management, 1*(3), 223–250.
70. Tyagi, M., Kumar, P., & Kumar, D. (2015). Assessment of critical enablers for flexible supply chain performance measurement system using fuzzy DEMATEL approach. *Global Journal of Flexible Systems Management, 16*(2), 115–132.
71. Unhelkar, B. (2017). *Big data strategies for agile business: Framework, practices, and transformation.* CRC Press.

72. Waller, M. A., & Fawcett, S. E. (2013). Data science, predictive analytics, and big data: A revolution that will transform supply chain design and management. *Journal of Business Logistics, 34*(2), 77–84.
73. Wang, Y., Kung, L. A., & Byrd, T. A. (2018). Big data analytics: Understanding its capabilities and potential benefits for healthcare organizations. *Technological Forecasting and Social Change, 126*, 3–13.
74. Weick, K. E., Sutcliffe, K. M., & Obstfeld, D. (2008). Organizing for high reliability: Processes of collective mindfulness. *Crisis Management, 3*(1), 81–123.
75. Wielki, J. (2013). Implementation of the Big Data concept in organizations—Possibilities, impediments and challenges. In *Proceedings of the 2013 Federated Conference on Computer Science and Information Systems* (pp. 985–989). Retrieved from https://www.researchgate.net/publication/257341462_Implementation_of_the_Big_Data_concept_in_organizations_-_Possibilities_impediments_and_challenges
76. Yadav, N. (2014). Total interpretive structural modelling (TISM) of strategic performance management for Indian telecom service providers. *International Journal of Productivity and Performance Management, 63*(4), 421–445.
77. Yang, M., Zhou, X., Zeng, J., & Xu, J. (2016). Challenges and solutions of information security issues in the age of big data. *China Communications, 13*(3), 193–202.
78. Zacharia, Z. G., Sanders, N. R., & Nix, N. W. (2011). The emerging role of the third-party logistics provider (3PL) as an orchestrator. *Journal of Business Logistics, 32*(1), 40–54.

Mahdi Nasrollahi is an Assistant Professor at Imam Khomeini International University (IKIU), Iran. He received his Ph.D. in Industrial Management from University of Tehran. His research interests include SCM, MADM and DEA. He has written several papers about Supply Chain Management, Multiple Criteria Decision Making and Mathematical Modeling. Mahdi Nasrollahi can be contacted at: m.nasrollahi@soc.ikiu.ac.ir.

Mohammad Reza Fathi is an Assistant Professor, College of Farabi at University of Tehran, Iran. He received his Ph.D. in Industrial Management from University of Tehran. His research interests include SCM, MADM and DEA. He has written several papers about Advanced Supply Chain Management, Multiple Criteria Decision Making and Mathematical Modeling. Mohammad Reza Fathi can be contacted at: Reza.fathi@ut.ac.ir.

Data Architecture for Big Data Service Operations Management (The New Vision of Data Architecture for the Future Human Society)

Fang Miao, Wenhui Yang, Yan Xie, and Wenjie Fan

Abstract Organizations collect big data and do analysis to improve their service operations, encounter some fundamental and common issues such as how to manage all kinds of data, how to meet the various of business requirements, and how to protect personal privacy and organization's interests. All these issues bring us have to re-understand data from the human civilizations. This chapter proposed a platform construction for data management and control called Data Architecture (DA), which can be used in big data service operation management and provide complicated data applications with data protection in the open Internet environment. The functions of DA are: (1) redefinition of the procedures of software engineering from business-oriented to data-oriented, to make the software engineering method more flexible, extensible and sustainable; (2) solving data sharing problems in an easy way for comprehensive data applications of cross-level, cross-sectoral, cross-industry, cross-region, cross-system and cross-business; (3) solving the data security problems in the open Internet environment to protect data and the data owner's privacy, as well as to grantee the data owner's interests. DA is simply described as one body with two wings. The one body is that the data must be combined with ownership. The one wing is that the data should be innately registered and the using of registered data information to manage the data. Another wing is that the data should be innately encrypted with the data owner's key to protect data with confirmation of data owner. To share data and make data applicable, DA also establishes data ownership authorization to use, data operation recording, illegal data using finding, data history tracing, data

F. Miao (✉) · W. Fan
Big Data Research Institute, Chengdu University, Chengdu, China
e-mail: miaofang@126.com

W. Fan
e-mail: 729009746@qq.com

W. Yang
Cyber Security College, Chengdu University of Technology, Chengdu, China
e-mail: ywhui@cdut.edu.cn

Y. Xie
Sichuan Provincial Big Data Center, Chengdu, China
e-mail: xiey521@126.com

A. Emrouznejad and V. Charles (eds.), *Big Data and Blockchain for Service Operations Management*, Studies in Big Data 98,
https://doi.org/10.1007/978-3-030-87304-2_4

value assessment, etc. Thus, data can be shared and used safely in open environment with ownership mandate under the support of DA. As for the use of big data for service operations management, DA can fully support it and meet further business application requirements.

1 Introduction

The tremendous volume of available data collected by firms has been transforming the service industry [17]. It brings big data values such as shaping the service industry, decision making assisted, offering personalized service, etc. in different industries such as finance, transportation, hospitality, tourism, education, governance and online platforms. Great opportunities also bring great challenges [5, 8].

The operation of modern service industry is much more different than before. It faces complex and comprehensive requirements and applications with cross-level, cross-sectoral, cross-industry, cross-region, cross-system and cross-business. Only big data can be fusional and can help the situations but the establish of a big data platform to suit the complex and comprehensive applications becomes a challenge [9].

The challenge is that before the supporting of the service operations management, the platform should support data management, data sharing, data security, data application services, and it should be sustainable and extensible. Simple stacking of technologies and traditional methods could not meet the above requirements. It needs to introduce a new construction called data architecture (DA) to solve these issues.

1.1 The Challenge and Requirements of Big Data Service Operations Management in a Complex and Synthetical Situation

The operations in different service industries become more and more complicated, need more and more quality data and sophistic methods to support. Data from different sectoral, systems, regions, owners and levels increased the difficulties of the management. It is not only the management of data quantity and quality, but also the management and services of data control, data sharing, data security, data ownership, data value, data benefits, data application and etc.

The service operations covered nearly all industries such as environmental protection, urban management, transportation, pension, medical treatment, community, education, energy, health care, social security, service industry, finance, logistics, e-commerce, enterprises etc. The service operation applications are no longer simple, local, unique and no exchange data with other systems. The new service operation applications are complicated, cross-region, open, multi-users, coordination, sharing data with other systems. This involves some new requirements, such as data ownership, data security, data benefit guarantee, data sharing methods and so on.

For example, to improve the service provision and accomplish efficiency gains in public service operation, the IT (information technology) governance of shared services in a service network involved multiple agencies, different objectives, resources, capabilities, processes and levels [13]. All these data need to be collected, managed, processed, shared and to support the service operation.

Further, the requirements of data security, the protection of personal privacy and the protection of business secrets in the open Internet environment are most urgent.

These are the most fundamental and common problems we must face. If these problems cannot be solved, they will greatly affect the effectiveness and efficiency of service operation.

1.2 Traditional Methods Could not Meet the Requirements of Big Data Service Operations Management

There is no good way to response the challenge mentioned above by using traditional technical methods or simply stack them together, even just simply using the new technology such as big data, cloud computing, Internet of Things, and artificial intelligent etc.

The different kind of information technologies are designed to solve different kind of information problems. From the point view of technology, we could not simply mix them together to solve a new kind of problem.

Because the traditional software engineering methods are business-oriented [31] and not following the rules of information technology itself. When we face the complex virtual world made by data, information and knowledge, we should re-think the methods and the architecture of software engineering.

In the past, the service operation system could be management through certain hardware, software and data support. A slightly more complex system may require enhanced hardware, software and data support. For example, hardware needs to upgrade from stand-alone to network, from network to cloud computing; software needs to upgrade from process-oriented to object-oriented, from service-oriented to lightweight, enterprise-level development tools and methods, to middleware, component-based, waterfall-based, agile, business middleware and other architectures; data needs to change from simple database to complex data warehouse. Then to support non-structured data and big data by Hadoop, No-SQL and other systems, as well as persistent databases, data desks and so on.

But we are facing new challenges: service operation system is becoming more and more complex, the tools and technologies we use are not suitable to solve the problems we face, and the cost of maintenance and upgrading is also increasing.

1.3 The New Data Architecture (DA) is Introduced to Solve the Problems

Apart from hardware architecture and software architecture, in order to solve the above problems, as well as to meet the needs of future complex information system or data systems, the Data Architecture (DA) is proposed and introduced. We can also understand it by the other two concepts: (1) Data-Oriented Architecture (DOA) [20], which is a software architecture and a series of theory, method, technique and product intending to build information system or data system treated data as the core; (2) Data-Oriented Security Architecture, or Data Ownership-based Security Architecture (DOSA) [18, 19, 21–23], which is emphasizing data ownership and data security based on DOA.

DA is intend to build a new data architecture to response the challenge of many fundamental and comprehensive problems concerned with data in the open Internet environment and for the future virtual data society. It must be data-oriented. It is of data ownership designated. It has all data innate registered, and all sensitive data innate encrypted to make the data manageable and controllable. It also considers the data use by ownership authorization, data procedure recording, data tracing to the source, data illegal use discovering, data value assessment, etc. It is a feasible frame and eco-system both for data application and for service operations.

The purposes of DA are (1) to change the method of software engineering from business-oriented to data-oriented, to meet the data technology in the future, and to make the software engineering method more flexible, extensible and sustainable; (2) to solve data sharing problems from different sources, different structures, different departments, different owners, different systems, etc. by using a unifying, ordinary and extensible architecture, for data comprehensive applications; (3) to solve data security and data sharing problems in the open Internet environment, to protect data and the data owner's privacy, as well as to grantee the data owner's interests.

DA can not only solve the problems mentioned above, it can also be more significant to human beings for the future. We need to re-understand data and to know more about data fundamental characteristics before we can know the DA better.

2 The Re-understanding of Data

Data, of which is a very technical concept but are the fundamental elements involved in our daily life and work, even affect the future of human beings. There are still a lot of characteristics, features and issues concerned with data which we even don't know well but cause obstacles, difficulties and problems in our application and business systems, services and operations.

We need to further understand and master data. We need to know what role the data will play in the future. We can re-recognize them from the development of human civilizations. Compared with material, we should understand the similarities and differences between data and material. In this way, we can understand how to face and solve the problems related to data in the future.

2.1 Civilizations, Data and the Virtual Data World

Everyone is thinking or talking about the future, and we know that technological progress is the fundamental driving force for the advancement of human civilizations [26, 32].

From the evolution of human civilization, we can see the future of human development. Chen [6] said that civilization is a unique way of survival and development created by mankind. Human being promote the evolution of civilization by exploring the laws of the outside world and innovating the story of human beings themselves. More and more people realize that data will play a very important role for the future. Brooks [4] said that in a world of increasing complexity, relying on data can reduce cognitive biases and "illuminate patterns of behavior we haven't yet noticed". Lohr [16] believed Big Data is transforming society, using the term "Data-ism" to describe the Big Data revolution. Harari [11] mentioned that all competing political or social structures can be seen as data processing systems: "Dataism declares that the universe consists of data flows, and the value of any phenomenon or entity is determined by its contribution to data processing".

Because of the development of computers and Internet, we have entered a new virtual, digital and data world from the real, physical and material world.

The real world, including the material world and human society, is composed of nature and humanities. The nature world created by the nature what we see are that of sun, moon, stars and our earth. There are mountains, plains, rivers and lakes in the earth. They are following the laws of the universe to create life in the earth repeatedly and endlessly. Everything grows ecologically and vigorously. While the humanity world are created by human beings and are composed of consciousness, spirit, thought, nationality, country, region, religion, politics, economy, public society, company, family, science, technology, engineering, culture, art, education, health, exchange, cooperation, system, law, finance, service, as well as infrastructure, buildings, devices, etc. For human beings to understand and to adapt, all the real world are reflected and expressed by the virtual world.

The virtual world created by human beings has experienced the development of murals, oracle bone inscriptions, printing and other traditional technologies. Until the twentieth century, the emergence and rapid development of computer and network technologies have realized the transmission, exchange and accumulation of ideas and knowledge in the form of numbers, words, data, information, knowledge, and promoted the progress and development of science and technology.

Human beings invented and established the method of fusing real world with virtual world by data to know, understand, adapt and control the real world. See Fig. 1. The two worlds are coexist and interacted. Human beings living in the real world cannot live without the virtual world composed of data.

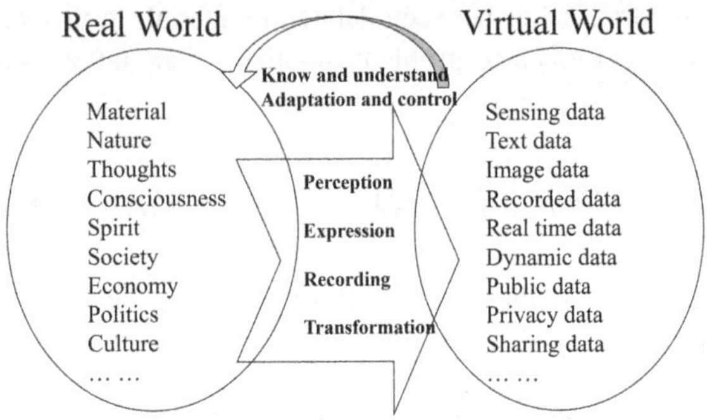

Fig. 1 The real world and the virtual world

Human has created a cyber space or virtual world by computer, Internet and related techniques. For example, computer technology provides information and data computing and storage servers, provides various terminals (desktop computers, mobile phones) for humans; wired, wireless Internet and Internet of Things form a communication network; hardware and software constitute various information system; and so on. People take advantages of the virtual world to communicate and interact with the real world for better cognition and life.

The contents of the virtual world are those data of static, real-time dynamic, constantly changing, visible and more invisible. These data constitute the entire contents of the virtual world, supporting the operation of the virtual world and the whole world.

People are entering an era of data. Therefore, people need to understand the data role, data value and data methods to know the rules and architecture of this new world. Let us recognize that data play an important role and significance to human civilization, social, economic, industry development and services operation. Data value and data method is the key and kernel for achieving human society development in the future. Researching and understanding data assets, value assessment, rights attribution, responsibility attribution, share characteristics, safety and security, protection of interests, the application system etc., are the new theory of cognitive point of view, ideas and methods.

Figure 2 shows the relations of digits, data, information, knowledge, wisdom and decision. Human begins invented the methods to perceive and understand the world through digits, and to express and transmit the contents of the world through data.

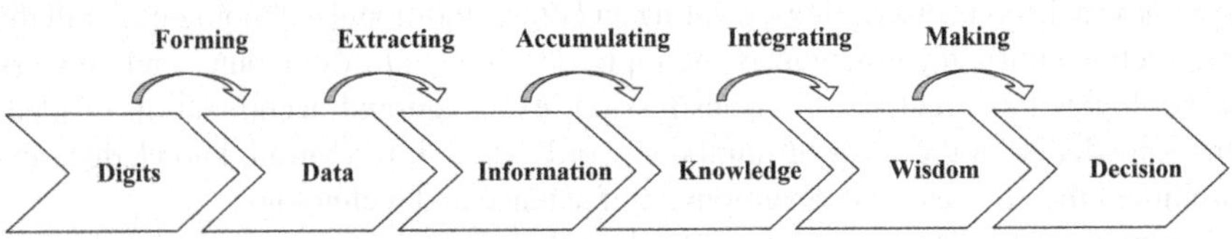

Fig. 2 Relation among digits, data, information, etc.

Table 1 Difference between information and data

	Information	Data
Concept	Extracted from data; a kind of data with value to people	The carrier of all information and knowledge; the contents of multi-media
Service object	People	People and computers
Quantity	Small; extract from data	Lager; could create various information
Application	Differ from person to person; utilitarianism	Generality; universality
Scope	For the relevant personnel	Benefit for more people
Technology	Acquisition, transmission, processing, storage, calculation, sharing, display, needs small resources and capacity	Acquisition, transmission, processing, storage, calculation, sharing, display, needs more resources and capacity
System	Small, relatively simple	Large, complex

There are many differences between information and data even if they are all look the same when they are processed in computer, stored in storage, or transmitted in network. The main difference between information and data is that information have meanings to certain people, but data are only waiting people to process them to become information [28], [35]. Table 1 roughly shows the main differences between information and data in concept, service object, quantity, application, scope, technology and system.

In terms of concepts of digit, data and information etc., we find that data has become the basic and fundamental elements for understanding and adapting to the world.

2.2 The Material Civilizations and the Data Civilizations

Human beings have now been experiencing four civilizations since they came out of primitive society [30]. Roughly we can divide the four civilizations into two parts: material civilizations and nonmaterial or data civilizations. See Fig. 3.

The first two civilizations of agriculture and industry based on the development and application of materials [1], supported by technologies of feeding, building etc. Now, human society are entering the two new civilizations of information and intelligence based on the development and application of data [15, 34] supported by technologies of computing, networking etc. These four civilizations form two worlds and one is the true or nature world made of materials and another is the virtual or artificial world (known as cyber space) made of data [2].

The first civilization we call it agriculture civilization or the first material civilization, lasted thousands of years. The main feature is that human beings **conquered**

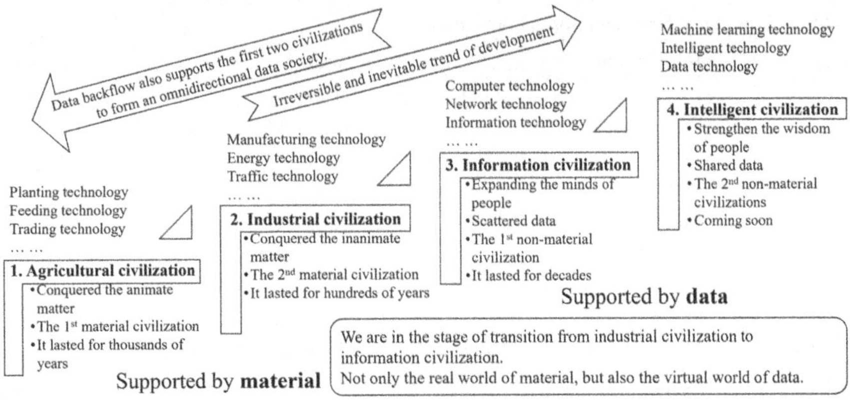

Fig. 3 The four civilizations human beings have been experiencing

living matter to solve basic food and clothing issues, by planting technology, feeding technology, trading technology, etc.

The second civilization we call it industrial civilization or the second material civilization, lasted hundreds of years. The main feature is that human beings **conquered nonliving matter** to solve basic residence and transportation issues, by manufacturing technology, energy technology, traffic technology, etc.

The third civilization we call it information civilization or the first nonmaterial or the first data civilization, lasted for decades. The main feature is that human beings **expanded the people's mind** to solve communication issues, as well as information service for eat, wear, residence and transportation, by computer technology, network technology, information technology, etc.

The fourth civilization we call it intelligent civilization or the second nonmaterial or the second data civilization, which is coming soon. The main feature is that human beings **strengthen people's wisdom** to solve human thought issues, as well as better service for eat, wear, residence, transportation and life, by machine learning technology, artificial intelligent technology, data technology, etc.

These four civilizations or two worlds are of interaction and fusion, coexist, influence, integrate and develop with each other, and bring human beings satisfied life and rich sprits. The latter data civilizations also support the former material civilizations, realize informationization and intelligence, so that they can play better roles. Some systems generated by the former material civilizations will also play certain roles in the latter data civilizations.

From the evolution of human civilizations we can see that data are vital important. Data are the bases of information, knowledge and wisdom. Data are the fundamentals of all nonmaterial. Data are the sources of human understanding of the world. Data are the contents of communication, the contents stored, processed and displayed in servers, cloud hosts and terminals. All technologies, businesses, functions and processes are data-oriented and data-centered. Material can carry economy, data can not only carry economy, but also carry human thoughts and culture.

At present, we are in the transition period from the material civilizations to the data civilizations. It is also the period of alternation, influence and change of the old and new civilizations. It is unavoidable that in the procedure of entering data civilization, the thought, the management system and the method system we are using are that of used in the traditional material civilization. This is bound to bring about many problems, produce many conflicts and face many subversions, if we don't realize the importance of data to the future, if we don't have any measure and scheme concerned about data.

Now we are facing the data world and data civilization, but are we ready to deal with it? We need to have a clear understanding of the data itself and the needs of the future society for the data and information systems.

Fang Miao et al. studied some fundamental topics about data and proposed an architecture of data-oriented and data ownership-based, in order to solve some bottleneck issues we faced in cyber space [18–24]. We hope it may be helpful for the future ordered cyber data society.

Data is so important to the future but we even don't know the characteristics and attributes of data well, such as data philosophy, data thinking, data theory, data rules, data assets, data ownership, data protection, data sharing, data transaction, data application, data method, data architecture, etc.

2.3 Data Characteristics in Future Cyber Society

Comparing the basic characteristics and basic elements with material, data bearing, assets, value, ownership, replication, transmission, sharing, protection, evaluation, genetics, elimination, processing, management, etc. should be studied. Data carrying capacity can carry information, knowledge, culture, wisdom and even human civilization. The asset attribute of the data is associated with its value attribute, and is also associated with data protection, data ownership, data value mining and other elements.

Unlike material, data is very easy to copy and transmit. It causes that data is difficult to control and manage, and the safety of data is also a big issue.

Table 2 shows the comparison of material and data. From which we notice that to make good use of data, we need to enhance data management and domination, as well as data assets security, data ownership, data sharing, data application, data profits, etc.

The Summary of Material Characteristics

Material is the essential factor to support agriculture civilization and industrial civilization, and form the core contents of real world.

Material has value, asset and ownership attributes. In general, you have it means you own it. It is not easy or needs costs of reproduction and dissemination. There is Property Law in many countries to make the material society be ordered.

Material cannot be shared at the same time, but can only be shared in time.

Table 2 Comparisons of material and data

Category	Material	Data
Civilization support	**Agriculture, Industrial**	Information, Intelligent, and Agriculture, Industrial
Carrying the world	The **real world**	The **virtual world**, Cyber space, and Intertwined with real world
Ownership attributes	Yes	**Yes**, but in general you have it does not mean you own it
Value attributes	Yes	**Yes**, data itself is valuable and also its derived information and knowledge are valuable
Asset attributes	Yes	**Yes**, virtual digital property protection and inherit are very urgent
Law	Property Law	**Not yet**. Data Asset Law should be needed and urgent
Duplicate	Not easy	**Easy**, almost zero cost, but not all data can be copied (classified)
Transmission	High cost	**Easy**, almost zero cost, but some data needs restrict to transmission
Sharing nature	Bad, time sharing	**Good**, time sharing and simultaneous, the key nature of data
Fulfill security	Perimeter protection, system protection	**Perimeter protection**, system protection (sharing constraints; Loopholes, Trojan horses, back doors, stealing)
Individual protection	Put into the safe	**Data encryption** (Management and use constraints)
Key keeping	Owner/credible bank	**Owner**?/Administrator (Believable? Encryption is still unsafe)

Material safety is perimeter protection, system monitoring and safe lock.

The Summary of Data Characteristics

There are different types of data such as big data; structured, unstructured, semi-structured data; time-stamped data; machine data; spatiotemporal data; open data; dark data; real time data; genomics data; operational data; high-dimensional data; unverified outdated data; translytic data; etc. [3]. All these data are the essential factors to support information civilization and intelligent civilization and form the core contents of virtual world.

Data have value, asset and ownership attributes too. But in general, you own them does not mean you have ownership of them. Because data are very easy to reproduce and disseminate with nearly no cost. Normally there are many copies of data in different people's hands. It makes the data ownership more complicated than material. There should be a Data Asset Law or Data Right Law concerned with the data ownership to make the data society to be ordered.

Data can be shared at the same time and in time. This is the embodiment of data value, and the data economics.

Data safety is perimeter protection, system monitoring and data themselves encrypted. But when data encrypted, who owns the decrypt key who is the data owner.

Various of data are not only the kind of valuable resources, but also the new cognitions of methodology from the viewpoint of data. We can see data bring us two revolutions. On the one hand, it is necessary to use data correctly to maximize its value and bring about a value revolution; on the other hand, it is necessary to stand on the data point of view and follow the data law to construct information systems or data systems to realize a method revolution.

We know in the true world, the security of materials including human beings is fundamental important for any other things. As well as in the virtual world (cyber space), the security of data is also fundamental important for any other things.

Data is easily obtained and disseminated in the open Internet environment, yet the behavior of infringement of the right to privacy and the copy of illegal data are difficult to be detected. This is an unequal contradiction. The consequence is more and more serious.

2.4 Basic Characteristics and Classification of Data

Data are the origin of non-material (i.e. information, knowledge, wisdom, etc.). The real world is inseparable from the virtual world made up of data. The development of human civilization has entered the era of data civilization. We need a new vision for the coming world, namely, data thinking and data architecture. We should follow data law to use data technology to deal with data matters, not follow what we need and from business point of view. And we should add data architecture to build information and data systems apart from only hardware architecture and software architecture. It comes to do researches from the perspective of human civilization and human future development, from the perspective of data philosophy and data science.

Conducting data foundation researches, we should explore the scientific and engineering essentials as well as systems of data rules, data theory, data methods, data technology, data engineering, data applications, data security, etc. We will involve researches in data ownership, data sovereignty, data assets, data value, data law, data society, data order, data architecture, data heritage, data ethics and other aspects.

Considering the future influence of data to human beings, we roughly give the framework of data study as following.

The role and impact of data on the development of society as a whole

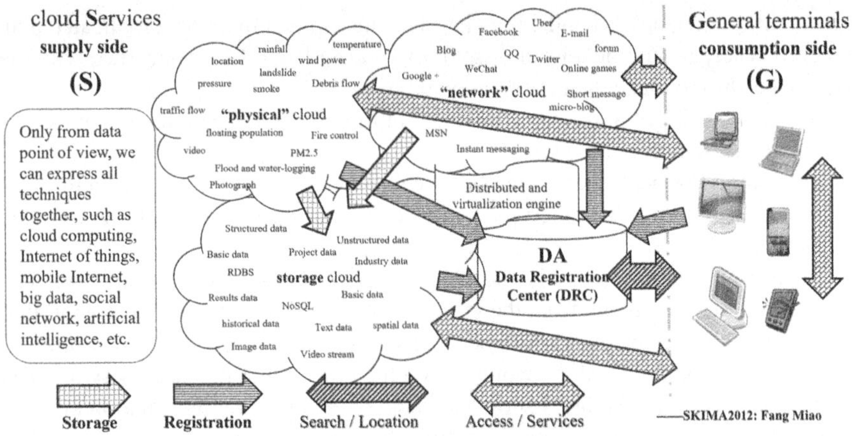

Fig. 4 Data cloud from the view of data

We should study data's position, role, influence and possible way of action in the development of human society. We should also study the past, present and future development of data to human civilization, in terms of economic, social, cultural, nature, science, as well as the philosophical significance, etc.

From the view of data, we can treat the virtual world as data cloud as shown in Fig. 4.

We divided the virtual world into two parts, one is supply side called cloud services (S, Service clouds), another is consumption side called general terminals which is facing human beings (G, General terminals), such as desktop, laptop, smart mobile phone, Web browser, PCs, etc.

People use G to communicate and interactive with S to get data and information, and to create various applications.

Three clouds compose S:

(1) "Physical" cloud known as Internet of Things (IoT), composed by physical sensing dynamic real time data;
(2) "Network" cloud is social network, composed by social network dynamic real time data;
(3) Storage cloud, composed by any types of statistic and historical data.

Data types classification

Data could be classified according to different perspectives.

(1) Structure: structured data, unstructured data and semi-structured data.
(2) Real-time nature: historical data, real-time data, static data, dynamic data, stock data and incremental data.
(3) Ownership: public data, private data, mixed data, proprietary data, cooperative data, third-party data, owner data, user data, manager data, agent data, etc.

(4) Security: protected data, shared data, open data, non-secret data, secret data, confidential data and top-secret data.
(5) Function: content-bearing data, system blood-type data, decision-support data, visual presentation data, etc.
(6) Presentation: tabular data, image data, video data, virtual reality data and augmented reality data, etc.

For different types of data, we should explore the general, universal processing method and processing procedure, and build a basic and well adapted methodology and technology system.

2.5 *Data Assets and Values*

(1) Data asset classification

Data assets are easy to replicate and disseminate, which brings new problems to the protection and publication of data assets. The characteristics of data assets should be studied comparison with real assets. Data assets are determined based on different forms. They can be digital currency with clear value and tradable, or they can be digital copyright works that generate value through widespread dissemination, or they can be network game equipment and have value within a certain range. Data assets can be other important data, value implicit, and ownership is clear.

The digital currency is a kind of data assets with only value attributes but without certain data characteristics. The digital copyright works are a kind of data assets with unique data characteristics and value attributes. The online game equipment and game currency are a kind of data assets which similar to digital currency, but only with exchange value in a certain range.

(2) Value evaluation and embodiment of data assets

The value of data assets varies by environment and location, by application scenario, varies from person to person, from time to time, from use to use, and is less likely to be valued than real assets. The value of data assets is more flexible, more complex, and more varied than real assets. To establish value assessment models of data assets in a various of application scenarios is necessary.

(3) Protection and transaction methods of different types of data assets

The protection of data assets is to enable data assets to be traded, applied, and valued on the premise that the interests of data asset owners are protected. Unlike real assets, the protection and trading of data assets must follow the characteristics of the data. For different kinds of data assets, the purpose of protection, the object of protection, the method of protection and the method of transaction are different. Data asset protection and transaction design should be classified.

For example, there are different demands for the protection and transaction of three kinds of data assets: digital currency has transaction characteristics, it needs to prevent theft, not allowed to reuse, and can record the use situation; digital copyright works are allowed to be widely used by others, but to protect the interests of owners and prevent unauthorized use; online game equipment and game currency are similar to digital currency, but are only valid within a specific range, and can be traded and prevented for theft. Legal issues of data assets.

The protection and use of data assets involve the construction of laws and systems. The protection and use of data assets need corresponding technology support. The data ownership is the first element of the data assets law. It needs the method of data confirmation, the method of data protection, the method of data ownership determination, the method of data authorization, and the supporting technical implementation. The data rights law should be the first law of the data legislation.

2.6 Data Method and Engineering

(1) Problems existing in traditional business-oriented software engineering

Traditional software engineering methods are business-oriented or demand-oriented methods, which design software according to business process and business function, adopting the thinking mode of industrialized production products [31]. It is difficult to realize the concept of software service, and it is difficult to reflect the role of data and the core position of data. It does not conform to the rules of software nature design. Business-oriented software engineering leads to complex design, long-term development process, difficult to adapt to business changes, difficult function expansion and high cost.

(2) An overview of Data-Oriented Software Engineering methods (DOSE)

The DOSE method is based on data as the core and data sharing as the concept, and develops software engineering methods to meet different business needs on the data resource pool. The data resource pool, through the data registration center, manages various types of data from different sources, different structures, massive amounts, and real-time dynamic changes, which can realize data sharing of different locations, different types, and different owners, while ensuring data security. The construction of the data resource pool is mainly the construction of the data registration center. It registers data innately in situ and in time for the historical data, incremental data, static data, dynamic data, structured data, unstructured data, etc. Processing data registration information in the background, to provide directories, indexes, search engines, and views to support the use of data.

The development of business functions relies on data access and data generation, depending on the data and information provided by the data registration center. Split the complete business process into many small processes-oriented to data, and then assemble small data-oriented processes into large business-oriented processes through user interface programs. DOSE is a development model of data big platform and fragmented application system.

(3) Features of DOSE methods

DOSE treats data as the kernel of the software design and development. Through the data registration center to build a data resource pool on which to develop business functions, can solve the problems of complex system data sharing and software system scalability.

In the development of business functions, there is no need to do the analysis of the needs of the entire business, and the development is more efficient and flexible.

(4) DOSE method implementation process

It needs to investigate the data types, data structures, distribution of inventory data and incremental data required for the development system. As for the requirements of data security, it needs data encryption and authorization methods based on ownership binding.

According to the data registration standard and registration principle, the distributed database system is used as the data registration center carrier, and the automatic registration of existing data is realized through the development of small programs to form data resources pool DRC for further software development.

Developing Data Access Interfaces (DAIs) and Application Programming Interfaces (APIs) in conjunction with DRC, DOSE provides application data access and programs access Software Development Kit (SDK) for subsequent programs development. DAIs especially for data accessing can be considered as a series of extended SQL statements, which no longer directly access data bases but indirect access them by pass through DRC. APIs concerned with data are mostly as same as common APIs, but they need more functions for managing and controlling of data.

Similarly to the traditional software development process to develop data-oriented business functions, the difference is that access to data through the DRC instead of direct access to data, and the registration of newly generated data in situ and in time.

Finally DOSE forms a virtuous and sustainable data-oriented application development ecosystem.

2.7 *Data Ownership and Confirmation*

As we talk about data ownership, we divided it into 4 types: data owner, data producer, data user and data agent show in Fig. 5.

Data owner has the right and domination of the data.

Data producer is authorized by the owner to create and collect the data.

Data user is authorized by the owner to use the data.

Data agent is authorized by the owner to own a kind of rights of the data, such as to manage data for the owner, to sell data for the owner, or to transmit data for the owner, etc.

The following (Table 3) are some examples about the above definitions of data ownership in different situations.

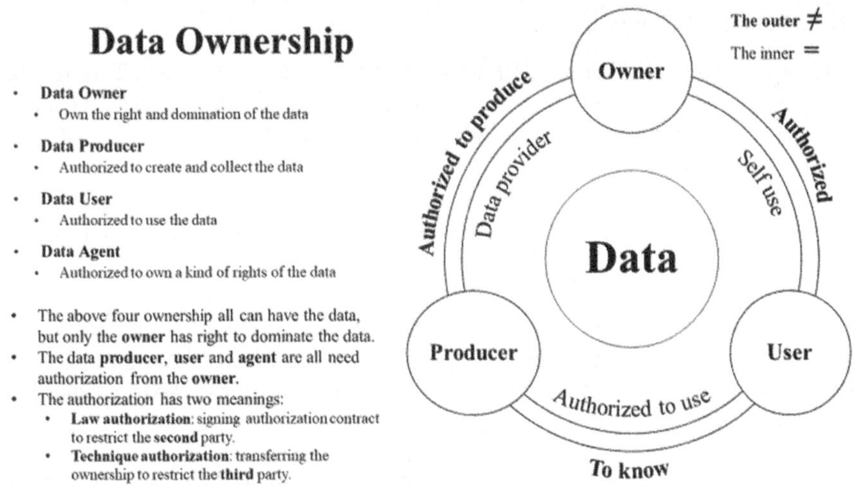

Fig. 5 Data ownership and description

Table 3 Different data ownership relations

Cases	Ownership relations	Examples
1	Data Owner = Data User = Data Producer = Data Owner	People uses digital camera to take pictures and self-appreciation; Scientific research institution creates data and makes analysis and decisions
2	Data Owner = Data User ≠ Data Producer ≠ Data Owner	Photo gallery takes digital pictures for customers; Government gets data from professional department to make decisions support
3	Data Owner ≠ Data User = Data Producer ≠ Data Owner	Hospital makes patient physical examination and gets data to prescribe prescription; Lawyer gathers client's information to make evidence
4	Data Owner = Data Producer ≠ Data User ≠ Data Owner	Digital rights author sells his/her digital works such as music, video, novel, article to live; Satellite's company collects the Earth remote sensing data for anyone who wants to use it
5	Data Owner ≠ Data Producer ≠ Data User ≠ Data Owner	Medical student studies patient physical examination data; Users download music and video works for entertainment from authorized Web sites

The above four kind of ownership people all have the data, but only owner has right to dominate the data. Data producer, user and agent all need authorization.

There are two kinds of authorizations to be used as ownership authorisation: (1) Law authorization, like normal situation to sign contract to get authorization. (2) Technical authorization is to make ownership to be authorized, is the key of this chapter.

(1) The basic relationship between people and data

For a long time, we have only seen data but not people. In fact, the generation and use of each data are related to people. Behind every data there are reflects of human responsibility, rights and obligations. The type of data ownership is different from the ownership of materials. Because of the reproducible and easy transmission of data, the type of data ownership is complex, diverse and poorly managed. We need to study and analysis the representative ownership and derivative ownership of data, as well as the types and classification methods of operable data ownership.

We divided data ownership into 4 types: data owner, data producer, data user and data agent. Data owner has the right and domination of the data. Data producer is authorized by the owner to create and collect the data. Data user is authorized by the owner to use the data. Data agent is authorized by the owner to own a kind of rights of the data, such as to manage data for the owner, to sell data for the owner, or to re-produce data for the owner, etc. The above four kind of ownership people all have the data, but only owner has right to dominate the data. Data producer, user and agent all need authorization.

There are basic issues of data ownership: (1) statement of people for data ownership; (2) method for binding people with data ownership; (3) the protection of people's data ownership; (4) the transfer of data ownership to other people; (5) the determination of people's right to use data; (6) the protection of people's right to use data; (7) the right of people to produce, process, manage, reproduce, and transmit data; (8) exploring the legal significance of research data ownership and the feasibility of legislation.

(2) The feasibility of real name authentication on the network

The relationship between people and data is reflected in the network and the identity of people. Therefore, it is necessary to study the feasibility of network real-name identity authentication. Common CA (Certificate Authority) digital certificate is one of the currently feasible methods.

(3) Application scenario of data ownership

Since data is related to various industries, the requirements for data ownership in different application scenarios of different industries are also different. We should study the role, adaptability and application scenarios of data ownership.

(4) The methods of determination and authorization of data ownership

There are possible ways to determine data ownership, including identification, registration, encryption, declaration, watermark, etc. with operability, convenience and feasibility.

The biggest feature of data compared with material is the almost zero-cost reproducibility and transferability, which lays the data sharing characteristics and can easily spread data, media, information, knowledge and wisdom through the network. Data authorization is the ownership authorization, that is, the right of data owner is transformed into the right of data users. It is necessary to study how to achieve this transformation by means of technology. Limiting the limitations and infeasibility of traditional platform password licensing, we should study on new data ownership authorization mechanism and method based on data ownership method. There are two kinds of authorizations to be used as ownership authorization: (1) Law authorization, is the same as the normal situation of signing contract to get authorization. This is to restrict the second party. This means that the default clause does not allow disclosure of the data to any other 3rd party. (2) Technical authorization, this is to transfer the ownership to the second party but restrict the 3rd party. This means even the 3rd party get the data but can't use it.

Data ownership authorization will be a regular activity in the future and may not be done by the data owner himself. Therefore, it is necessary to introduce an agent mechanism to help data owner to achieve the authority of the data, etc. The agent mechanism of data ownership, and agent authorization mechanism should be also studied.

2.8 Data Opening, Sharing and Security Grading Classification (Data Security Hierarchy Grading System and Application)

The traditional cyberspace security methods are territory-defending approach, focused on border security, system security and data transmission security. Data are almost in the naked state in storage, in systems and in application terminals. Hackers could steal data through exploiting system bugs and vulnerability, using Trojans and backdoors. Some organizations even develop traitors to steal and sell internal data. Data security in the open Internet environment is facing even more difficulties and challenges. In addition, data has not played its roles far away from it should be because the lack of data ownership, the lack of data security responsibility, the lack of data ordered rules and the lack of an efficient mechanism of data architecture.

(1) Data security hierarchy grading system

Data openness, sharing and security are different grads and have different requirements. They associate with the responsibilities, rights and obligations of the data owner and data user. The basic classification requirements of data for government, enterprises and individuals should be established before we consider the data protection or data open.

The reasonable and feasible data security grading system should be compatible with existing security level protection and security grading protection. It should be including data grading principles, grading standards, grading definitions, grading conversions, grading management, etc.

For data security features, we classified data from low to high into 3 large grades include 6 small grades. Grades 0–1 are open grades. Among them grade 0 is totally open level, no data needs to manage; grade 1 is general opening level, it needs registration and management, but no need to encryption. Grades 2–3 are share grades; both need registration and management, as well as encryption. Among them grade 2 is public level, or a sharing level in large scale; grade 3 is a share level, or a sharing level in small scale. Grades 4–5 are protection level, can only be used in the inner network environment. Among them grade 4 is a general protection level and needs encryption and registration, can only be used in inner environment; grade 5 is totally protected level, it needs encryption but do not manageable. Table 4 shows the detail of the data security hierarchy classification.

For the data in grade 2–4 and above, we can realize secured data sharing, secured data transactions and secured data applications.

(2) Application scenario of data security grading

The examples of application scenarios and adaptations of different data security gradings for government, enterprises and individuals are showed in Table 5. In practice, the data grade level can be decided by the data owner according to the importance of the data.

Table 4 Data security hierarchy and classification

			DAC			DRC		
Large grade	Grade level	Small grade	Encryption	Sharable	Publicly	Manageable		Contrast with classification protection
Open	0	Totally open	No	Yes	Yes	No, not need		Non-confidential
	1	General open situation	No	Yes	Yes	Yes	Need, innate register	
Share	2	Sharable in large scale	Yes	Yes	Yes	Yes		Secret class
	3	Sharable in small scale	Yes	Yes	No	Yes		
Protect	4	Protection for inner use	Yes	No	No	Yes		
	5	Protection in inner management	Yes	No	No	No, can not		Confidential class

- The grade 1 presents open application; registration, no encryption, recording sharing
- The grade 2–4 presents security application; registration, encryption, authority to share
- The grade 0 is totally open, no need to manage
- The grade 5 is for inner security use, cannot manage

Table 5 Data security hierarchy and examples

Grade	Description	Personal example	Enterprise example	Government example
0	Totally open No need for management	Statements	Advertising	Public notice
1	General open situation Manageable, to know application situation	Name	Basic profile	General bulletin data
2	Sharable in large range Emphasis on interests, need to be authorized	ID number	Operating data	General statistical data
3	Sharable in small scale Emphasis on protection, need to be authorized	Family address	Technical documents	Administrative data
4	Protection for inner use For internal use only	Economic situation	Core technical data	Internal data
5	Protection in inner management Can't manage by the system	Privacy	Core technical secrets	Secret data

2.9 Data Manageable Method and Technology System

(1) The necessity, feasibility and basic conditions of data manageable

Because of the data's reproducible and easy transmission, it brings great difficulties to data management and control. Traditional management and control methods for structured data and unstructured data can no longer meet the requirements of data society in open environment for data management and control. Only when we manage and control data comprehensively can we truly realize data sharing, data security and data value. From the birth of the first electronic computer, all kinds of information technology and database technology have been developed, and the development of security technology, cloud computing, big data and artificial intelligence technology have brought basic conditions for the manageable and controllable data.

The basic conditions for data manageable and controllable, are under the premise of not restricting the reproducibility and easy transferability of data. For example, to study how to ensure that the ownership of data is not lost, basic conditions such as data ownership, data value and data asset attributes are not lost. The necessary and sufficient conditions for data manageable should also be studied.

(2) Data manageable technical method

Conventional methods including distributed database management technology and file management technology cannot adapt to data security, sharing, and application issues for different locations, different structures and different owners. In the open environment, we need to protect the data, share the data, and embody the value of the data, and realize the controllable technical method of data transaction and application. It is necessary to study the technical methods of data manageable and controllable under the premise of ownership confirmation, ownership authorization, and ownership benefit protection. It is also necessary to study data recording, statistical analysis, violation discovery, tracking and traceability, etc., combined with big data, cloud computing, artificial intelligence, etc.

(3) Data manageable technical architecture

In the past, the data under the local condition can be controlled by only one or two software, and it has not adapted to the data manageable data security and sharing requirements under the open environment. Therefore, it is necessary to have corresponding technical architecture to achieve such data manageable and controllable. It is necessary to study the basic principles, main components, operation modes and relationship of this technology framework. It is necessary to study the unified manageable and controllable technical framework of different types of data, different safety grade data and different ownership types data.

2.10 Requirements in Data Platform

(1) Information System and Data System

In the current information society and information civilization era, all industries are inseparable from information systems built with information technology to meet their business needs.

With the widespread use and application of Internet, there is an increasing demand for integrated services, collaborative services, and personalized services. Single-purpose and small-scale traditional information systems are no longer sufficient. A new generation of data systems with data as the core needs to be built.

Data and information are both nonmaterial. Data can generate information. Information is also a kind of data. The technologies of information collection, transmission, storage, processing, calculation, analysis, and display are equally applicable to data. But the amount of data is much larger than the amount of information. It needs Big Data technologies to deal with data of 5/V features [7].

On the one hand, due to the growing demand for multi-party collaborative applications, multi-system data sharing, complex comprehensive analysis and personalized

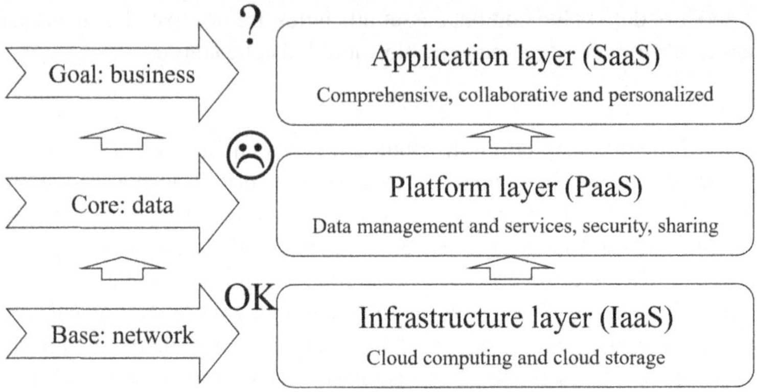

Fig. 6 The simplified model for a comprehensive data system

application services, on the other hand, the development of computer networks, cloud computing, big data technology and artificial intelligence technology, so that we have requirements also have the ability to process a large amount of data, not a small amount of information.

Looking to the future, we need more data systems to meet our growing complex, comprehensive, collaborative, and personalized application needs.

(2) Application Scenarios and Requirements of Data Systems

The application of data system goes beyond the application scenario of traditional information system with single requirement and small scope.

Smart cities, financial applications, government information resources opening and sharing, health and medical systems, tourism and traveling services, industrial Internet application, Internet of Things application, ..., more and more scenarios need data system but not only information system.

Figure 6 shows a simplified model for a comprehensive data system.

Thanks to the rapid development of cloud computing technology, network communication technology, the infrastructure layer including cloud storage, cloud services, and wireless communication networks, have become more powerful and easy to use. It fully supports the above two layers: platform and application layers.

The comprehensive, collaborative and personalized application layer is our goal for business, but it depends entirely on the platform layer. The core of the platform layer is data. Data management and services with security and sharing attributes are the bottleneck and pain point of the entire data system now.

The requirements of data systems in future virtual world (cyber society) compared with existing information systems:

- Data systems are ecological and sustainable.
- Data systems are open and autonomous.
- All types of data are manageable and extensible.

- The various applications on the system are manageable and growing, similar to the app store to provide APP services.
- Data are safe and protected.
- Data are shareable and controllable.
- The data system is supported by the data platform. The software development on the data platform is superimposed, easy and scalable.

(3) The Key Points of Data Platform

The key of data platform layer is data sharing and security. Data sharing needs to be solved by technology and mechanism. Data security needs to be solved by technology and mechanism too.

Traditional data management methods cannot adapt to the unified management and control of structured, unstructured, static, dynamic, changeable, real-time, different places, different structures and different owner's data, which leads to the data cannot be shared and secured.

Traditional data sharing methods:

(1) Open data interfaces on each system, other systems access data through interfaces;
(2) Establish a public data exchange platform, each system shares data by accessing the data exchange platform;
(3) Establish a catalogue on each system to provide data access; and so on. However, these methods are complex and costly, and they cannot reflect the updated data in time.

Traditional security methods are territory-defending approach, focused on border security, system security and data transmission security. Data are almost in the naked state in storage, in systems and in application terminals. Hackers could steal data through exploiting system bugs and vulnerability, using Trojans and backdoors. Some organizations even develop traitors to steal and sell internal data. Even if the data is encrypted, the decryption key is not in the hands of the data owner and is still not secure. Data security faces much more difficulties and challenges in open Internet environment.

We need to consider the data sharing and data security issues of the data platform by following data rules with a new data architecture perspective.

3 The Concept of Data Architecture (DA)

With the increasing demand of business and service operation, the corresponding information system becomes more and more complex. We know that information system includes three basic objective elements: hardware, software and data. We need to deal with increasingly complex systems through building architectures in these three aspects.

Hardware Architecture (HA) refers to the identification of a system's physical components and their interrelationships [25].

Software Architecture (SA) refers to organization of a system as a composition of components; global control structures, the protocols for communication, synchronization, and data access; the assignment of functionality to design elements; the composition of design elements; physical distribution; scaling and performance; and dimensions of evolution [10].

Data Architecture (DA) are proposed recently but there are different definitions. Tupper [33] thought it should be the blueprint for organizing and structuring of information for services, service providers, and the consumers of that data. Kimball and Ross [14] discussed the enterprise data warehouse bus architecture and associated matrix for establishing the data warehouse's integration strategy. WH Inmon described data architecture as the information paradigm [12]. In this chapter, DA is intend to face the future data society to give a simple architecture to manage all kinds of data and supporting all kinds of applications. It can be descripted as a data-oriented nature of data innate registration for data management and data innate encryption to protect the data from being free to use. The premise is data ownership binding and data registration. To share data and make data applicable, the mechanism of data ownership authorization, the recording procedure of data usage, the mechanism of finding the violations use of data, as well as data history tracing, are also established.

3.1 The Description of DA

For the future cyber data society, we need a data architecture to support data-oriented software engineering and sustainable software development, to support comprehensive, collaborative and personalized applications and service operations. This data architecture should be simple, protecting data and data sharing, supporting applications based on both open data and protected data.

Figure 7 shows the basic principle of data architecture for the future cyber data society.

The basic idea of this data architecture (DA) is data-oriented and data ownership-based. The brief description is one body with two wings and generalized K-V structure.

Data and ownership are binding together as one body. Data ownership can be divided as data owner, data producer, data user and data agent. It is important to be aware of the ownership of each data.

One wing is the innate registration of the data to produce the data attribute, we call it the generalized Key of the data. It supports at least four modules to make data shareable: automatic classification, automatic directory generation, automatic index generation, and automatic view generation.

Another wing is the innate encryption of the data by the owner's public key to protect the data, we call it the generalized Value of the data. It supports at least four modules to make protected data to be used and to find violations use of data: the

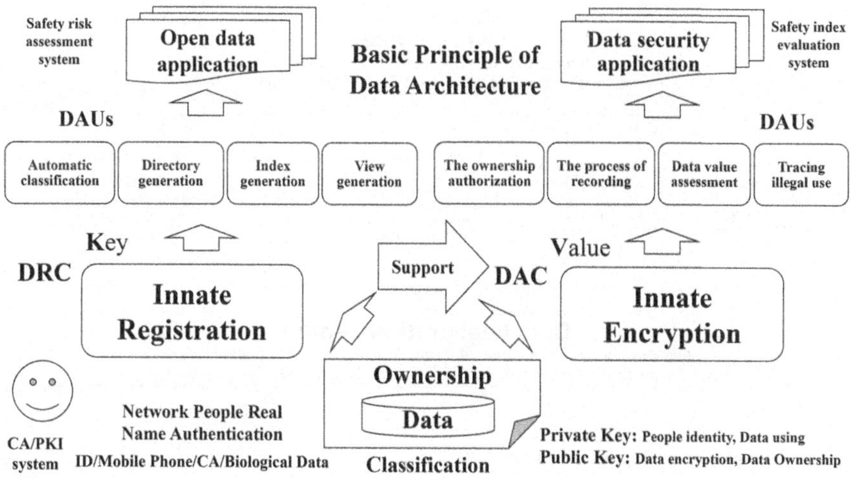

Fig. 7 The basic principle of data architecture

ownership authorization, the process of recording, data value assessment and tracing illegal use.

These two wings support the application of open data and the application of protected data respectively. But the former needs only one wing (left side), the latter needs both wings.

We combined data ownership with CA/PKI (Certificate Authorization/Public Key Infrastructure) to make the data ownership confirmation and data protection, as well as to support the architecture for people identity and authorization, data encryption, ownership binding, ownership authorization, and data use.

The DRC stands for Data Register and recording Center; DAC stands for Data ownership Authorization Center; DAUs stands for Data Application Units.

3.2 *Data Platform Supported by the DA*

Figure 8 shows a data platform for general purposes based on and supported by DA.

The main component of the data platform DRC, which composed by register principal and standard, register engine, directory creation engine, classification management, configuration engine and maintain engine.

There are some important modular such as Open CA Management, Encryption and Decryption, Data Directories, Data Vision, Authorization and Transaction, They support the CA basic functions.

The interface to applications are Data Services, which are DAI, API, SDK and Data Service Management (DSM).

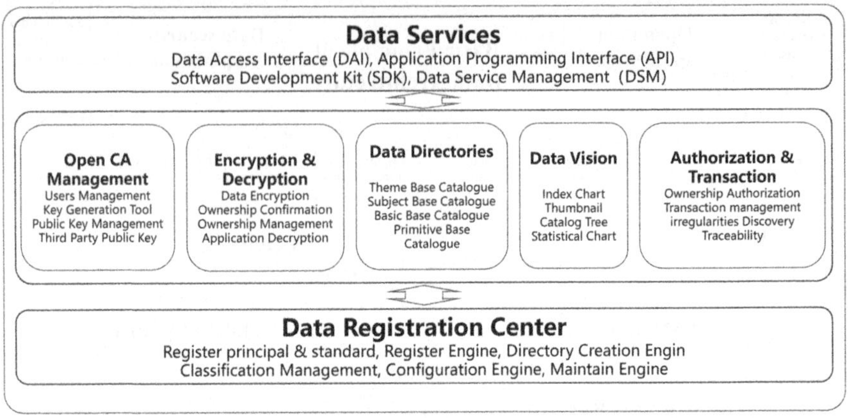

Fig. 8 Data platform supported by DA

3.3 *The Preliminary Application of DA*

We have made some application attempts in terms of the DA such as:

- a tourism platform;
- a government information resources sharing and application system;
- a system for personal privacy protection and get service;
- a data trading platform;
- a data acquisition and management system of IoT;
- a digital copyright works protection and trading systems;
- a big data service to civil people;
- a system supporting mobile phone data safely stored in open clouds; and so on.

The above framework design, prototype or demonstration of DA applications are show in Sect. 4. All these show the possibility and feasibility.

3.4 *The Significance of the DA*

From the study of the data architecture, we should think from epistemology and methodology to our human society.

On the one hand, we should learn to recognize the present from the perspective of the future, and look back from the future, not just forward. We should realize that human life and survival depend on material, human consciousness, thought and sprit depend on data. Material and data are the core elements of productivity. The development of data productivity can promote the development of appropriate production relations. Material and data support the development of human civilization from the progress of human science and technology.

On the other hand, data are different from material. They have their own laws. It is necessary to build a technical system according to data rules, to support data system by data architecture, and to support data society by corresponding technical and legal means.

The significance and roles of the DA played are:

- To interpret the whole world from the point of view of data and human being;
- Combing the relations among people, data and material;
- To construct data value system and data thought method system;
- To make data manageable and controlled;
- To solve "data, security, application" problem in an open, comprehensive and collaborative information system;
- Making data ownership feasible in theory and technology;
- Promoting the legislation of the "data right law";
- Providing basic technical support for the construction of an "orderly cyber data society".

3.5 *The Key Characters of DA*

(1) Data security and data application

On traditional knowledge, data security and data applications, is a paradox and inverse relationship. In general, an application system is to find a balance between security and application to achieve a certain protection effect and a certain application effect. In accordance with the data-oriented security architecture, data security is to better application, data application is under data security. To solve this contradiction, the first thing to consider is data own safety to protect data from not being free to use. While making protected data can be accessed and used for others that involves the concept of data ownership.

(2) Data ownership, data rights and data authorization

To understand data assets and data value assessment, data must have ownership attribute. Because data are very easy to copy and transfer, and can be owned by many people, so data ownership attribute is not easy to determine. We classify data ownership for four cases: data owner, data producer, data user and data agent, can cover almost all types of data ownership. To confirm data ownership, we used data encryption approach based on data ownership. General practice is based on the data owner's public key to encrypt data, so the data ownership is determined, and the data is protected. If others want to use the data, the data ownership authorization should be carried on. Generally, there are two methods of data ownership authorizations, the contract licensing and technology authorization. The former constrains the second party, while the latter allows the second party but constrains the third party. The technology authorization is ownership authorization, that is, using the user's public key to encrypt data to ensure only the user to use the data.

(3) Innate data encryption and registration

To achieve maximum protection of data, as well as be able to understand the condition of being protected data, we proposed a method of data innate encryption and data innate registration. Data registration is for its application, data encryption is to determine the ownership and to protect the data. To use related software to encrypt data automatically and to register data automatically when the data generated. In this way, we classified data into "data state" and "application state" two states. Data in "data state" is forever in an encrypted state, whether it is in a stored procedure, or in the course of transmission. Data in "application state" is in a decrypted state, appear only in the application process, while the application terminal is in a safe environment.

(4) Data recording, data discovery, and data provenance

To prevent the second party (data is authorized to use) the misuse of data, and/or to know data usage state, the data logging module is needed to record the data usage, the data authorization procedure, and so on. Also, data exception notice module, data source tracing module are needed to find data misuse or abnormalities, to trace the data source and to identify offenders.

(5) Data security application ecological system

DA's "one body two wings" builds data security application ecological system; break the traditional software engineering method and data security protection system.

DA determine data ownership and protect data via data encryption, build data big platform via data registration, achieve fragments of application in growth via shared collaborative mechanism. It can be vividly expressed as that a "plentiful and luxuriant" application "forest" is growing on the "fertile and rich" data "soil earth", under the protection of data and benefits of data owner, shown in Fig. 9.

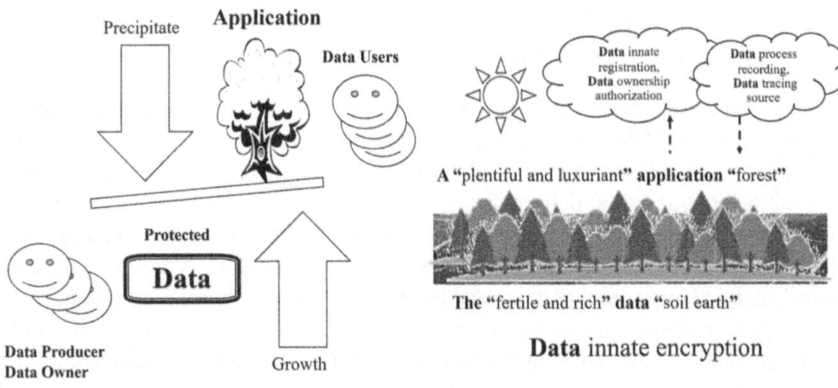

Fig. 9 DA builds Data-Security-Application Ecosystem

(6) Data security application and open platform

Most applications are centralized management, at least physically decentralized and logically centralized. For complex huge information systems, such as smart city applications, complex application scenarios involving multiple systems, multi platforms, multi-agent, multi data sources, and multiple applications, etc.

DA can form data application ecosystem in single and centralized platform, can also build multivariate, multi-center, hierarchical distributed, multi-platform data sharing ecosystem and fragmented applications based on data routing.

Blockchain technology have characteristics of decentralized, no third-party supervision, distributed ledger, tamper-proof, etc. It can be applied to a certain anonymous application scenario like virtual coins application without central management. DA's trust is based on data itself rather than on business, process, code, software, and hardware. It is more suitable to decentralized, anonymized applications. DA could construct a fully open platform for data application ecosystem of self-organization, self-management, full openness, no center or multi-center, transaction self-trust, data self-trust, data security protection and data security application.

3.6 Ten Steps Make the Data Architecture Work

We need 10 steps to make the data architecture work, shows in Fig. 10.

(1) Authentication

Everyone who is active in cyberspace needs to have a unique digital identity. Whether it is a data manager, a data server or a data user, or a legal entity representing an organization, a digital identity is required. This digital identity is CA. The CA provides each person with a digital certificate representing the identity, i.e., the public and private keys.

Figure 11 shows the open CA management platform for DA. The public key management depends on DRC.

It can provide everyone individually a PKI key generator to create a pair of public and private keys in his/her local terminal. The private key is just safely reserved in his/her local devices and no longer be handed in by a third party.

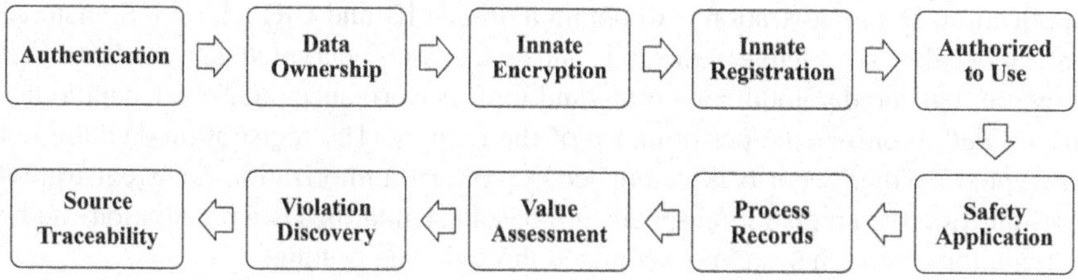

Fig. 10 The 10 steps to form the data application eco-system

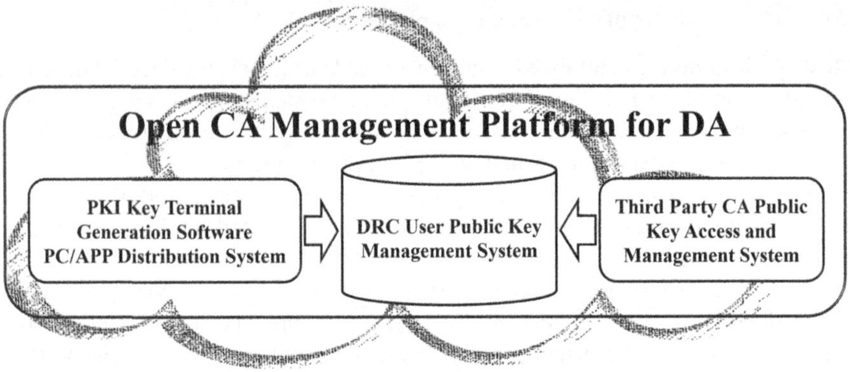

Fig. 11 Open CA certificate public key management platform for DA

The open CA platform can also handle any public key packages from any third party.

(2) Data Ownership

Each data has its own rights or ownership. Each person holding data is the data holder, but the data holder is not necessarily the data owner. For a single data, there are four different data holders: data owner, who owns all the rights and domination of the data; data producer, who is authorized by the owner to create and collect the data; data user, who is authorized by the owner to use the data; data agent, who is authorized by the owner to own a kind of rights of the data.

(3) Innate Encryption

The data are encrypted as soon as the data are produced. Data are kept encrypted whether they are stored or transferred. Encryption is related to data ownership. The encrypted key is the public key of the data owner or the public key of the authorized data user. Mass data are encrypted by using a symmetric key, and then the public key is used to encrypt the symmetric key.

(4) Innate Registration

Data must also be registered as soon as the data are produced. Open data do not require encryption, but they must be registered. The purpose of registration is for application. Data registration is to obtain a unique ID and URI address. Registered data are visible, searchable, indexable, and analyzable. Encrypted data can be used if they are registered. Cloud computing and high-performance database technologies are needed to ensure the performance of the registry. The registration data are not encrypted and their security is guaranteed by platform authorization. Data registration uses uniform data registration specifications and standards. Data registration can be autonomous for each system or federated through data routing.

(5) Authorized to Use

Users need data owner authorization to use the data. Authorization of legitimate data users requires a legal-level contractual license (contract authorization) and a technical-level ownership transfer authorization (technical authorization). The contract authorization, indicating that the authorized data should not be disclosed to any third party, is to bind the second party; the technical authorization is to encrypt the data with the user's public key and put it in the designated location, which is to allow the second party to bind the third party. Data replication is necessary. The consumption of time and space needs to be supported by cloud computing and fast hardware. Current conditions are now available.

(6) Safety Application

Data are only decrypted in applications. On the application terminal, the application software decrypts the data encrypted by the data user's public key by using his private key. The security of the application environment is required to ensure the security of applications and data. There should be no loopholes and Trojans in the application environment. Use technology to make decrypted data unreadable to other people or hardware and software devices.

(7) Process Records

Data authorization and application processes should be logged. The time stamps and spatial location information of the data activity should be recorded. Similar to the blockchain, the DRC records data events throughout the data lifecycle. This will also bring some time and space costs, but cloud computing is affordable.

(8) Value Assessment

Data are valuable only when they are used. Data can be easily copied and transferred for convenience. The data value assessment model is used to analyze the registered and recorded data information and calculate the data value. Data are assets and data owners should be rewarded for the value of data.

(9) Violation Discovery

There may be irregularities in the use of data. According to the data registration and recording information, the network crawler technology is used to compare and discover illegal data. It will become normal to patrol the network under the rule of data legal application.

(10) Source Traceability

Once the illegal outflow data are found, the registration and record information and the log file of the system are analyzed by big data, the traceability and location can be traced and the offender can be found.

4 DA Application Cases Study

DA of one body with two wings, constructs a reasonable and feasible mechanism in an open environment of Internet, coping with the situation of the new era. The following are some application cases study.

4.1 The Tourism Platform Based on DA

Figure 12 shows the tourism application platform framework based on DA [27]. The tourism application requirements include restaurant recommend, transportation recommend, hotel recommend, social activities, culture activities, economics activities. They need various data from governments, enterprises and individual tourists. DA can provide the platform to manage all kinds of data and to support all kinds requirements under the data sharing and data protection mechanism.

There are different kinds and various of data sources from tourists, enterprises, government and Web sites. The DA supports DOSA platform with DRC to provide many functions such as data collection, data registration, data management, data indexing, data search engine, data protection, data mining, data interests guarantee, etc. The platform then supports many kinds of applications for different people, such as restaurant recommendation, transportation recommendation, hotel recommendation, social activities, culture activities, economics activities, etc.

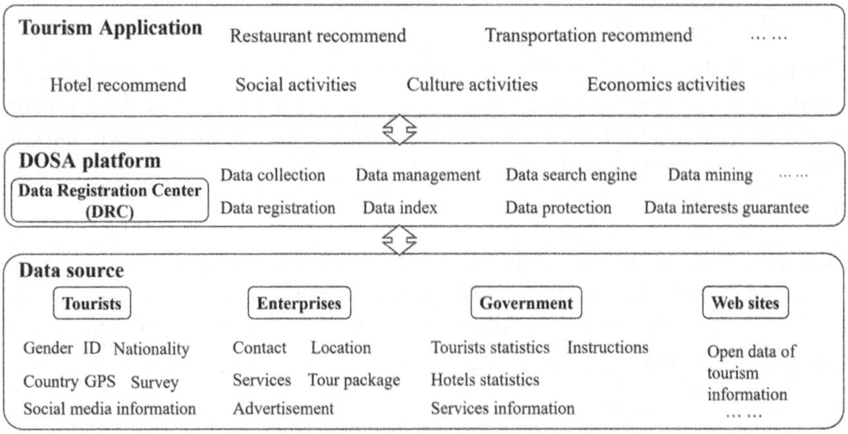

Fig. 12 The tourism platform framework based on DA

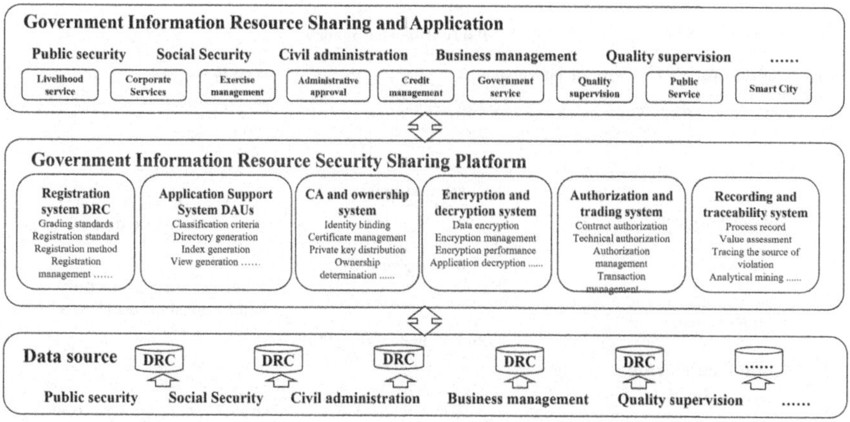

Fig. 13 The framework of government information resource security sharing platform based on DA

4.2 *Government Information Resources Sharing and Application System Under DA*

Government information resources sharing is concerned with different departments and is difficult to coordinate. Using DA, we can establish a sharing platform with registration system of DRCs, application supporting DAUs, CA and ownership system, authorization and trading system, as well as recording and tracing system. Then we can establish a sustainable system to connect data sources from all departments to government applications and services for all departments, shown in Fig. 13.

The framework can manage data sources from different departments such as public security, social security, civil administration, business management, quality supervision, etc. The platform then can support different applications for different departments, such as livelihood service, corporate service, exercise management, administrative approval, credit management, government service, quality supervision public service, smart city, etc.

4.3 *Personal Privacy Protection Based on DA*

The personal privacy should be protected, but we need to share them to others in a certain extend to get some kind of services. We often use our privacy data to get some service from the second party. For example, we share our personal phone number and family address to purchase goods from online stores. The privacy data we only allow the second party (Party B) to know but do not want any third party to know. The problem is that we don’t know and have no way to protect our privacy data if

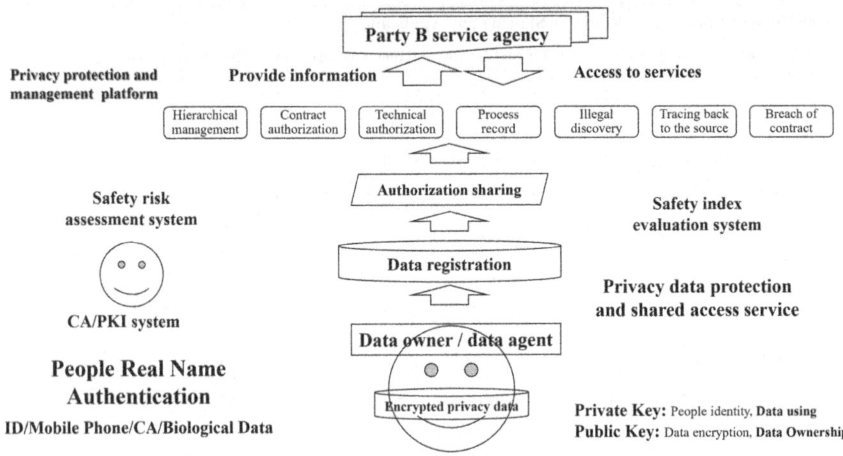

Fig. 14 Personal privacy protection based on DA

it is shared to the second party. Figure 14 shows the concept to protect our privacy under DA.

In this privacy protection and sharing system, the private data is encrypted by the owner's public key. It means that the person owns the data which were protected by the encryption with his own public key. When the person wants to get service from Party B, then he authorizes his private data to be used only by Party B. The authorization is to use Party B's public key to encrypt the data and then send to Party B. The contract authorization refers to the restriction signed with Party B that Party B shall not disclose the data to any third party. The technical authorization refers to restriction with any other third party who cannot decryption the data without the Party B's private key. If Party B leaks the data to any third party, we can find out who, when, where and how the data are leaked to whom through the process of recording, illegal discovery and traceability.

4.4 Data Trading Platform Based on DA

Data trading needs to know data owner and to protect the trading data. Figure 15 shows the procedures of data trading in the platform based on DA.

In this platform, data seller (data provider) is the agent of data owner, data buyer is the data user. There are 13 steps to finish one data transaction. (1) data should be registered; (2) using the platform to publish the data; (3) the platform gives the data pricing basis; (4) data buyer enquiring the data's information; (5) determine the data he wants; (6) the buyer apply the data authorization to the platform; (7) the buyer can negotiation the data price; (8) the data seller make sure the buyer's identity authority; (9) notify the buyer for payment; (10) the buyer makes payment;

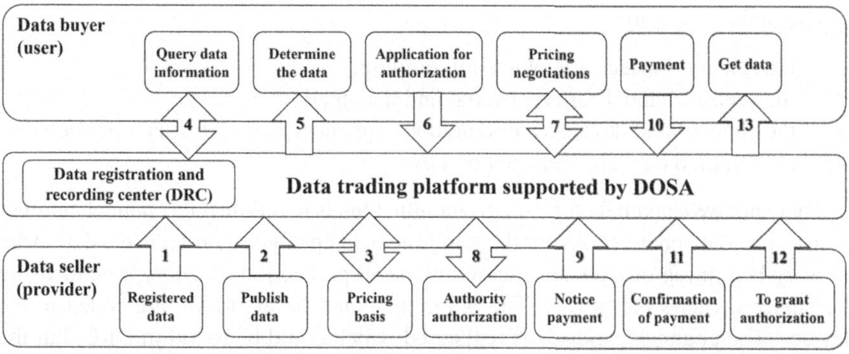

Fig. 15 Data trading platform based on DA

(11) confirmation the payment; (12) to grant the authorization, it means that the use of contract authorization and technical authorization for the data, like in Sect. 4.3, using the buyer's public key to encrypt the data, and then send the encrypted data to the buyer; (13) the buyer get the data.

4.5 Data Acquisition and Management System of IoT Based on DA

The security of IoT is now more and more important. It concerned with the safety of real time data acquisition. DA can provide the mechanism of protection real time data acquisition. Figure 16 shows the structure of safety data acquisition and management based on DA.

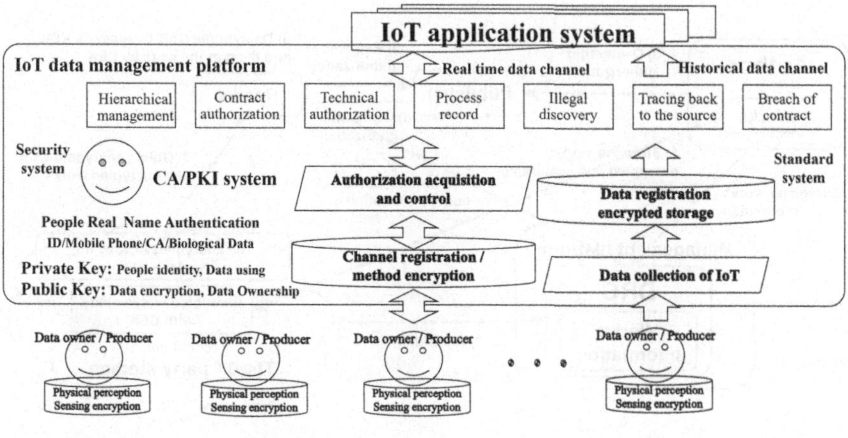

Fig. 16 Data acquisition and management system of IoT based on DA

The main points are:

(1) the sensing devices are ownership binding;
(2) the access channels are registered and managed;
(3) the channel and device access methods are encrypted by owner's public key;
(4) all obtained data are managed by DA.

There are two channels for data protection. One is real time data channel, another is historical data channel. As for the real time data channel, it protects the data from the very beginning of data sensing and physical perception by encryption of using the device owner's public key. The public key and the algorithm are coded in the devices. The channel's address is registered which could be found in DRC, but the accessing methods are encrypted by the owner's public key. Only the device owner can read or access the device. If some other people want to access the device, it needs to authorize the ownership of the sensing data and device accessing methods. As for the historical data channel, the protection methods are similar to Sect. 4.3.

4.6 Digital Copyright Works Protection and Trading Systems Based on DA

We established a digital copyright works protection and trading platform based on DA [21]. The main idea is to classify author as data owner, publisher as data agent, reader as data user, and then follow the rules of DA to protect the works and authorize it to reader and trade it by publisher.

Figure 17 shows the steps of the procedures of how a copyright work is protected and transacted. (1) The author has his work encrypted with a symmetric key which we call it CEK (Content Encryption Key), and then has the CEK encrypted with the author's public key which we call it KEK (Key Encryption Key). Then all encrypted

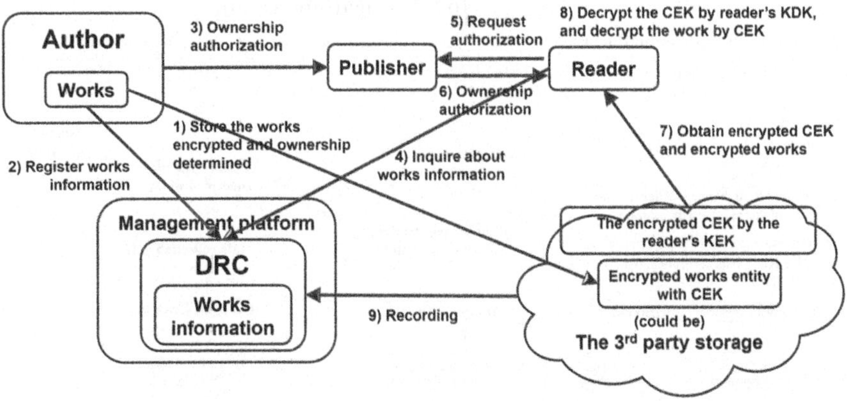

Fig. 17 Schematic diagram of digital copyright works distribution process based on DA

works and encrypted CEKs are stored into a 3rd party public cloud. (2) The platform automatically registers the information of the works and the CEKs in DRC. (3) The author signs the contract of the works authorization with the publisher, and changes the works ownership authorization by encryption of the CEK with the publisher's public key (KEK). (Note: here omitted some procedures such as: got data back from cloud and decrypted them by the author; restored the encrypted one into the cloud; registered the information in DRC; destroyed the original data; etc.) (4) Reader inquires the works information in DRC. (5) If the reader found a work he is interested, he may request the authorization of it to the publisher. (6) The publisher and the reader sign an agreement on authorization of the work, and the publisher changes the work ownership authorization by encryption of the CEK with the reader's public key (KEK). (Note: here also omitted some procedures). (7) The reader obtains the encrypted CEK and the encrypted work from the cloud. (8) The reader decrypts the CEK by his privacy key which we call it KDK (Key Decryption Key), and then decrypts the work by CEK. (9) The platform records all procedures in DRC automatically.

There are other procedures and modulars for illegal leak discovery and tracing the source, etc.

4.7 *Big Data Service to Civil People Based on DA*

Figure 18 shows a diagram of big data services for civil people which we call it as "data home". It wants to change the data owner becoming data user, to use his/her own data supported by DA.

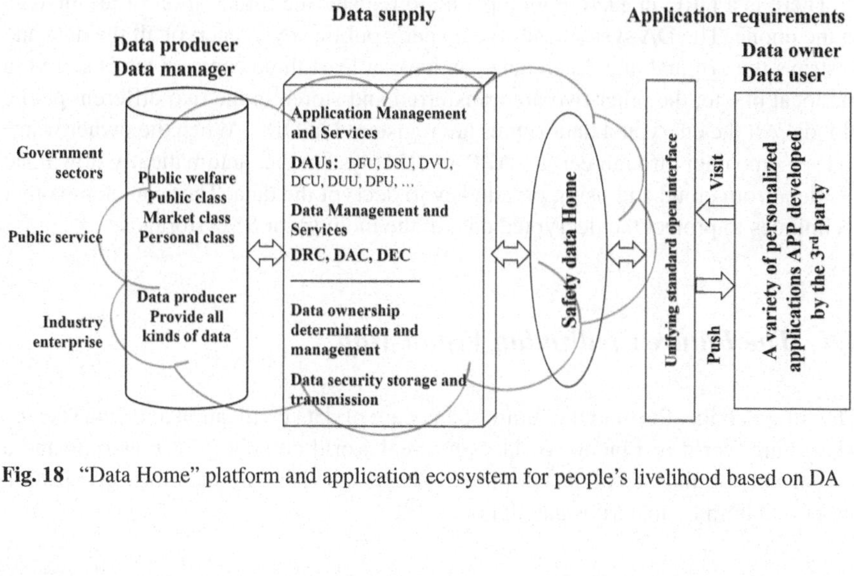

Fig. 18 "Data Home" platform and application ecosystem for people's livelihood based on DA

Now civil people have many personal data such as ID data, health data, training data, bank data, credits data, hydroelectric data, traveling data, consuming data, etc. in governments, enterprises and industrial departments, but not in people own hands. These data include important information and privacy for each people. We use DA mechanism and framework to get the data back to each people explained in APP developed by the third party to instruct people to live a better life.

The DA provides data management and services based on DRC, DAC and DEC (Data Exception processing Center) to deal with data ownership determination, data security storage, data transmission and data management. The DA also provides application management and services based on DAUs including DFU (Data Function Unit), DSU (Data Service Unit), DVU (Data Vision Unit), DCU (Data Connection Unit), DUU (Data Using Unit), DPU (Data Processing Unit), as well as DAI (Data Access Interface), API and SDK, etc. All these established a safety data home in the cloud. Through a unifying and standard open interface, each user can get his own data by visiting the safety home or by pushing to his terminals. In the user's terminal, the 3rd party can develop variety of applications to interpret the data and make instructions to people's daily life and work.

4.8 Security Mobile Phone Based on DA

With the development of 5G technology and commercial application, our smartphone will become a cloud-based data display and operation terminal. All data can be stored in clouds. Supported by DA, we can encrypt all private personal data and store them in any public clouds. This can make our phone leak-proof, theft-proof, extortion-proof and loss-proof. Figure 19 shows the case.

There is a DRC in each mobile phone to register the information of all the data in the phone. The DA system uses the owner's public key to encrypt all the data and destroys the original one. Each data can have at least three copies. One is stored in the local phone; the other two are transferred and stored in the two different public clouds. All the encrypted data copies are registered in DRC. When the owner wants to use some data through certain APP, the DA system can automatically download the data from cloud and using private key to decrypt the data. There are some other techniques to protect the decrypted data in the memory or SSD storages.

4.9 The Future Civilization Vision Map

The future civilization and the future society are of data civilization and data society. The future world is a blend world with a real world composed of materials and a virtual world composed of data. Figure 20 shows the vision for the future civilization supported both by materials and data.

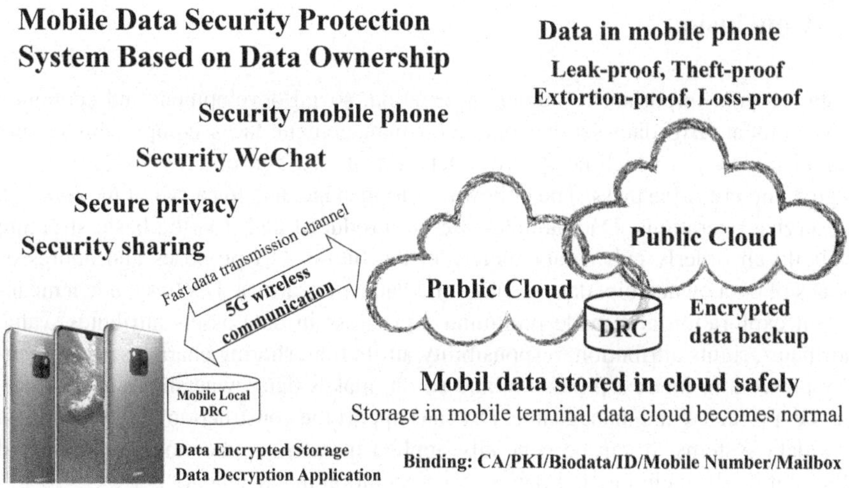

Fig. 19 Security mobile phone based on DA

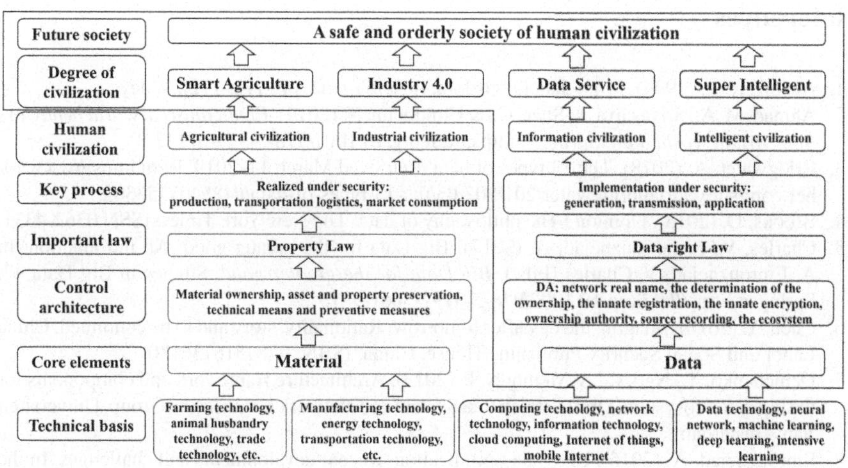

Fig. 20 The influence of data on human future civilizations

Materials and data are both core elements of our combined real and virtual worlds. The technical basis supports them to form the control architecture of our society. Associated with important laws such as property law and data right law accordingly and separately, the two worlds are implemented with secure and order. The four civilizations make people living a lovely and comfortable life in the future.

5 Conclusion

Data are vital important to human civilization, social development and economic development. Big data service operation management faces comprehensive and complex situation which needs cross-department, cross-level, cross-system, cross-region and cross-business. The only way is to manage data to carry out the function of service operations. Data architecture is introduced and it is the basic structure to build an orderly cyber data society in the future. DA proposes and addresses issues of data ownership, data security and data applications. DA has made a meaningful exploration and made preliminary progress in data assets attributes, value attributes, rights attribution, responsibility attribution, sharing characteristics, security protection, interest protection and so on, makes data ownership can be determined, protected and manageable. DA can support the construction of data platform and data systems. It can be generally applied in many application situations and different fields, including big data service operation management.

References

1. Ackoff, R. L. (1999). From data to wisdom. *Ackoff's best* (pp. 170–172). Wiley.
2. Ahmad, M. A., Srivastava, J., Shen, C., & Contractor, N. (2014). *Predicting real world behaviors from virtual world data*. Springer. https://doi.org/10.1007/978-3-319-07142-8
3. Bridgwater, A. (2018). The 13 types of data. Retrieved March 11, 2019, from https://www.forbes.com/sites/adrianbridgwater/2018/07/05/the-13-types-of-data/#485407553362.
4. Brooks, D. (2013). Opinion | The philosophy of data. The New York Times. ISSN 0362-4331.
5. Charles, V., & Emrouznejad, A. (2017). Big Data for the greater good: An introduction. In A. Emrouznejad, V. Charles (Eds.), *Big Data for the greater good*. Studies in Big Data 42. Springer. https://doi.org/10.1007/978-3-319-93061-9_1.
6. Chen, Y. (2018). Imaging the day after tomorrow: Rationality, story and to be continued. China Labor and Social Security Publishing House, China. ISBN: 9787516736180.
7. Demchenko, Y., Ngo, C., & Membrey, P. (2013). Architecture framework and components for the big data ecosystem, Draft version 0.2. System and Network Engineering Group, Universiteit Van Amsterdam.
8. Emrouznejad, A. (2016). Big Data optimization: Recent developments and challenges. In the series of "Studies in Big Data". Springer. ISBN: 978-3-319-30263-8.
9. Emrouznejad, A., & Marra, M. (2017). Big data: Who, what and where? Social, cognitive and journals map of big data publication. In A. Emrouznejad (Ed.), *Big Data optimization: Recent developments and challenges*. Studies in Big Data 17. Springer. https://doi.org/10.1007/978-3-319-30265-2_1.
10. Garlan, D., Morrison, R., Balasubramaniam, D., Oquendo, F., Warboys, B., Greenwood, R. M. et al. (1998) Software architecture.
11. Harari, Y. N. (2016). Homo Deus: A brief history of tomorrow. Harvill Secker.
12. Inmon, & William, H. (1989). Data architecture: the information paradigm. QED Information Sciences.
13. Janssen, M., & Joha, A. (2007). Understanding it governance for the operation of shared services in public service networks. *International Journal of Networking & Virtual Organisations, 4*(1), 20–34.
14. Kimball, R., & Ross, M. (2016). Data architecture. The Kimball Group Reader. Wiley.

15. Kitchin, R. (2014). The data revolution: Big Data, data infrastructures & their consequences. SAGE Publications, Social Science. ISBN: 1473908264, 9781473908260.
16. Lohr, S. (2015). *Data-ism: The revolution transforming decision making, consumer behavior, and almost everything else*. HarperCollins Publishers.
17. Cohen, M. C. (2017). Big data and service operations. *Production and Operations Management*. https://doi.org/10.1111/poms.12832
18. Miao, F., Fan, W., Yang, W., & Xie, Y. (2019). The study of data-oriented and ownership-based security architecture in open Internet environment (pp. 121–129). ICCSP 2019, January 19–21, Kuala Lumpur, Malaysia, ACM. ISBN 978-1-4503-6618-2/19/01. https://doi.org/10.1145/3309074.3309093
19. Miao, F., Yang, W., Fan, W., Xie, Y., Guo, Q., You, Y., Liu, Z., & Liu, L. (2018). Digital copyright works management system based on DOSA. In *Proceeding CSAE '18, Proceedings of the 2nd International Conference on Computer Science and Application Engineering*, Article No. 179, Hohhot, China. ACM. ISBN 978-1-4503-6512-3/18/10, https://doi.org/10.1145/3207677.3278047.
20. Miao, F., Yang, W., Li, Y., Yu, B., Zhou, X., Song, C., & Wang, G. (2014). A data-oriented information technology system. CN Patent 201410341092.X, PCT/CN2015/080013.
21. Miao, F., Yang, W., Xie, Y., Fan, W., & Wang, J. (2018). The study of ownership-based data security application. In *Proceedings of the 7th International Conference on Advanced Technologies (ICAT'18)* (pp. 393–399). Antalya, Turkey. E-ISBN: 978-605-68537-1-5.
22. Miao, F., Yang, W., Ye, A., & Chen, H. (2016). The architecture for data security and application in Digital Earth platform. In *IOP Conference Series: Earth and Environmental Science* (Vol. 46, no. 1, p. 012050).
23. Miao, F., Yang, W., Ye, A., et al. (2015). A data-oriented security architecture method and system. CN Patent 201511025657.4, Public No. CN105450669B.
24. Miao, F., Yang, W., Xie, Y., Fan, W. (2019). Consideration and research on data architecture for the future cyber society. In *IEEE SmartWorld, Ubiquitous Intelligence & Computing, Advanced & Trusted Computing, Scalable Computing & Communications, Cloud & Big Data Computing, Internet of People and Smart City Innovation (SmartWorld/SCALCOM/UIC/ATC/CBDCom/IOP/SCI)*, Leicester, United Kingdom (pp. 1671–1676). https://doi.org/10.1109/SmartWorld-UIC-ATC-SCALCOM-IOP-SCI.2019.00298.
25. Miller, F. P., Vandome, A. F., & Mcbrewster, J. (2010). Hardware architecture. *Astronomy & Astrophysics*, 69–76.
26. Mumford, L. (1967). Technics and human development: Myth of the machine (Vol. 1). New York: Harcourt Brace Jovanovich Publishers.
27. Panpeng, V., Miao, F., Phaphuangwittayakul, A., Rattanadamrongaksorn, T. (2021). Preliminary study and implementation of Chiang Mai Tourism platform based on DOSA. In X.S. Yang, S. Sherratt, N. Dey, & A. Joshi (Eds.), *Proceedings of Fifth International Congress on Information and Communication Technology*. Advances in Intelligent Systems and Computing (Vol. 1184). Springer. https://doi.org/10.1007/978-981-15-5859-7_51.
28. Pras, A., & Schoenwaelder, J. (2003). On the difference between information models and data models. RFC Editor.
29. Radnor, Z., & Bateman, N. (2016). Debate: The development of a new discipline—Public service operations management. *Public Money & Management, 36*(4), 246–248.
30. Sass, S. L. (1998). *The substance of civilization: Materials and human history from the stone age to the age of silicon*. Arcade Publishing.
31. Sun, Y., & Liu, X. (2010). Business-oriented software process improvement based on CMMI using QFD. In *Information and software technology*, 52, 1 (Jan. 2010) (pp. 79–91). https://doi.org/10.1016/j.infsof.2009.08.003.
32. Targowski, A. (2008). Information technology and societal development. Information Science Reference— Imprint of. IGI Publishing Hershey, PA. ISBN:1605660051 9781605660059.
33. Tupper, C. (2011). Data architecture.
34. Walker, S. J. (2014). Big Data: A revolution that will transform how we live, work, and think. *International Journal of Advertising, 33*(1), 181–183. https://doi.org/10.2501/IJA-33-1-181-183

35. Wohl, M. (2003). Turning data into usable information: in today's world, gathering data is the easy part; putting it into usable information is a lot harder. (fw focus: technology). Franchising World.

Professor Dr. Fang Miao was born in Beijing City in 1958. He received the Ph.D. degree in engineering from Chengdu University of Technology (Former Chengdu Institute of Geology) in 1990. He is a professor and Ph.D. supervisor, as well as the director of Research Institute of Big Data at Chengdu University. His research interests include Big Data, Geophysics, Spatial Information Technology, Computer Science and Technology, etc.

Associate Professor Ms. Wenhui Yang was born in Shanghai City in 1969. She received the B.S. degree in engineering from Chengdu University of Technology (Former Chengdu Institute of Geology) in 1991. She is an associate professor of Information Science and Technology College at Chengdu University of Technology. Her research interests include Big Data, Computer Applications, etc.

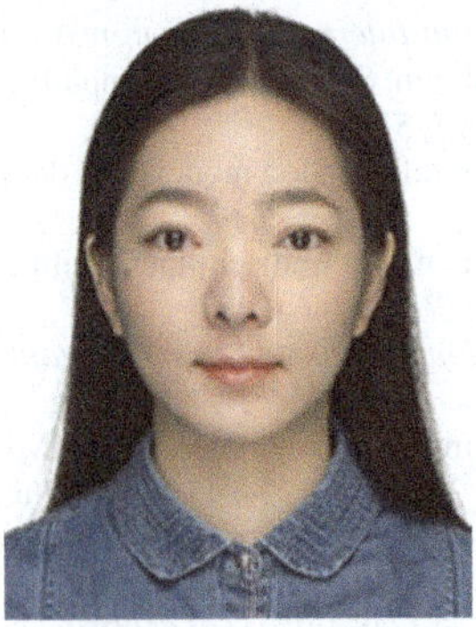

Senior Engineer, Ms. Yan Xie was born in Changling County, Jiling Province in 1986. She received the Ph.D. degree in engineering from Chengdu University of Technology in 2020. She is a senior researcher in Sichuan Provincial Big Data Center. Her research interests include Big Data, Artificial Intelligence, Spatial Information Technology, etc.

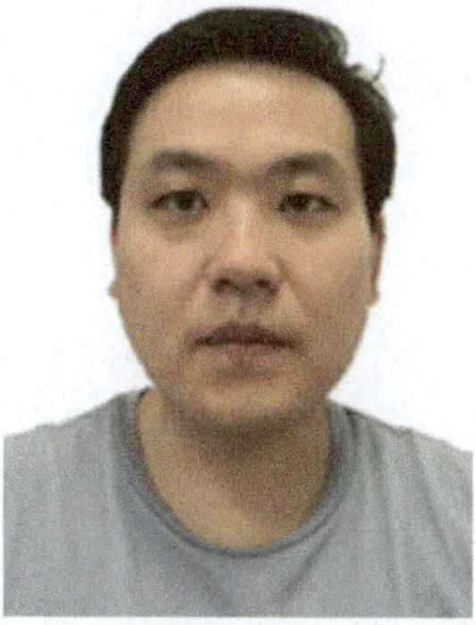

Associate Professor Mr. Wenjie Fan was born in Anyue City, Sichuan Province in 1978. He received the M.S. degree in engineering from Sichuan Normal University in 2012. He is an associated professor of Computer College at Chengdu University. His research interests include Big Data, Computer Applications, etc.

Big Data for Educational Service Management

Santosh Kumar Ray, Mohammed M. Alani, and Amir Ahmad

Abstract With over 3.7 billion people using the Internet in 2018, the amount of data being generated has exceeded 2.5 quintillion bytes per day. This rapid increase in the generation of data has pushed the applications of big data to new heights. Like any other field, education must also have its share of applications. Large amounts of data generated in the education sector can be harvested for useful information that can improve the delivery of education inside and outside of classroom. This chapter presents a survey of the state-of-the-art applications of big data analytics in the field of educational services. Educational services consist of all the tasks and subtasks that support the education process and help it achieve its goals. The utilization of big data is expected to increase the satisfaction of all stakeholders of a higher education institute. Educational services are not limited to teaching and learning processes in the classroom, rather they include efficient registration and administrative process, library services, research activities, extracurricular activities, and overall development of faculty and students. This chapter explains how big data can help in improving the overall education services. It also describes the challenges that institutions face while implementing big data-based solutions. Finally. This chapter suggests the possible directions for future research that can be explored to maximize the benefits of big data in improving educational services.

S. K. Ray
Department of IT, Al Khawarizmi International College, Al Ain, UAE

M. M. Alani (✉)
School of Information Technology Administration and Security, Seneca College, Toronto, Canada
e-mail: m@alani.me

A. Ahmad
College of Information Technology, UAE University, Al Ain, UAE

A. Emrouznejad and V. Charles (eds.), *Big Data and Blockchain for Service Operations Management*, Studies in Big Data 98,
https://doi.org/10.1007/978-3-030-87304-2_5

1 Introduction

According to classical three-sector theory [1], the economy can be divided into three sectors: Primary sector that involves the extraction and production of raw material with minimal processing, the secondary sector that produces finished or usable products, and tertiary sector which includes virtually everything which is not in the first two sectors. The tertiary sector, also called the service sector, is concerned with the people and processes involved in the primary and secondary sectors. The service sector usually does not have a tangible product to offer, but it helps to increase the productivity in the other two sectors. In contrast to a typical producer and consumer relationship in primary and secondary sectors, the service sector involves several stakeholders. The stakeholders in the service industry may be internal or external to the organization, maybe a producer or consumer, or may be direct or indirect beneficiary of a service. The main roles of these stakeholders are to define, deploy and monitor the services. Based on the activities involved in it and the definition of the service sector, it can be safely assumed that the education industry falls into the service sector category. Different references identify the stakeholders in education differently [2, 3]. In this chapter, we will focus on internal stakeholders of a higher education institution which are: learners, educators, educational researchers, course developers, learning institutions, and education administrators. Some external stakeholders can be alumni, prospective students, parents of the students, vendors, government, and employers. However, in this chapter, we will stick to the six internal stakeholders mentioned above. The two central tasks in the higher education environment are conducting research and imparting education. These two central tasks are comprised of several processes such as content creation, delivery of content to the learners, access, and management of the content, video conferencing, etc. However, there are several supporting activities to achieve the objectives of these central tasks. Some of these activities may include the creation of education portfolios of learners, professional development of faculties and administrative staffs, management of social media and online community, preparing a marketing strategy, organizing fundraising services, dashboard management, asset management, library services, etc. The developments in the field of information and communication technology have enabled the universities to record both the activities of the stakeholders and the content created during these activities. Accumulation of the data generated by the stakeholders over the years results into a huge volume of data. As seen from Fig. 1, this enormous volume of data generated through educational services, if analyzed properly with the help of big data analytics, can be used to improve the university services ranging from the student learning experience, student retention to faculty acquisition and retention.

Big data has become a very popular term across the academia and industry alike. However, due to its complex and evolving nature, researchers and practitioners of big data have different opinions on the precise definition of big data [4]. Most commonly agreed definition of big data describes it in terms of three V's namely, Volume, Velocity, and Variety [5, 6]. The first V, Volume, indicates that big data consists

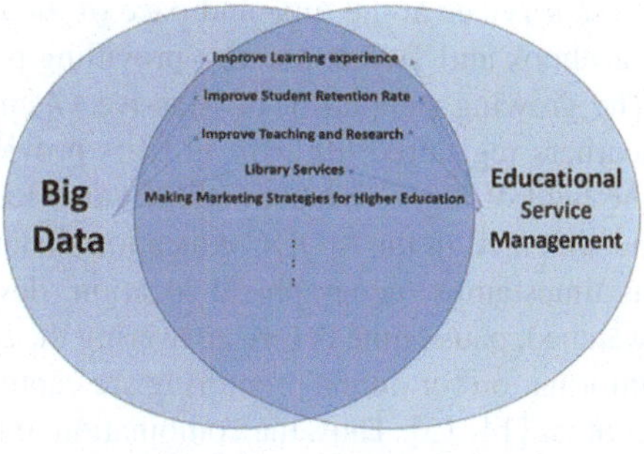

Fig. 1 Use of Big Data to improve educational services

of a very large volume of data usually ranging from few terabytes to hundreds of petabytes. The velocity of data is defined as the rate of generation and processing of data in real-time. For example, Walmart generates more than 2.5 PB of data per hour through user transactions [7], which is a very high speed by standards of traditional data warehouses and traditional database processing techniques. Variety indicates that big data consists of data of multiple formats generated from distributed and heterogeneous sources. In contrast to relational databases, this data is highly unstructured. Over time, the practitioners of big data kept adding more Vs to describe the nature of big data in a holistic way, for example Charles and Gherman [8] conducted a literature survey and found that big data can be described using using 4 Vs, fourth being Veracity. A search on the Internet reveals the use of up to 42 Vs to describe big data [9]. Due to the limitation of space, we cannot describe all Vs in this chapter. However, we will describe the next two Vs added (in chronological order) to characterize big data: Value and Veracity. Unlike the original three Vs which deal with quantitative aspects of big data, the next two Vs deal with qualitative aspects of big data. The big data does not make any sense unless some meaningful information can be extracted from it, which can add some value to the organization. Veracity helps to judge the quality of the big data. It helps to determine the reliability and accuracy of the content generated by devices and users.

The service industry has become the backbone of the modern world economy. According to the World Bank Report 2017, the service industry contributes 65.1% of the world GDP [10]. In some places like Macao, China this percentage is as high as 92.5%. Big companies and governments are investing a lot in developing infrastructures to boost the service sector. The development of the service sector infrastructure requires the installment and availability of a larger number of sensors, mobile devices, satellites, and other computer-based systems. In addition, companies are providing cloud-based services to their customers. The extensive use of these devices is generating a large amount of data at very high speed. The education sector, being a part of the service sector, is also witnessing the generation of a large volume of data at a rate never seen before. Higher education institutions are adopting cloud-based technology in delivering lectures, registering students, marketing for their services, monitoring the security of their campuses, and other tasks [11]. The availability of these services anytime anywhere helps the stakeholders to interact with these services at the time and pace of their choice. Social media platforms such as Facebook and Twitter are also providing platforms to support mobile learning [12]. The growing popularity of Massive Open Online Courses (MOOCs) (81 million learners registered till 2017 [13]) is providing another great source of big data in the field of higher education. All video lectures, activities of learners are properly documented in the MOOC education system. Several data and metadata types such as timestamps, geographical location, device identifier, the fraction of the video watched, pause time before answering the question, facial recognition and biometric information for online proctoring are captured in MOOCs and other kinds of online courses [14, 15]. Thus, the combination of traditional education and MOOC courses can generate a massive dataset that can be mined to improve customer services in the education sector.

Daniel and Butson [16] proposed a big data analytics framework for higher education that divides big data analytics into four major components: institutional analytics, information technology analytics, academic/program analytics, and learning analytics. This framework classifies data into six categories: administrative data, research data, student data, teaching and learning data, department data, and curriculum data. Linking and analysis of these related data can provide meaningful insights for higher education institutions. Linking of data can facilitate the accessibility, sharing, and enrichment of data on the Web [17]. Analysis of the educational datasets proposed in this framework can be done by using the combination of two overlapping techniques namely Educational Data Mining (EDM) and Learning Analytics (LA). These techniques develop a capacity for quantitative research in response to the growing need for evidence-based analysis related to education policy and practice [18]. Educational data mining is an emerging research field that uses data mining techniques to analyze data generated in an educational context with an objective to understand issues related to the learning process [19]. Educational data mining combines theoretical concepts such psychopedagogy, cognitive psychology, psychometrics with practical techniques such as statistical analysis, machine learning, data mining, social network analysis (SNA), and visual data analytics to understand learning, administrative and research issues in the education sector. Figure 2 shows the education data mining and its relation with other areas. Education data mining provides meaningful information from raw data generated in the educational context with an objective to improve the learning experience of students. Some of the models used in educational data mining are: multiple linear regression model, multilayer perceptron (MLP) network, radial basis function (RBF) network, and support vector machine [20]. A review of the developments in the field of educational data mining can be found in [19]. The Society for Learning Analytics Research [21] defines LA as: "... *the measurement, collection, analysis and reporting of data about learners and their contexts, for purposes of understanding and optimizing learning and the environments in which it occurs*." The objective of LA is to analyze the large dataset and provide feedback that has an impact directly on students, instructors, and details of the learning process. For example, using LA techniques, educators can determine that poor grade in a specific course, say chemistry, can be attributed to poor understanding of one concept in the prerequisite course in the previous semester [22].

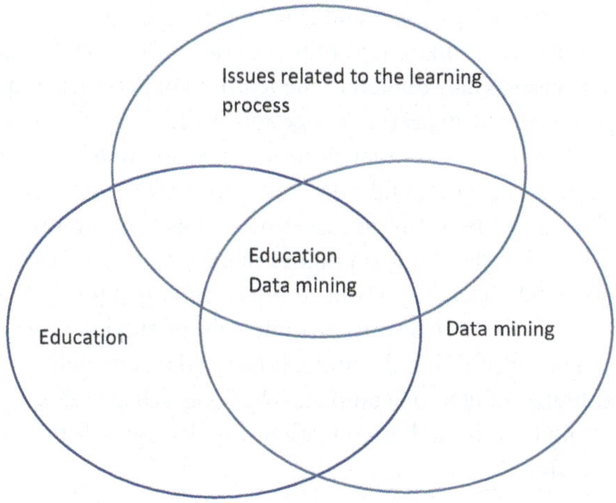

Fig. 2 Education data mining

Information based on analyzing the student activities and interaction among students can be used to provide a personal recommendation to students learning or customization in the course material [23, 24]. Although LA and EDM share many attributes and have some similar goals and interests, there are some major differences between them as described in [25]. LA has an origin in Semantic Web, intelligent curriculum, and systemic interventions while EDM has origin in educational software, student modeling, and predicting course outcomes. Researchers in EDM focus more on automated methods for discovery in educational data while researchers in LA focus more on human-led methods. Researchers in LA use statistics, visualization, social network analysis, sentiment analysis, influence analytics, discourse analysis, concept analysis, and sense-making models more frequently while researchers in EDM rely more on classification, clustering, Bayesian modeling, relationship mining and discovery with models. LA focuses on the description of data and results while EDM focuses more on procedures and techniques. In the specific context of higher education, a detailed description of applications and tools of EDM and LA in handling big data can be found in [26].

Higher education institutions are under tremendous pressure to provide high-quality educational experience while keeping minimal costs [27, 28]. This compels researchers to focus on the improvement of all components of educational services. Some of the components and activities may be more significant and more frequent while some other components and activities may not be that frequent or significant. The unique contribution of this chapter is to introduce the reader about the usage of big data in improving day to day activities irrespective of the significance or frequency of the activities. We have attempted to cover teaching–learning, research, student recruitment, student retention, library services, marketing services, sports, and other activities in this chapter and how these activities can be improved with the help of big data techniques. Based on the background information provided in this section, the rest of the chapter discusses how big data can be used to improve some of the processes and activities related to central tasks in higher education institutions. The remaining of the sections of the chapter are organized as follow: Sect. 2 discusses issues related to the learning experience and role of big data in developing personalized experience. Section 3 discusses the performance of the student, the dropout rate, and other issues related to student retention and the role of big data in early detection and mitigation of these issues. Section 4 discusses the role of big data in the two central tasks that is teaching and research. Section 5 discusses the role of big data in formulating marketing strategies for the higher education sector. Section 6 discusses library services and big data, and how these two can benefit each other. Section 7 discusses some miscellaneous issues of campuses where big data can be helpful. For example, how big data can help to tackle health and safety issues at higher education campuses. Section 8 discusses challenges in the implementation of big data in higher education. Finally, Sect. 9 provides the conclusion and future research directions.

2 Big Data to Improve the Learning Experience

The learning experience in the context of higher education is a broad concept encompassing various kinds of emotions a learner goes through while interacting with living and non-living entities of a higher education institution. The students' learning experience may include their experience with teachers in classes, experience with university's service units and administration, interaction with peers, interaction with equipment in department laboratories, interaction with library resources, cleanliness and safety of the college campus, etc. Thus, the focal service provided in any university or college is the learning experience. However, the traditional system of collecting feedback from students to measure learning experience has serious limitations. The ubiquitous presence of technology and the availability of big data can help to improve the learning experience and overcome the limitations of traditional feedback methods. For example, in the field of medical education, it is very risky for students to work with live patients. On the other hand, working with plastic dummies does not provide an enriching experience to the students. Applications, for example, human simulators prepared using Artificial Intelligence and virtual reality will create an almost-real environment where medical students can have an enriching learning experience. To improve the accuracy of the measurement of the learning experience, E-portfolios of the students can be created which would contain all academic records of the students such as assignments, projects, online quizzes, and their evaluations. Once stored on the server of the university or the data centers of university, these portfolios can be accessed anytime anywhere by all stakeholders. An educational cloud created in this way will have a lot of data. A critical analysis of this data may give key insights into the students' learning experiences and help in identifying a learning curve for the student to assure that the student has achieved the intended learning outcomes. Lessons learned from this kind of analysis can be used to customize the programs to meet the requirements of individual students. This customized learning experience will not only increase the genuine interest of the student in the course but will also create a healthy learning environment where students with different needs and capabilities are attended to.

The twenty-first century has witnessed the integration of ICT into teaching and learning in massive proportions. Currently, educators are using several online and offline tools to create high quality, easily understandable content for learners. One of the most adopted ICT tools in higher education institutions during the first decade of this century was web-based Learning Management Systems (LMS). LMSs such as Blackboard and Moodle helped educators in bringing learning content and resources together besides other administrative jobs such as assessment of student's work etc. However, due to fast-changing requirements of the higher education sector, LMSs are proving to have limitations in their monitoring capabilities of students' activities. This has pushed the emergence of distributed heterogeneous tools used by all the stakeholders of the learning process [23]. These distributed homogenous tools are embedding data mining techniques to collect the required data, analyze them, and suggest the appropriate actions. Big data can also help in several ways, such as tracking the time taken by students to learn a given concept. This will be an indicator of the level of difficulty of the concepts provided in the study material and will help in determining the learning ability of the students. For example, Researcher Paulo Blikstein, in [29], examined a sample of college students in a computer programming class to see how they solve a modeling assignment. He used NetLogo software [30] to maintain logs of all user actions, from button clicks and keystrokes to code changes, error messages, and use of different variables. He found that error rates progressed in an "inverse parabolic shape" as students tried things and made a lot of mistakes initially, and then progressed through problem-solving until they had developed the correct model. Researchers are also investigating various techniques to analyze the students' moods instead of simply analyzing the computer commands used by the students. This can help to assess students' interest in the course more deeply. Moridis and Economides, in [31], proposed a formula-based method and a neural network-based method to automatically collect the affect-state of the student during learning.

A typical large size class, especially in e-learning environments, consists of students from different academic backgrounds whose educational requirements are quite different and sometimes conflicting too. Offering the same learning path of content to all of them can negatively affect the overall performance of the class. It has been observed that personalized learning can make learning activities more effective by suiting the learning process to learners' needs and enhancing learner motivation [32, 33]. EDM techniques can be used to create a customized learning environment in which students can be provided personalized learning paths which can optimize their performance [34]. Some data mining techniques such as clustering, association rules [35], and feature selection [36] have been applied in developing personalized learning systems with an aim to improve individual learning performance. Due to their inherent strength such as displaying results in user-understandable formats, ability to analyze both continuous and discontinuous variables efficiently, and flexibility with type and scale of databases, decision trees have been popular in designing personalized learning contents [37, 38]. Designing the personalized learning content path requires an accurate estimation of the learning abilities of students at various stages. Researchers have considered this issue as well and have used statistical techniques such as Gaussian approximation method to estimate the learning ability of students in a typical web-based learning environment [39]. Once the learning abilities of the individual learners have been determined, big data can help in recommending customized textual/video contents according to the learning abilities of the learners. Then learners can read/watch the recommended content anytime and at their own paces. Once they become familiar with the fundamental concepts of the recommended content, physical classes can be used as a discussion-only environment to provide a deeper understanding of the contents.

3 Big Data to Improve Student Retention Rate

Governments across the world are reducing the funding of higher education institutions. To fill the gap between decreasing revenue and increasing expenses, institutions are under tremendous pressure to increase new enrollments and retain existing students at higher rates. One of the several measures adopted by the institutions is to predict the possible dropout students at the early stages so that the necessary remedial measures can be pre-planned. Big data analytics and related technologies can help in predicting possible students' dropout by extracting and analyzing various parameters such as cumulative grades in pre-requisite courses, marks obtained in earlier assessments, students' participation in in-class and out-of-class activities, multi-modal skills, etc. Once a group of possible dropout students is identified, the student retention rate of the institute can be improved by taking proactive measures such as one-to-one tutoring, peer-tutoring, customized student counseling, the arrangement of remedial classes, etc. [20, 40]. In order to predict students' dropout at early stages in e-learning courses, Lykourentzou et al. [41] applied a combination of three machine learning techniques named; feed-forward neural networks, support vector machines, and probabilistic ensemble simplified fuzzy ARTMAP on detailed students' logs from a learning management system. Dekker et al. [42] used various classification algorithms to predict students' dropout and identify factors of success based on the first-year enrollment data itself.

In addition to the detection of possible students' dropout, researchers used big data techniques to measure satisfaction, motivation, and performance of the students by analyzing their activities during lectures and exams. Kizilcec et al. [43], classified learners according to their interactions with course content (video lectures and assessment) in learning activities in MOOCs. The study analyzed behaviors and engagement patterns of students in three different computer science courses and compared clusters based on learners' characteristics and behavior. Giesbers et al. [44], investigated the relationship between students' interaction with synchronous tools in online education and performance on a final exam. Another objective of this research was to determine whether actual usage of synchronous tools increases the motivation to participate in online courses that support these tools. Abdous et al. [40], and He [45], examined the relation between interactions within a Live Video Streaming (LVS) environment and students' final grades in order to predict their performance, discover behavior patterns in LVSs that help increase performance, and understanding ways in which students are engaged in online activities. Dejaeger et al. [46], studied how retention of students can be increased by enhancing students' satisfaction. The study investigated students' satisfaction using several class and training variables such as the perceived usefulness of training, perceived training efficiency, etc.

Realizing the limitations of statistical models in the accurate prediction of student performances, Xing et al. [47], used learning analytics, educational data mining, and human computer interaction (HCI) theory to develop a model to predict the final performance of the student. They used data of 122 students from an online Math course to generate categories of student performance. Zacharis et al. [48] used a multivariate model to predict the "at-risk" students. The study looked into 29 variables used in a Moodle-based learning environment and found only 14 variables to be significant. However, the results suggested that only four variables—reading and posting messages, Content creation contribution, Quiz efforts and Number of files viewed—provide an 81.3% accuracy in prediction. Niemi and Gitin [49], used some non-academic parameters (e.g. age, gender, race, marital status, military status, previous college education, estimated family financial contribution, and the number of transfer credits from another university) besides the academic variables (final exam score, discussion participation, project scores, and other assignment scores) to predict the possible rate of students dropout. They studied and applied the learning analytics to a database of 14,791 students enrolled in a fully online program. They observed that married people or people working with the military have a lower probability to drop the course. The use of textbooks and e-books play a very crucial role in improving the performance of the students. Junco and Clem [15] used linear regression analysis technique to link the usage of digital textbooks with the performance of students. They stressed that this easily traceable variable is a much stronger indicator of performance than other variables including the past performance of the student. Mouri et al. [50] also used Bayesian Network to analyze approximately 330,000 logs from 99 first-year students to establish the relationship between the usage of e-books and the performance of students.

Studies mentioned above clearly show that, despite published previous work, improving retention rate using big data is still an under-explored area. More research needs to be conducted to identify the high-impact factors that increase the probability of dropping out, and then to find solutions to mitigate the risks of dropping out.

4 Big Data to Improve Teaching and Research

Big data techniques can be useful in identifying academic resources to increase awareness of the instructors. Deep analysis of textual and video data available through MOOCs can provide many insights for instructors that can help them create a better learning environment for students. For example, after analyzing the video data and performance of the students, researchers discovered that the presentation of the instructor's face in video lectures influences attrition and achievement rates and they found heterogeneous effects on attrition [51].

Assessment is a critical part of the teaching and learning process. Easy availability of gadgets, in addition to advances in technology, have dramatically increased the complexity of designing and evaluating assessment components of a course [52]. Gibson et al. [53], discuss four challenges in educational assessment posed by the technical advances: how to deal with change over time and time-based data; how digital performance space's relationships interact with learner actions, communications and products; how layers of interpretation are formed from translations of atomistic data into meaningful larger units; and how to represent the dynamics of interactions between and among learners who are being assessed by their interactions in digital performance spaces. Traditional methods of assessments cannot provide efficient solutions to these challenges. However, the use of EDM in the assessment of learning can result in faster progress as EDM can provide real-time continuous assessment [54]. Instead of conducting a periodic exam with a fixed set of questions for all students, big data can be used to create customized tests meeting the assessment requirements of the knowledge level of the student. This can enable the instructor to find out the precise weakness of each student and to prepare a study plan tailored to the needs of individual students. Romero et al. [55], used association rules mining to improve quizzes and courses. They used different objective and subjective rule evaluation measures to select the most interesting and useful rules. Based on the selected rules, the proposed system provided feedback to the instructors to improve quizzes and courses. Data mining methods such as clustering, classification, and association analysis have been used to study how well the questions in the test and the corresponding elaborated feedback were designed or tailored towards the individual needs of the students [56]. Kizilcec et al. [57] conducted research to study how the level of transparency about the peer grading process affects learners' trust in peer grading. They concluded that a fair and transparent peer grading procedure can promote resilience in trust of learners who received a lower than the expected grade. However, the downside of peer pressure was studied by Rogers and Feller [58] who found that exposure to exemplary peer performance causes attrition, due to the upward social comparison that undermines motivation and expected success.

Big data can assist faculty and research scholars to conduct research in a more efficient and organized way. Availability of large data increases the confidence and accuracy of the results of research methodologies. Researchers can leverage the EDM and LA techniques to explore and research large data made available by a MOOC. For example, an in-depth analysis of the demography of students enrolled in MOOCs can provide researchers with heterogeneous samples of people from traditionally under-represented demographic and socio-cultural groups in more narrowly obtained educational datasets. The researcher can leverage this data to conduct large-scale field experiments and evaluate multiple theory at a minimal cost [59]. For example, article recommender, a big data-based software, can analyze the research interests of the research scholars on their past research and reading habits. Article recommender software then recommends interesting articles to researchers by offering articles related to the area of their research. This will save efforts and time of researchers in finding articles by keyword search through search engines whose output depends on the right formulation of a search query, or through citations from other articles. By using citations from other articles, the researchers may be at risk of missing some more important articles just because the authors of the article did not cite them [60]. Big data-based software can also scan millions of articles and journals and books to assist research in generating and validating new hypotheses. Big data techniques can analyze the research interest of millions of researchers and recommend possible research collaboration possibilities.

5 Big Data in Making Marketing Strategies for Higher Education

Until the beginning of the twenty-first century, marketing of services was not considered a big concern for the higher education institutions. Most of the institutions were easily getting students because there was comparatively a smaller number of higher education institutions, and the demand for university graduates was high. But, the supply of university graduates has increased over time due to the increased number of higher education institutions and online courses (MOOC). Under these changed circumstances, marketing has become a necessity for higher education institutions. Good marketing plan, combined with high-quality teaching and learning can help universities compete at a global scale [61]. Big data can help in preparing and measuring the effectiveness of marketing strategies. Białoń [62] points out that effective marketing strategy of a higher education institution must include: aligning marketing strategy with the mission of the institute, and correct choice of the targeted segments of the market and marketing tools. Research also noted that corporate marketing strategies are not always suitable for educational services because educational services are more complex with many different elements that interfere with the client's decision-making process in comparison to a corporate product. The study recommends relationship marketing and spiritual marketing strategies for the education services where concepts of creating a better world are promoted and a long-term relationship is maintained with the clients (e.g. students and employers). Big data can provide several inputs to prepare a precise and dynamic strategy. Big data can suggest what kind of knowledge and skills are in demand (by analyzing the demands of local and global industry from their recruitment websites, job advertisements), the demand for which skill is less present but growing fast. Big data can be used in analyzing logs of visits on the institution's website and social media pages (most probably these are potential students or potential employer). There are several information dissemination channels such as advertising in the press, SMS, educational fairs, open day in colleges, in addition to social media. Not all these channels are equally effective. Analysis of how effective an individual channel is, overall and in a specific geographical area, will help in refining marketing strategies, reduce advertisement budget, and make it more effective. Data analytics techniques can help to prepare visually pleasing infographics that will make interpretation of facts easier for administration and students.

6 Big Data and Library Services

Over the years of their existence, libraries have collected a huge volume of data in printed as well as electronic forms. Due to developments in technology, hard copies of books can be easily converted into electronic format resulting in larger volumes of data. In electronic form, textual contents of books can be considered as unstructured data where big data techniques might not have much role to play. However, if we structure this data into the form of fields such as authors, co-authors, publishers, their geographic locations, publication and circulation histories, named entities in the text, dates, etc., then we will have a large set of data that will provide us much relevant information. Similarly, possible analysis of borrowing patterns of students can provide a good insight into the learning experience. It can provide information about what topics are in demand or likely to grow in demand so that resources related to those topics can be provided rapidly, or even pre-emptively. Big data and library services share a mutual benefit relationship. There have been several studies in recent years on the role of big data in the context of library services [63, 64]. Under the project "Big Cities meet Big Data" in Finland, Zhan and Widén [65] conducted empirical research to examine the role of public libraries in managing big data and bringing big data to citizens. Based on an online survey, the study identified eight possible roles for libraries in the context of big data namely educator, marketer, advocator, advisor, developer, organization server, data organizer and data container. It classified these eight roles into two groups: service-oriented roles (first six) and system-oriented roles (last two). The service-oriented roles include helping library users to use big data for their benefits. System-oriented roles are related to sustainable development.

Libraries, as resource hubs, can help to popularize big data by providing knowledge about big data to masses. A library in a higher education environment, maintains a warehouse of literary resources, does indexing and abstracting of documents and manuscripts, and keeps records of transactions of books and other resources. With the rise of Web 4.0, Noah [66] visions library 4.0 with massive data along with other features such as being intelligence-based, cutting edge displays, augmented reality, context awareness, etc. As library 4.0 will be intelligence-based and have massive data, big data techniques will play a big role in making these libraries more effective. For example, research data management is a critical task in a higher education environment. Universities produce a lot of research data and documents. These research data and documents are stored in ad hoc and scattered manner in most cases. One of the requirements of high-quality research, as sought by the research community and funding agencies, is to carefully maintain the data so that the research findings can be reproduced and verified. Big data techniques can help libraries to organize and retrieve research data and documents. Also, consider the scenario when libraries of several institutions are linked together. In that case, with the help of big data, university libraries can recommend the possible research topics in a specific field in real-time by analyzing the interest of scholars and the work of the researchers. Big data can also be used to compute return-on-investment on library resources by providing evidence of usage.

The Sheridan Libraries at Johns Hopkins University [67] created a model for data management services to provide dedicated and customized data management planning support for research grants. Contextualized as a case study in a specific library setting, this model aimed to understand the social, technical, and financial dimensions of a cross-disciplinary data management service and support the development of effective and sustainable data management services in academic libraries. They analyzed several documents from the University website, data repository website and other sources. They also conducted semi-structured interviews with pertinent personnel. Their investigation revealed that human, financial and technological considerations are crucial to the success of data management services. They also concluded that the success of such models could be achieved by ensuring big data-based services.

Many libraries are using big data tools to deliver relevant contents to their users. Brooklyn Public Library [68] was using services of consultants to extract and manipulate data from a software system used at the library, and prepare reports. They used to spend approximately two weeks in a month to prepare these manually made reports. Also, accessing data was time-consuming and complicated. To solve these issues, they invested in Tableau big data tool for better serving to patrons in an efficient way [69]. Now, they use Tableau to prepare data visualization reports for all staffs working at the library to understand data in a better way. Fast processing of data and visually better reports are helping library management to make informed business decisions.

7 Big Data and Other Services in Education Systems

As discussed in Sect. 1, there are various stakeholders in the higher education sector. Each of these stakeholders produces and consumes data. Most of the studies related to the use of big data in higher education cover data related to only a few stakeholders, mostly students, and teachers. Daniel and Butson [16] proposed a big data analytics framework for higher education that identifies and includes stakeholders which were usually not covered in big data analytics literature. Their proposed framework divides big data analytics into four major components: institutional analytics, information technology analytics, academic/program analytics, and learning analytics. The proposed framework classifies data into six categories: administrative data, research data, student data, teaching and learning data, department data, and curriculum data.

In previous sections, we covered many services and data related to these kinds of analytics, especially learning analytics. In this section, we will discuss a few other services in brief.

Higher education institutions keep collecting operational data from data warehouses and data dashboards on a regular basis. Effective analysis of this data can help university administration in taking better decision to make improvements at the institutional level. For example, some universities have several campuses spread across the country or the globe. Various campuses of these big universities have different requirements according to the demography of students and faculties. Data and text analytics can help administrators to make effective decisions. Institutional analytics can be further broken into assessment policy analytics, instructional analytics, and structural analytics. Information technology analytics integrates data from various systems such as student information system, learning management system, alumni systems, and systems managing learning experiences outside the classroom to develop models to identify obstacles to student access and usability. Academic/program analytics encapsulates all the activities in higher education affecting the administration, research, resource allocation, and management [70]. Higher education institutions organize several activities throughout the year. These activities include art and cultural activities, sports activities, academic events, charity events among others. These activities are helpful in improving the institution's reputation, but they require fundraising. However, investors will support and fund these activities only when the institution is doing or likely to do well in these activities. Big data and artificial intelligence tools can help institutions to do well in these activities by analyzing fine details. For example, the University of Florida and Duke University use an artificial intelligence application named HomeCourt [71] to track statistics for baseball players. Similarly, big data can help to manage vendors in a better way. Vendors can be informed in advance what services are expected from them, and how these services will be delivered and measured.

Health and safety are critical components in higher education institutions especially, with higher institutions with the residential facility. Being one of the important stakeholders of the education sector, parents are more concerned about the well-being and safety of their wards. If the institution is making proper arrangements of health and safety of the students, it will improve its brand values. This will help them not only to attract new students, it will also help them to increase the retention rate of students. The evolution of the Internet of Things (IoT) can help institutions to closely monitor the student's activities. For example, the use of radio-frequency identification devices (RFIDs) can help to record the purchase of food items and time spent in gyms. In fact, Oral Roberts University in Oklahoma requires students to wear fitbits so that data related to their exercise, eat, sleep, weight, calorie burns can be sent to professors for further analysis [72]. With the development of technology, it will be possible to monitor eye movements, breathing patterns, or heartbeat patterns in real-time to determine levels of anxiety of students. However, the amount of data generated in tracking these activities will be huge. Big data techniques can be used to analyze this data in a meaningful way, preferably in real-time. This will help administrators to communicate to hospitals and parents quickly in case of an emergency.

Similarly, cheaply available IoT enabled devices could help in monitoring activities of students, staffs, and visitors in the institution's campus. Big data can help in managing the security of the campus by quickly analyzing video contents of surveillance cameras installed on campuses. This will also help in maintaining discipline on campuses. Racial discrimination is a big issue in many educational establishments around the world. Big data can help in identifying racial discrimination. To achieve this objective, Baker et al. [73] planted messages in discussion forums across 126 MOOCs (1,008 messages in total, eight per course) and randomly assigned learner names to be provocative of different races and genders. They found evidence of discrimination in the behavior of instructors and students. For example, instructors wrote more replies for white male names than for white female, Indian, and Chinese names.

8 Challenges in Implementing Big Data

Though big data provides several benefits to higher education institutions, implementing big data throughout the institution is not a trivial task. One of the foremost challenges is the cost associated with the collection and storage of data. Collection and storage of data is not only time consuming, rather it requires a lot of resources in terms of hardware and software. Collection of data may involve people from a different background who may not be aware of the correct format or significance of data. This can compromise the quality of data collected. Data cleansing process will also result in the loss of some crucial data. Secondly, devices and formats of data may not be interoperable, and integration of data and devices can be a challenging task. It is important to clean the data to get accurate results. Dealing with different types of legacy data stroreges make this problem difficult. It is difficult to convince the stakeholders that Big data analytics will be useful to them. Cost benefit analysis may be one of the solutions for it.

Privacy and confidentiality of the data is another challenge. Though most of higher education institutions have disaster recovery plans, password protection techniques, and research ethics, their ability to cope with the challenges of big-data-based environments is not always on par with the challenges [74, 75]. Therefore, students, parents, faculties and administrative staffs may not agree to be monitored. In order to convince the stakeholders about the privacy and confidentiality of the data collected from them, an ethical framework need to be developed. This referential framework may consist of various theories of philosophical ethics and can be used to explain the concept of evolution of privacy in the age of the digital revolution to the stakeholders [76]. This framework can also explain how privacy is linked to big data and how privacy concerns can be handled with the help of computation on encrypted data [77]. Thirdly, various departments and sections in a typical higher education environment work as a separate entity. They use separate modules or entirely different software. This separation of data is a hindrance in providing a holistic view of hidden patterns if all data were analyzed together. In several instances, people involved in analyzing the data have no knowledge of the domain and those who are domain experts are not aware of the data analysis techniques [78]. As a result, data analysts might not be aware of what outcome to be produced, or they may exclude some critical parameters. Because of the rapid change of the technology, solutions often cannot handle the growth of new data types and new educational issues.

9 Conclusion and Future Directions

Big data, being one of the breakthrough technological development in recent times, is affecting every aspect of our life. Like any other business sector, the service sector, including higher education, can benefit from using advanced analytics techniques such as text analytics, machine learning, data mining, and natural language processing. The advanced computing techniques make it possible to link the otherwise standalone databases, which in turn allows researchers and business users to analyze previously inaccessible or unusable data to gain new insights resulting in better and faster decisions [79, 80]. In this chapter, we discussed how the two central tasks of a higher education institution, namely research and imparting education, can benefit from advanced data analytics techniques. The massive data collected from the interaction between the learners and instructors through computing devices can be linked together and analyzed for predicting the performance of the students. The feedback thus received can be used to improve the teaching–learning process. We also demonstrated that the use of big data is not limited to teaching and learning activities only. Rather, the big data can be used to formulate the marketing strategy, attract and recruit new students, improve student retention, improving library services, and so on. Despite the several benefits provided by big data, it is not an easy task to implement big data in higher education. In this chapter, we highlighted some challenges, such as privacy concerns, related to the implementation of big data in a higher education institution.

The use of big data analytics in higher education is becoming popular. Many universities including Marist College, North Carolina State University, Purdue University, Austin Peay State University, Georgia State University and John Hopkins University, to name a few, are adopting big data analytics to enhance learning experience, supporting research, improving student-university interaction and getting a lot of benefit out of it [81]. For example, the University of California saved approximately $500 million over four years by using IBM smarter analytics [82]. As discussed earlier, big data can benefit the higher education institutions by increasing the student retention rate, among other benefits. A case study done by Kellen et al. [83] suggested that the increase of one percent in the graduation rate of students not only helps those students to get good career path but also helps the institution to increase revenue by approximately one million dollars. Though the adoption of big data is now limited to a few top higher education institutions [84], we believe that in the near feature this trend will be followed by medium and small higher education institutions. In order to achieve that, the researchers need to develop affordable technologies.

The linking of standalone databases through big data techniques will deliver one system accessible to all stakeholders. This will make the higher education system more transparent. Everyone will be aware of who is responsible for a specific task or subtasks, who is responsible for the accuracy of the process and information, and so on. This will not only reduce the inconsistency of data format and the probability of error but also increase the efficiency of the system. The linking of the datasets and establishing a transparent system will require to develop and build infrastructure, tools and human capacity [85]. In order to reap the full benefits of research in big data, all stakeholders need to provide adequate professional training [86]. Therefore, a lot of research needs to be conducted to develop infrastructure for the efficient collection, cleaning, analysis, and distribution of large datasets.

Traditionally education in universities have been provided as a complete package (equivalent to enterprise software suits) with very little flexibility and customizability, in the form of elective courses only, and personalization for the learners. Like a customer may not be interested in the entire package in a software suite, a learner may not be interested in all the courses of a complete educational program such as a bachelor's or a master's degree. The learner may be interested in filling the gap between his/her existing skills and skills required to get a new job or meet the needs of a specific role in an organization. This gap can be filled by something called education as a service (EaaS). The concept of EaaS is similar to the delivery of Software as a service (SaaS). Most of the major software corporations, for example, Adobe Corporation, has changed the mode of delivery of their software products to be completely through the cloud. Instead of selling the complete software suits, many modules of that suit remain unused, companies are selling individual modules. This model of delivery of software, product or services, called the a-la-carte model, is being adopted in other areas of businesses. Higher education sector may also start adopting this business model. In fact, Educause-NMC's 2016 higher education report [87] suggests that many higher education institutions may start selling courses instead of complete programs in the next few years. Companies like Intel Corporation have already started the cloud-based infrastructure required to facilitate EaaS [88]. Big data, by analyzing large activities over months or years, can help to precisely assess the skills of a person and determine exact requirements for improving his/her skills. Researchers in the field of big data and data analytics need to develop techniques to make this analysis at the minimum cost. Only then education institutions will be able to use these techniques to analyze the data generated by their stakeholders. If the university is offering a degree curriculum, based on the analysis by big data, the student can get equivalent credit in some courses of the curriculum so that he/she can focus on a deep understanding of more useful courses.

Many countries like the USA and European countries are already creating value in higher education using big data. However, research in applications of big data in higher education in developing and the underdeveloped country is still a challenge. One of the reasons behind this lag is, perhaps, lower penetration rates of the Internet in these countries. According to a new report by the UN International Telecommunication Union (ITU) [89], less than a quarter of all people will be online by 2020 in the 47 least-developed-countries. We believe that once governments of these countries are able to improve Internet penetration, big data techniques can accelerate the development and research in the field of education in under-developed countries. As big data techniques require a lot of computational resources, providing high-end computational resources at a large scale is not possible in these countries. Therefore, it is both a challenge and an opportunity for the computer and data analytics researchers to develop low-cost high capacity devices, software, and applications to promote education and research in poor countries. Recently, Google launched a lightweight and AI-powered internet search app named Google Go which lets user access information even on unstable connections [90]. Similar research and innovations in the field of data analytics and AI will help the population across the globe to take complete benefit of technological developments.

References

1. Simeon, R. (2013). *Working in the global economy: How to develop and manage your career across borders.* Routledge Taylor & Francis Group.
2. Kuzu, Ö. H., Gökbel, H., & Güleş, H. K. (2013). Developing sustainable relations with internal and external stakeholders in universities: Vision and mission views. *Procedia - Social and Behavioral Sciences, 103*, 281–289.
3. Marshall, S. J., & Marshall, S. J. (2018). Internal and External Stakeholders in Higher Education. In *Shaping the university of the future* (pp. 77–102). Springer Singapore.
4. Ishikiriyama, C. S., & Carlos Francisco Simoes, G. (2017). Big Data: A global overview. In A. Emrouznejad, & V. Charles, (Eds.), *Big Data for the greater good* (pp. 35–50). Springer.
5. Laney, D. (2001). 3D data management: Controlling data volume, velocity and variety.
6. Furht, B., & Villanustre, F. (2016). Introduction to big data. In *Big Data technologies and applications* (pp. 3–11). Cham: Springer International Publishing.
7. Oussous, A., Benjelloun, F.-Z., Ait Lahcen, A., & Belfkih, S. (2018). Big Data technologies: A survey," *J. King Saud Univ. - Comput. Inf. Sci.*, vol. 30, no. 4, pp. 431–448, Oct. 2018.
8. Charles, V., & Gherman, T. (2013). Achieving Competitive Advantage Through Big Data. Strategic Implications. *Middle-East Journal of Scientific Research, 16*(8), 1069–1074.
9. Shafer, T. (2017). The 42 V's of Big Data and Data Science. Elder Research, Inc. Retrieved April 9, 2019, from https://www.kdnuggets.com/2017/04/42-vs-big-data-data-science.html.
10. World Development Indicators|The World Bank. (2017). The World Bank. Retrieved April 9, 2019, from http://wdi.worldbank.org/table/4.2.
11. Al-Samarraie, H., & Saeed, N. (2018). A systematic review of cloud computing tools for collaborative learning: Opportunities and challenges to the blended-learning environment. *Computers & Education, 124*, 77–91.
12. Ray, S. K., & Saeed, M. (2015). Mobile learning using social media platforms: An empirical analysis of users' behaviours. *International Journal of Mobile Learning and Organisation, 9*(3), 258–270.
13. Shah, D. (2018). By The Numbers: MOOCS in 2017—Class Central. Retrieved April 9, 2019, fromhttps://www.classcentral.com/report/mooc-stats-2017.
14. Atif, Y., Mathew, S. S., & Lakas, A. (2015). Building a smart campus to support ubiquitous learning. *Journal of Ambient Intelligence and Humanized Computing, 6*(2), 223–238.
15. Junco, R., & Clem, C. (2015). Predicting course outcomes with digital textbook usage data. *Internet and Higher Education, 27*, 54–63.
16. Daniel, B.K., & Butson, R. (2013). Technology Enhanced Analytics (TEA) in Higher Education. In *International conference on educational technologies (ICEduTech)* (pp. 89–96).

17. Rajabi, E., & Beheshti, A. (2016). Interlinking Big Data to Web of Data. In Ali, E. (Ed.), *Big Data optimization: Recent developments and challenges* (pp. 133–145). Springer.
18. Besbes, R., & Besbes, S. (2016). Cognitive dashboard for teachers professional development. In *Qatar foundation annual research conference* (pp. 1–4).
19. Romero, C., & Ventura, S. (2010). Educational data mining: A review of the state of the art. *IEEE Transactions on Systems, Man, and Cybernetics, Part C,* 40(6), 601–618.
20. Huang, S., & Fang, N. (2013). Predicting student academic performance in an engineering dynamics course: A comparison of four types of predictive mathematical models. *Computers & Education, 61,* 133–145.
21. LACE SoLAR Flare (OU, 2015) | Society for Learning Analytics Research (SoLAR). Retrieved April 9, 2019, from https://www.solaresearch.org/events/flare/lace-solar-flare-ou-2015/.
22. M. Murphy, S. Redding, and J. S. Twyman (2016). Handbook on personalized learning for states, districts, and schools. In Center on Innovations in Learning, Temple University. 1301 Cecil B Moore Avenue Ritter Annex 422, Philadelphia, PA 19122. Tel: 215-204-3364; Fax: 215-204-5130; Web site: http://www.centeril.org.
23. Romero-Zaldivar, V.-A., Pardo, A., Burgos, D., & Delgado Kloos, C. (2012). Monitoring student progress using virtual appliances: A case study. *Computers & Education, 58*(4), 1058–1067.
24. Kop, R. (2010). The Design and Development of a Personal Learning Environment: Researching the Learning Experience. In *European Distance and E-learning Network annual Conference 2010.*
25. Baker, R. S. J. d, & Siemens, G. (2012). Learning analytics and educational data mining. In *Proc. 2nd Int. Conf. Learn. Anal. Knowl. - LAK '12* (p. 252).
26. Santosh, R., & Mohammed, S. (2018). Applications of educational data mining and learning analytics in handling Big Data in higher education. In Mohammed, M., & Alani, O. A., Tawfik, H., Saeed, M. (Eds.), *Applications of Big Data analytics: Trends, issues, and challenges* (pp. 135–160). Cham: Springer.
27. The Cisco Connected Learning Experience Strategies for Higher Education. (2014).
28. Altbach, P. G., Reisberg, L., & Rumbley, L. E. (2010). *Trends in global higher education: Tracking an academic revolution.* UNESCO Publishing.
29. Blikstein, P. (2011). Using learning analytics to assess students' behavior in open-ended programming tasks. In *Proceedings of the 1st international conference on learning analytics and knowledge - LAK '11* (p. 110).
30. NetLogo. Retrieved April 9, 2019, from https://ccl.northwestern.edu/netlogo/.
31. Moridis, C. N., & Economides, A. A. (2009). Prediction of student's mood during an online test using formula-based and neural network-based method. *Computers & Education, 53*(3), 644–652.
32. Jeong, H.-Y., Choi, C.-R., & Song, Y.-J. (2012). Personalized learning course planner with E-learning DSS using user profile. *Expert Systems with Applications, 39*(3), 2567–2577.
33. Arrigo, M., Kukulska-Hulme, A., Arnedillo-Sánchez, I., & Kismihok, G. (2012). Meta-analyses from a collaborative project in mobile lifelong learning. British Educational Research Journal, 1–26.
34. Lin, C. F., Yeh, Y., Hung, Y. H., & Chang, R. I. (2013). Data mining for providing a personalized learning path in creativity: An application of decision trees. *Computers & Education, 68,* 199–210.
35. Hsu, M.-H. (2008). A personalized English learning recommender system for ESL students. *Expert Systems with Applications, 34*(1), 683–688.
36. Méndez, J. R., Fdez-Riverola, F., Iglesias, E. L., Díaz, F., & Corchado, J. M. (2006). *Tracking concept drift at feature selection stage in spamhunting: an anti-spam instance-based reasoning system* (pp. 504–518). Springer.
37. Khaing, K. T. (2010). Enhanced features ranking and selection using recursive feature elimination (RFE) and k-nearest neighbor algorithms in support vector machine for intrusion detection system.

38. Chien, C.-F., Wang, W.-C., & Cheng, J.-C. (2007). Data mining for yield enhancement in semiconductor manufacturing and an empirical study. *Expert Systems with Applications, 33*(1), 192–198.
39. Lee, Y.-J. (2012). Developing an efficient computational method that estimates the ability of students in a Web-based learning environment. *Computers & Education, 58*(1), 579–589.
40. Abdous, M., He, W., & Yen, C.-J. (2012). Using Data Mining for Predicting Relationships between Online Question Theme and Final Grade.
41. Lykourentzou, I., Giannoukos, I., Nikolopoulos, V., Mpardis, G., & Loumos, V. (2009). Dropout prediction in e-learning courses through the combination of machine learning techniques. *Computers & Education, 53*(3), 950–965.
42. Dekker, G. W., Pechenizkiy, M., & Vleeshouwers, J. M. (2009). Predicting students drop out: A case study. In *The 2nd international conference on educational data mining* (pp. 41–50).
43. Kizilcec, R. F., Piech, C., & Schneider, E. (2013). Deconstructing disengagement. In *Proceedings of the third international conference on learning analytics and knowledge—LAK '13.* (p. 170).
44. Giesbers, B., Rienties, B., Tempelaar, D., & Gijselaers, W. (2013). Investigating the relations between motivation, tool use, participation, and performance in an e-learning course using web-videoconferencing. *Computers in Human Behavior, 29*(1), 285–292.
45. He, W. (2013). Examining students' online interaction in a live video streaming environment using data mining and text mining. *Computers in Human Behavior, 29*(1), 90–102.
46. Dejaeger, K., Goethals, F., Giangreco, A., Mola, L., & Baesens, B. (2012). Gaining insight into student satisfaction using comprehensible data mining techniques. *European Journal of Operational Research, 218*(2), 548–562.
47. Xing, W., Guo, R., Petakovic, E., & Goggins, S. (2015). Participation-based student final performance prediction model through interpretable Genetic Programming: Integrating learning analytics, educational data mining and theory. *Computers in Human Behavior, 47*, 168–181.
48. Zacharis, N. Z. (2015). A multivariate approach to predicting student outcomes in web-enabled blended learning courses. *Internet and Higher Education, 27*, 44–53.
49. Niemi, D., & Gitin, E. (2012). Using Big Data to predict student dropouts: Technology affordances for research. In *IADIS International conference on cognition and exploratory learning in digital age* (pp. 261–264).
50. Mouri, K., Okubo, F., Shimada, A., & Ogata, H. (2016). Bayesian network for predicting students' final grade using e-book logs in university education. In *2016 IEEE 16th international conference on advanced learning technologies (ICALT)* (pp. 85–89).
51. Kizilcec, R. F., Bailenson, J. N., & Gomez, C. J. (2015). The instructor's face in video instruction: Evidence from two large-scale field studies. *Journal of Educational Psychology, 107*(3), 724–739.
52. De Freitas, S. (2018). *Education in computer generated environmentS*. Routledge.
53. Gibson, D. C., Webb, M., & Ifenthaler, D. (2015). Challenges of Big Data in educational assessment. In *12th International conference on cognition and exploratory learning in digital age* (pp. 92–100).
54. Shute, V. J. (2008). Focus on formative feedback. *Review of Educational Research, 78*(1), 153–189.
55. Romero, C., Zafra, A., Luna, J. M., & Ventura, S. (2013). Association rule mining using genetic programming to provide feedback to instructors from multiple-choice quiz data. *Expert Systems, 30*(2), 162–172.
56. Pechenizkiy, M., Calders, T., Vasilyeva, E., & De Bra, P. (2008). Mining the student assessment data: Lessons drawn from a small scale case study. In *International conference on educational data mining* (pp. 187–191).
57. Kizilcec, R. F., Pérez-Sanagustín, M., & Maldonado, J. J. (2016). Recommending self-regulated learning strategies does not improve performance in a MOOC. In *Proceedings of the third (2016) ACM conference on learning @ Scale - L@S '16* (pp. 101–104).
58. Rogers, T., & Feller, A. (2016). Discouraged by Peer excellence: exposure to exemplary peer performance causes quitting. *Psychological Science, 27*(3), 365–374.

59. Kizilcec, R. F., & Brooks, C. (2017). Diverse Big Data and randomized field experiments in MOOCs. In *Handbook of learning analytics* (pp. 211–222).
60. Musto, C., Semeraro, G., De Gemmis, M., & Lops, P. (2015). Word embedding techniques for content-based recommender systems: An empirical evaluation. *CEUR Workshop Proceedings, 1441*, 448–456.
61. Ng, I. C. L., & Forbes, J. (2009). Education as service: The understanding of university experience through the service logic. *Journal of Marketing for Higher Education, 19*(1), 38–64.
62. Białoń, L. (2015). Creating marketing strategies for higher education institutions. *Minib, 18*(4), 129–146.
63. Teets, M., & Goldner, M. (2013). Libraries' role in curating and exposing Big Data. *Future Internet, 5*(3), 429–438.
64. Hoy, M. B. (2014). Big Data: An introduction for librarians. *Medical Reference Services Quarterly, 33*(3), 320–326.
65. Zhan, M., & Widén, G. (2018). Public libraries: Roles in Big Data. *The Electronic Library, 36*(1), 133–145.
66. Noh, Y. (2015). Imagining library 4.0: Creating a model for future libraries. *The Journal of Academic Librarianship, 41*(6), 786–797.
67. Shen, Y., & Varvel, V. E. (2013). Developing data management services at the johns hopkins university. *The Journal of Academic Librarianship, 39*(6), 552–557.
68. Brooklyn Public Library. Retrieved April 9, 2019, from https://www.bklynlibrary.org/. [Accessed: 09-Apr-2019].
69. Brooklyn Public Library saves time, money, and headcount with Tableau | Tableau Software. Retrieved April 9, 2019, from https://www.tableau.com/solutions/customer/brooklyn-public-library-saves-time-money-and-headcount-tableau.
70. Tulasi, B. (2013). Significance of Big Data and analytics in higher education. *International Journal of Computers and Applications, 68*(14), 21–23.
71. HomeCourt. Retrieved April 9, 2019, from https://www.homecourt.ai/.
72. ORU students log 3 billion steps on Fitbit devices—Washington Times. Retrieved April 9, 2019, from https://www.washingtontimes.com/news/2017/mar/20/oru-students-log-3-billion-steps-on-fitbit-devices/
73. Baker, R., Dee, T., Evans, B., & John, J. (2018). Bias in online classes: Evidence from a field experiment (CEPA working paper no. 18-03). Stanford Center for Education Policy Analysis, (18).
74. Slade, S., & Prinsloo, P. (2013). Learning analytics: Ethical issues and dilemmas. *American Behavioral Scientist, 57*(10), 1510–1529.
75. Jones, S. J. (2012). Technology review: The possibilities of learning analytics to improve learner-centered decision-making. *Community College Enterprise, 18*(1), 89–92.
76. Charles, V., Tavana, M., & Gherman, T. (2015). The right to be forgotten—Is privacy sold out in the big data age? *International Journal of Enterprise Information Systems, 7*(4), 283–298.
77. Nielsen, K. (2017). Big Data and Sensitive Data. In Emrouznejad, A., & V. Charles (Eds.), *Big Data for the greater good* (pp. 183–204). Springer.
78. Macfadyen, L. P., & Dawson, S. (2010). Mining LMS data to develop an 'early warning system' for educators: A proof of concept. *Computers & Education, 54*(2), 588–599.
79. A. Emrouznejad and M. Marra, "Big data: Who, what and where? Social, cognitive and journals map of big data publication," in *Big Data Optimization: Recent Developments and Challenges*, A. Emrouznejad, Ed. Springer, 2017, pp. 1–16.
80. V. Charles and A. Emrouznejad, "Big Data for the Greater Good: An Introduction," in *Big Data for the Greater Good*, A. Emrouznejad and V. Charles, Eds. Springer, 2017, pp. 1–18.
81. C. Boulton, "Big data on campus | CIO," 2016. [Online]. Available: https://www.cio.com/article/3121250/education/big-data-on-campus.html. [Accessed: 21-Dec-2018].
82. Smarter Education with IBM Business Analytics. Smarter Analytics, 2012. Retrieved December 21, 2018, from ftp://ftp.software.ibm.com/la/documents/gb/mx/Business_Analytics_for_Higher_Education.pdf.

83. Kellen, V., Recktenwald, A., & Burr, S. (2013). Applying Big Data in higher education: A case study | cutter consortium. Retrieved December 21, 2018, from https://www.cutter.com/article/applying-big-data-higher-education-case-study-400836.
84. Macfadyen, L. P. (2017). Overcoming barriers to educational analytics: How systems thinking and pragmatism can help. *Educational Technology, 57*, 31–39.
85. Daniel, B. K. (2017). Big Data and data science: A critical review of issues for educational research. *British Journal of Educational Technology, 50*(1), 101–113.
86. Daniel, B. K. (2017). *Big data and learning analytics in higher education: Current theory and practice*. Springer.
87. "NMC Horizon Report > 2016 Higher Education Edition | The New Media Consortium." [Online]. Retrieved April 9, 2019, from https://www.nmc.org/publication/nmc-horizon-report-2016-higher-education-edition/.
88. R. C. Fogel (2010). The Education Cloud: Delivering Education as a Service. https://www.k12blueprint.com/sites/default/files/ITDM_education_cloud_final.pdf.
89. Least Developed Countries on track to achieve SDG 9.c on universal and affordable Internet access by 2020. Retrieved April 9, 2019, from https://www.itu.int/en/mediacentre/Pages/2018-PR02.aspx.
90. Protalinski, E. (2019). Google Go launches globally on Android. Retrieved August 24, 2019, from https://venturebeat.com/2019/08/20/google-go-launches-globally-on-android/.

Santosh Ray is currently working as an associate professor and Head of Information Technology & Research in Department of Information Technology, Khawarizmi International College, Al Ain, UAE. He has teaching and research experience of over 18 years in the field of Information technology. His research area includes Information Retrieval, Natural Language Processing, Social Media, Mobile Learning, and Educational Data Mining. He is a member of British Computer Society and IEEE CS.

Mohammed M. Alani received his Ph.D. in 2007. He has authored 4 books and many papers in well-known international venues. He has served in many educational leadership positions in various countries in the Middle East. He's currently working as a Cybersecurity Professor in Seneca College of Applied Arts and Sciences, teaching in several Universities and Colleges in Ontario, Canada, and providing consultancies to large corporations. His major fields of interest include cybersecurity, networking, and big data.

Amir Ahmad has a Ph.D. from University of Manchester, UK in Computer Science. He is currently working as an Associate Professor in College of Information Technology, UAE University, Al Ain, UAE. His research areas are machine learning, data mining and nanotechnology.

A Novel Big Data Approach for Text Supported Service Operations Management

Lukas Povoda, Radim Burget, Martin Rajnoha, and Peter Brezany

Abstract Operations management is being applied across all industry sectors and offers more accurate decision making, cost saving and a path to lean production. All depend on proper information, which is a valuable input for operations management and can represent the key to gain competitive advantage. Thanks to the information, we can improve and optimize processes and react to problems before they even appear. Unfortunately, getting this knowledge is often a challenging task. It is estimated that only a small portion of publicly available data can be considered as valuable. Considering the size of today's data storage, it is out of possibilities for anyone to read them all and operatively react to the changing environment and competition. In case of the text data, they are often written in many different languages, use different expert domains or customer segments, who have their own specific language style. To gain valuable information from this data, it is necessary to know each particular language (not necessarily human, but also machine generated) and grammar. This chapter presents the latest advances in artificial intelligence for the text data analysis and operations management. First, we provide the state of the art of the text processing approaches, then we discuss selected use-cases from the field of operations management and how the latest methods can help to solve those problems. The last part of the chapter outlines some ideas for further improvement of the current approaches and how to effectively analyse data in a multilingual environment and decrease memory demands.

L. Povoda (✉) · R. Burget · M. Rajnoha
Department of Telecommunications, Faculty of Electrical Engineering and Communication, Brno University of Technology, Brno, Czech Republic

R. Burget
e-mail: burgetrm@feec.vutbr.cz

M. Rajnoha
e-mail: martin.rajnoha@vutbr.cz

P. Brezany
Faculty of Computer Science, University of Vienna, Vienna, Wien 1010, Austria
e-mail: Peter.Brezany@univie.ac.at

SIX Research Group, Brno University of Technology, Brno, Czech Republic

A. Emrouznejad and V. Charles (eds.), *Big Data and Blockchain for Service Operations Management*, Studies in Big Data 98,
https://doi.org/10.1007/978-3-030-87304-2_6

List of Acronyms

BERT	Bidirectional Encoder Representations from Transformers
BoW	Bag of Words
CBOW	Continuous Bag of Words
CNN	Convolutional Neural Network
ELMo	Embeddings from Language Models
FLM	Fast Large Margin
FP	Fingerprint
GloVe	Global Vectors for Word Representation
GPT	Penerative Pre-training Model
GPT-2	Penerative Pre-training Model, version 2
GPU	Graphics Processing Unit
k-NN	k-Nearest Neighbour
LSTM	Long Short-term Memory
RCK	Recurring Kernel
RNN	Recurrent Neural Network
SNAP	Stanford Network Analysis Project
SOTA	State of the Art
SVM	Support Vector Machine
TCN	Temporal Convolutional Network
TF–IDF	Term Frequency—Inverse Document Frequency
TPU	Tensor Processing Unit
ULMFit	Universal Language Model Fine-tuning
UTF-8	8-bit Unicode Transformation Format
VGG	Visual Geometry Group

1 Introduction

Operations management is defined as a process of designing and controlling subprocesses of production and redesigning business operations in the production of goods or services. It is responsible for development and use of the resources that are necessary for a company to deliver goods and services to its customers. To do so, information is the key. In order to manage these processes, it is absolutely necessary to have appropriate information. That information is usually in the form of numerical values, images, videos, or text data and only a small portion of all the data available is structured.

This work focuses mainly on textual and unstructured (e.g. natural language) form of data and how to extract knowledge from that kind of data to utilize gained knowledge for operations management. It is estimated that approximately 80% of all the data stored in digital space have a textual form [9]. A large part of this data is publicly available, but large amounts of data are being stored privately in corporate data warehouses or cloud storages. Companies often store these data in hope to utilize this data in future. However, only a small portion (statistics says approx. 5%) of all the data can be considered as a valuable source of information [9]. Processing and analysing these billions and trillions of records (generally referred to as Big Data [27], the global overview in article [10]) is impossible by reading and understanding by humans. Furthermore, as described in [37], it is quite a challenging task even for computers. However, the knowledge stored in them is often necessary for good operative management, especially in the case of fast changing environment and strong competition. However, automated text analysis can be quite challenging, since data sources are often written in plenty of different languages, data suppose some level of common sense, particular domain knowledge or customer segment that has its specific language style, etc.

This chapter compares different methodologies for text analysis and proposes a novel method where the key advantage is in no need of preliminary knowledge about the language and grammar. The method alternates current method based on the language modelling (such as embedding models) with lower demands on hardware in the training process and lower demands on memory when the trained model is used.

For operations management having relevant and accurate information is the key. It uses the information extracted from the data and transforms the data into knowledge that is consequently used for quality management, location strategies management, supply chain management, accountability or process optimization. So far, these principles are already deployed in many industry fields. For example, some traditional areas are engineering, information technology, marketing, human resources management, quality assurance, manufacturing, legal sector and others. This is supported by the current trend—increased interest in data mining (especially text mining) and latest advances in the field that brought new possibilities. To name a few of them, for example text sentiment analysis, fact checking, automatic question answering and other. Solving these problems can lead to effective interaction with customers and feedback evaluation from millions of them, quality optimization, controlling and scheduling capacities efficiently, which would enable to save the costs and achieve lean production.

There is increasing need to effectively analyse data that are gathered from international environment. For example, companies with a worldwide presence have to process different languages. Here the computer systems need knowledge of the particular language and grammar for text understanding. In traditional methods, this is commonly achieved by using combination of (1) preprocessing algorithms (like stemming or lemmatization, using different dictionaries for synonyms replacement, grouping words, stop words, spell check), (2) knowledge of the language structure and morphology and (3) optimization methods to achieve higher accuracy. For Big Data problems it means high memory and computational demands, it needs to be configured for each language and usually the systems are trained for each language separately and thus it is not possible to combine the knowledge from different languages. Also, to create and keep those sources of information up-to-date, it is a resources and time demanding task. In the world almost 7000 actively used languages exist worldwide and there are also countless computer-generated logs—structured, semi-structured or unstructured that can often become a valuable source of information as well [16]. Moreover, languages are evolving in time, for example words and abbreviations that would not have any meaning few years ago, like ":-)", LOL, IMHO, FYI, YouTuber, hater, etc. So, they add a language dependency of the input text in the analytical system. However, removing those language dependent part reduces success of such solution rapidly. Therefore, more research has recently been pursuing a different approach to this issue.

The rest of the chapter is focused on information retrieval from text data, which is organised as follows. First, background about the used methodologies and objectives of this chapter are given. Then different approaches are discussed, including statistical, big data, embeddings, word level and character level for automated multilingual data analysis. We show that the automated methods can compete with statistical and embedding based methods in regards to accuracy but can significantly save memory and computation resources.

1.1 Background

In recent years, there has been significant progress in operations management supported by computer systems. This is thanks to increased computational power together with significant progress in artificial intelligence methods. One of such a technology is the so-called deep learning, which allowed significant progress which supports expansion of knowledge mining from various data sources, including natural language texts. Extracting knowledge from text data is often referred to as the data mining, knowledge extraction, or question answering. It is being addressed by a wide range of research institutions all over the world as well as technological companies including for example Google, Facebook, Amazon or Apple, or different startups. Many of these companies invest a lot of effort and money to data mining research and developing a specialized hardware that can further accelerate computations (e.g. Google Tensor Processing Units). Operations management is strongly connected to text data mining. To use knowledge contained in the text data, we need to understand the true meaning of the text. This is often a very challenging task even for humans. In many cases, this task can achieve higher complexity than for example image analysis. Meaning of a text can be changed using irony or sarcasm, texts can be written in different styles (e.g. politely, in everyday language, slang), different languages can use different phrases to achieve the same meaning (e.g. phraseologisms).

Using the methods of data mining to support operations management can lead to significant advantage in global competition. In the text mining field, one of the most common problems is sentiment analysis (also known as opinion mining), text classification or text regression, which can bring new external inputs for operations management. The common use cases are for example: prioritization of dissatisfied customers on support services or helpdesks to achieve higher customer support quality, obtaining indirect feedback from social media and public discussions to better market adaptation, and prediction of future market behaviour.

To automate text analysis, we usually need to prepare a few things first. First, knowledge of the particular grammar and specifics of the language are the keys to do the proper text analysis. Different methods have implemented this knowledge in different ways: (1) methods based on traditional approach use advanced text preprocessing to extract function vectors, which are created based on the knowledge of a particular language, its syntax and grammar, (2) methods based on deep learning usually process raw input text data (without preprocessing) to obtain this language knowledge themselves. Embedding models, as representatives of different solutions, fall also in the second category. These models are supporting current state-of-the-art methods by converting the raw input into an embedded vector, which is then processed by—usually—a deep neural network that gives an answer about the text (for example text classification).

Embedding models represent language knowledge by using encoding, usually of each word, into a vector, which represents the relationship between other words. The latest approaches allow not just embedding on the word level, but rather on the sentence level. As the article [15] explains, these embedding models can boost the accuracy and achieve high precision even with the simplest machine learning algorithms. To the end of 2018, also new types of deep contextual word embeddings were introduced. These models utilize ideas of deep learning and bidirectional recurrent neural networks. Embeddings from Language Models (ELMo) was introduced in [23] and it models complex language characteristics. This model provided 6–20% relative error reduction, significantly faster training and lower dataset size requirements. The extended solution was proposed by Google in the article [5] where the model called BERT (Bidirectional Encoder Representations from Transformers) was introduced. It obtains new state-of-the-art results on eleven natural language processing tasks and outperforms human performance by 2.0%. One of the joint disadvantages of all embedding based methods is that the embeddings need to be pretrained on huge datasets which also demands large computational resources. This needs to be done for each language. This is a problem especially for the minority languages or if we need to analyse machine generated logs.

To avoid the need for a pre-trained embedding model, some other approaches were introduced recently. The deep learning approach with recurrent neural networks (RNN) was quite promising because text data can be considered as time series of words (i.e. vectors) and it is more like humans reading text data. Further work has shown, that character level convolutional neural networks perform even better than RNNs [35]. This idea was further elaborated by Facebook AI Research team, where the so-called Very Deep convolutional neural network (CNN) was designed [4] and FastText [2]. Thanks to these ideas, drawbacks of RNN like exploding gradient were eliminated. In 2017, Google DeepMind was trying to disprove this opinion. In the article [34], they compare the performance of Discriminative LSTM (Long Short-Term Memory) and Generative LSTM, however, the accuracy did not achieve the accuracy of character-level CNN networks. This indicates that CNN can outperform RNN with higher memory efficiency. To the end of 2018, the idea of CNN as a promising successor of RNNs was enhanced by the publication of temporal convolutional networks (TCN). This claim was supported by a wide range of experiments on different data sets [1].

1.2 Research Objectives

In order to automate analysis of texts, we first need to know the language it is written in, know the grammar and have this knowledge in the form of a program or pretrained model on some dataset. Unfortunately, the number of actively used languages worldwide is more than 7000 and there are countless machine languages and log formats and only a few of the languages have already such a model prepared. It is estimated there are more than 280 actively used languages in Europe (according to Eurobarometer 2012), in the United States there are more than 330 actively used languages and in India there are more than 450 different languages. More than half of the world's population uses just 23 languages [21].

Before the deep learning methods for text analysis started to be widely used, ones of the most favourite were statistics-based approaches. We call them in this text as the "traditional" methods. These methods use complex text preprocessing like stemming (reducing each word to its stem), lemmatization (determining the lemma of each word), removing stop words (words with little information that can only confuse text analyser) etc. All these methods lead to strong language dependency of such a model. Libraries to perform these tasks are publicly available only for common languages such as Chinese, Spanish, English, etc.

The problem occurs if we want to process a language, which is not so well known. After preprocessing, relatively simple statistical approach was commonly used to evaluate the statistical distribution of words in the evaluated document and provided conclusions about the text. Unfortunately, those methods do not take the word order into account (e.g. beer free, vs. free beer). Using optimization methods especially in simpler cases, we can achieve accurate and computationally efficient solution. Big Data approach is preferred for more complex tasks where cluster computing or massive parallelism using graphics processing unit (GPU) or tensor processing unit (TPU) can be used. Advantage of the big data approach is that it does not need so much preprocessing work and just from a large amount of data we can learn data automatically. Due to significant computational demands the optimization methods (like forward attribute selection, backward elimination) are usually not feasible.

Therefore, the main objective is to create a universal re-usable method for decision making support. To achieve this, knowledge of the language and grammar should be extracted automatically. Therefore, different language models which deal with problems of language and grammar encoding are currently arising. However, these models have high hardware demands for training such model and high demands on memory (and also computation) to encode the input text.

2 Data Processing

This section discusses four basic approaches how text data can be preprocessed, i.e. transformed to a form of vectors that are suitable for machine learning algorithms. In particular it discusses traditional approaches, big-data approaches, methods utilizing embeddings and introduces methods independent on a language. It discusses also advantages and disadvantages of those methods.

Traditional methods based on statistical analysis use text preprocessing where the grammar and language knowledge is contained. On input, the text is transformed into a vector of features which is the statistical representation of words in the given text document. The length of such a vector is defined by the total number of unique words (tokens) in the given language, or words that were in the training set. In such solutions, words or tokens are the smallest units. Therefore, languages based on characters (not on words—for example the Chinese language) cannot be directly processed with such a solution. Change comes with the idea of character-based input.

To cover all languages, this can be a significant barrier. Language dependency in different parts of preprocessing means that every part must be manually modified and adapted to every language such a system has to handle. For operations management, this may represent considerable costs. Therefore, there is a need to eliminate all language dependent parts and maintain achieved performance.

2.1 Traditional Preprocessing Approaches

The traditional methods are ones of the most demanding regarding knowledge of the language and handmade rules for preprocessing. It usually consists of several preprocessing steps: tokenization, spell check, lemmatization (or stemming), synonyms replacement and removing stop words. Tokenization ensures correct splitting of sentences into words, while special cases (like numbers, names, emails, emoticons etc.) should be processed. The spell checker ensures that every word is valid and known and does not contain typing errors. In cases an unknown word is detected, the word is replaced with the closest possible known word from the dictionary. In preprocessing, this operation brings significant memory and computation demands (e.g. English is estimated to have more than 170 thousand words) since similarity is usually estimated between the unknown word and the whole set of known words. After this step, it is usually attempted to reduce the number of used words, i.e. dimensionality of the machine learning problem. It is done by mapping words into a single form (e.g. small, smaller, smallest have relatively similar meaning and just small can be used), finding synonyms or antonyms and removing unimportant words. This is done by using lemmatization, which uses dictionaries to search for the root form of every word. Unfortunately to create lemma from every word is not a simple task and the rules for transforming a word to its lemma can be very complex and time consuming. Alternative to lemmatization is stemming which is unfortunately less accurate and can change or confuse some words. Stop words usually refer to the most common words in a language with little information and potential to confuse classifiers during information extraction. Then synonyms are replaced and words without meaning are removed. As it is clear, every operation named above is language dependent, which means they have to be manually developed for each particular language.

A simple and efficient baseline for statistical word representation is Bag-of-Words (BoW) which has successfully proved interesting accuracy [11]. According to this model, text (sentence or document) is represented as a set of words, i.e. it neglects grammar and word order, but it includes histogram about number of word occurrences and takes word's statistical importance into account. This approach is called Term Frequency—Inverse Document Frequency (TF–IDF) [30] and is well known and still widely used. Each value in the feature vector increases with the number of occurrences of a given word in the document and decreases with the frequency of occurrence of that word throughout the whole corpus. This makes it possible to detect words that are more common in texts.

One of the biggest disadvantages of this approach is that it completely losses word order information (e.g. "free beer" or "beer free" sentences are indistinguishable from each other). Here the idea of creation of word pairs (bi-grams or tri-grams) can help to capture the meaning with a broader context. For example, negation can be captured using this technique. Unfortunately, *n*-grams increase significantly the dimensionality of the feature vector (e.g. $170{,}000^2$ in the case of English and bi-grams). It is quite memory demanding because the input vector can contain hundreds or thousands of millions features based on a given language. To reduce this dimensionality, traditional methods use optimization methods mainly based on feature selection to decrease the size of the input vector. For multiclass classification complex optimization techniques were proposed in article [24, 25] like for example sequential elimination of attributes or optimization with token groups. Even optimization methods can bring another language dependency especially when they are based on word as the basic unit.

TF-IDF and Bag of Words methods using bi-grams (or even *n*-grams) give usually worse performance than modern embedding methods. The article [35] compares state-of-the-art methods on well-known data sets including various *n*-grams. Traditional methods still perform well when there is no big training dataset available, they are about to be deployed to a domain specific problem (i.e. language, where there is not yet pretrained embedding) or text variability is not very high, e.g. text server logs.

2.2 Traditional Methods in Big Data Environment

TF-IDF with Bag of Words can be alternatively trained in distributed environments. This can partially solve problems with big volumes of data, training time and memory limits. Using the Big Data approach, more complex language analysis operations are possible and therefore the preprocessing parts could be significantly simpler. By distributing the task on multiple processors, it is also possible to finish the training or analysis in a reasonable time. However, complexity of optimization methods is increasing significantly. Therefore, in the context of Big Data, many optimization techniques are commonly not being used because of the fact that training a single model is often a matter of several days of computations even with several processors. Forward feature selection or Backward feature elimination have complexity quadratic to problem dimensionality which for e.g. English and 2-g means $170{,}000^2$ or 3-g $170{,}000^3 = 4.9e15$ iterations. For this reason, specialized big data optimization approaches were developed.

In the context of big data approach, it was shown that Support Vector Machine (a machine learning algorithm) performs quite well with advanced preprocessing rules. As the article (Povoda, Burget and Dutta, 2016) proves, this machine learning algorithm can be trained by using big amount of data to achieve similar accuracy as classifiers with complex text preprocessing described in Sect. 2.1. The advantage is, that the preprocessing rules could be significantly simpler. On the other hand we need a lot of labelled training samples which must be created for each particular case and each particular language. Accuracy of such a method evaluated on public datasets is currently worse than the state-of-the-art methods published in [4, 34, 14].

2.3 *Word Embeddings*

In the Sect. 2.1, all the preprocessing was based on manually created rules. This is unfortunately not only significantly time consuming; it is also almost impossible to take all the possible relationships between each particular word into account. The problem with the manual creation of preprocessing parts is partially solved by Big Data approach (see Sect. 2.2). Thus, we can save time and costs with automatically created rules. Unfortunately, on the other hand large dataset of labelled training data is required, it is quite time and resource demanding to train a model and such a created model does not generalize well for the cases it hasn't yet seen before.

The idea of embedding methods is based on language modelling. The language modelling is basically the unsupervised learning task that tries to generate the next word. Using this principle during training we can relatively easily describe each word by a feature vector representation. Such feature representation is not only significantly smaller in regards of dimensionality when compared to vectors described in Sects. 2.1 and 2.2. It also expresses relationships between words, can search for synonyms, antonyms etc. One of the biggest advantages is that it can be trained from unlabelled data (i.e. time-consuming data labelling is not needed here). Once the embedding is trained, it can be transferred also to other domains, where the same language is used. Unfortunately, to train the model is usually quite time consuming and still needs a lot of training data—although only unlabelled are sufficient.

The embedding methods can be divided into the methods with local context (like Continuous Bag of Words—CBOW [20] or Skip-grams [8]) and context dependent methods (like ELMo [23], BERT [5], Universal Language Model Fine-tuning – ULMFit [7], GPT or GPT-2 [28]).

CBOW is a method that gets surrounding words and tries to predict a vector, which best describes the word which is in the middle of the surrounding words. Skip-grams (or Word2Vec[1]) has opposite philosophy, which gets a single word and tries to describe the context of the word (i.e. surrounding words). CBOW is less computationally demanding than the skip-grams and has slightly better accuracy for the frequent words. Skip-grams is considered to work well with small amount of the training data and represents better even rare words or phrases.

The context dependent models can, in contrast to e.g. Skip-Grams, take into account word context and order. ELMo uses bi-directional recurrent neural networks, in particular LSTM networks. Their main advantage is that they need significantly less training data than the previous approaches. BERT is built upon the ELMO, but instead of LSTMs it uses the so-called Transformer [5]. The transformers are attention-based models with positional encodings to represent words positions. This is mainly what makes the transformers to deal better with long term memory than LSTM. It also extends to both left and right contexts of sentences which can be used simultaneously (so called masked language modelling). It differs from ELMo and all the other methods because it can work on sub-word level while the other methods work on word level only.

Universal Language Model Fine-tuning (ULMFit) [7] is another language model introduced in Jan 2018. It uses a 3-layer LSTM network. It reduced errors by 18–24% on the majority of datasets. It also performs well on smaller data sets—with just 100 samples it matched performance on 100-times more data.

GPT and its direct successor GPT-2 [28] are language models based on transformers. They use in contrast to ELMo and BERT only one directional sequence modelling network. Especially GPT-2 is considered a big success. According to current reports, it helped to achieve SOTA results on 7 out of 8 tested language modelling datasets in a zero-shot transfer setting without any task-specific fine-tuning. Its advantage should be expected especially on small training datasets and with long term dependency tasks. Both GPT and GPT-2 use byte pair encoding with UTF-8 sequences and have 12 decoding blocks of transformers. Architecture of GPT and GPT-2 is quite similar—only some normalization layers were added to input of each sub-block. Recently, an unsupervised language representation learning method named as "XLNet" was introduced in [33]. It captures more important dependencies between prediction targets, such as (New, York), which e.g. BERT omits and proves state of the art on many benchmarks [33].

[1] Word2vec is a neural network that processes text, on the output is a set of feature vectors.

2.4 Convolutional Feature Extraction

In the previous approaches, a part of data processing is visible: input raw data are processed to a feature vector from which the searched information is extracted. The knowledge of language and grammar is used in the first part as a prerequisite for text understanding, which is actually a statistical extraction of possible meanings. Current state-of-the-art methods use language models known as embedding models (see article [14]). An embedding model converts each word from the text into a vector that represents the given word with possible meanings, relationships to other words, contexts, and so on. Such representation is made using large volume of samples in which the word was present. Therefore, embedding models are always focused on a particular language and represent a language-dependent part of the system.

The main idea of different researches based on Deep Learning was that Deep Learning is able to use Big Data to train a robust model. Utilizing specific artificial neural architectures, which have succeeded in computer vision tasks [29], a different approach was proposed in the article [35]. This approach was later adopted by Google [34] and Facebook [4] research teams. The idea of processing raw input data using one-hot based text encoding was used in different approaches, even in new Deep-Learning-based embedding models like ELMo [23]. This approach generates a matrix of ones and zeros, which can be regarded as raw representation of the text. This is done by reading the text letter by letter and encode each letter by a vector. See Fig. 1 where the word "nature" is represented in such format. Thanks to this approach, the deep neural network has access to the original raw input data. If the structure of the deep neural network were designed optimally, this network would be capable of extracting the information on its own—without any preprocessing.

The main idea behind the Deep Learning approaches heads to the machine understanding of textual data. Language models can represent language and grammar of the text. According to OpenAI article [28], current findings in this area suggest a promising path to language processing systems which learn to perform tasks from their naturally occurring demonstrations.

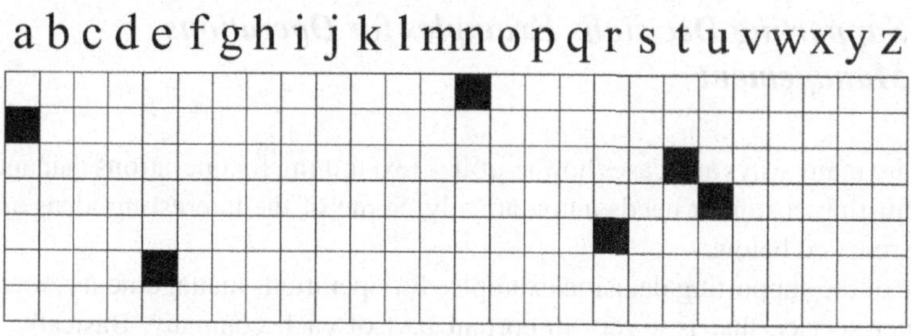

Fig. 1 Word "nature" represented as a raw input encoded with one-hot encoding

3 Supporting Decisions with Extracted Knowledge

Significant progress in information technologies development and latest artificial intelligence technologies almost force companies to store and analyse any kind of data. Of course, data mining for knowledge extraction has been used for years, but text mining or knowledge mining from natural language texts was a very tentative topic due to its complexity. The latest methods based on deep neural networks bring new perspectives for the companies even for more complex tasks such as text sentiment analysis, fact checking, automatic question answering, and other. These tasks can bring a new external input and their analysis and supporting decisions for operations management thus can lead to costs saving, lean production and advantage in global competition in general.

Knowledge extraction and text mining vary from case to case, but for solving more complex tasks, already mentioned deep learning methods have to be used. The common problem is insufficient amount of training data [13]. Even if there is plenty of data stored for years, the model training requires enormous amount of labelled data with searched information. This data preparation is basically a problem for many companies because it includes data selection, preprocessing (cleaning, sampling), transformation (decomposition, aggregation) and labelling. However, a public dataset or an automatically labelled dataset can be used for some cases. There are already prepared and published datasets with textual reviews with numerical value of the review that can be used as basement data for the model training. The Yelp dataset contains user reviews and user data for use in personal, educational and academic purposes. This dataset contains more than 4.7 million reviews rated with stars from 1 to 5. However, the database can contain some duplicated examples which have to be removed before training. Another similar dataset is Amazon reviews database obtained by Stanford Network Analysis Project (SNAP) published in [19] which contains more than 82 million product reviews and 5-star rating alongside user typed text.

3.1 Supporting Decisions Examples for Operations Management

There are many ways and cases how to utilize text mining for operations management and adjusting customer needs automatically. Some of the interesting ideas and use cases are listed below.

One of the supporting decision examples for operations management is regarding customer service that is a very important part of each company. Basically, profit and reputation directly depend on the quality of the company customer service that of course demands human resources and significantly affects the company's budget. There are obviously more communication channels e.g. voice (phone), on-site support or live chat, social networks and email communication that are text-based.

Knowledge extraction and text mining techniques together with other signals (e.g. voice) can be used for automatic customer type classification. Quality of the classification model then depends only on the internal data collected within the company. There are different points of view on the customer classification and they can be divided according to their mood or if they are ideal customers or ones that never buy the product or service. But in general, with some modifications there are three types of customers: care, sell, retence.

A human is not able to process such an amount of information—historical behaviour of the customers, an enormous number of customers—to make some predictions. It is very valuable to know the mood and the type of the client before the customer service operator starts to interact with the client. The operator can adapt and change the way of negotiating and talking with the client according to its mood and type immediately. Chatbots are widely used nowadays for the first contact with the customers. It is possible to fit and adjust an automatic answer for the specific customer behaviour. Prioritization of client requests according to their types predicted by classification model is a powerful tool to achieve higher customer care quality and early care, or to make a new offer for the customer before his leaving. Marketing campaigns and advertising may be optimised and better targeted according to customer types.

3.2 *Sentiment Analysis for Operations Management*

Classification is a general method and its usage is not only about estimating types of customers, but a popular classification problem is deciding whether a text is negative or positive. Social networks, blog posts, forums, reviews etc. are places where everybody can write anything and anytime. For example, the attitude of the people interacting with the company is actually important for the company and it has to be known whether their experiences are negative or positive. Any comments that mentioned a company, not strictly only on the company channels but also concurrent channels can be processed and analysed. Further analysing the positive comments should lead to focus and keep the positive things up to date. On the other hand, negative comments should be analysed immediately and recommendation and improvements should be proposed to avoid these dissatisfactions.

This kind of classification may bring also a low-level automatic feedback valuable for operations management. Automatic feedback statistics can be done on the social network posts level or after publishing a blog about new feature etc. Number of negative and positive comments (reactions) can be then analysed and some basic statistics about post or feature accomplishment is created automatically, see example in Fig. 2.

An advantage of this binary classification into negative or positive class is primarily in independence on the internal datasets. External dataset such as already mentioned Yelp or Amazon reviews might be used for training classification model (considering 1 star as a negative class and 5 stars as a positive class).

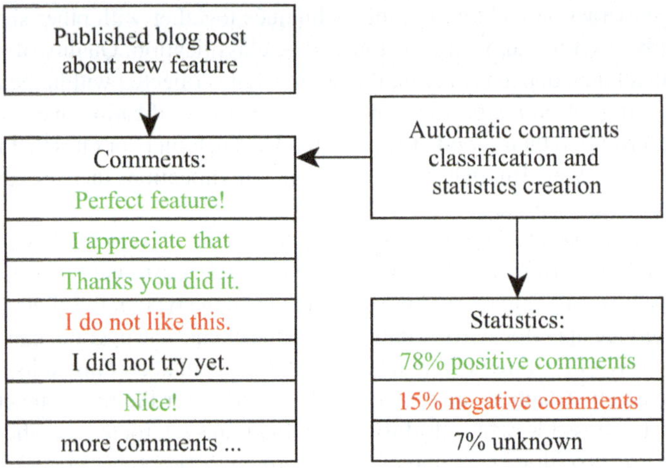

Fig. 2 An example of comments classifying and analysing on the blog post about new feature

3.3 Text Similarities for Operations Management

Blog posts and newsletters are great tools how to present and inform customers or potential clients about product advantages, features etc. But what about competitions overview and seeing how the products or features are similar or not? The amount of text to be read and remembered by humans is enormous for making any analysis. Knowledge extraction and text mining may significantly help with similarity measurements of two articles. Graph convolutional networks are used in [17] for matching long text documents. There can be basically three most important cases how to evaluate a post similarity. The first one is the best case that says the blog post does not match to any other competitions blog posts. This means the product or feature mentioned in the blog post is a big competition advantage and probably no one else disposes of this feature. The second case is that articles partially match. This says competitors have probably the same feature which means the companies are up to date and there is no significant advantage of any of them. The last case is that articles highly match. This case may warn about some copyright or intellectual property violations in the worst case (the competitor copies the article content or traces the developing process). Crawler or scraper techniques allow to make the whole process more automatic. Example of automatic similarity matching is in Fig. 3.

From another point of view, the concurrence blog post analysis may help for future features or product direction. Future development fulfilling all customer needs and requirements is a dream of every operations management. Planning the development and direction of the product according to customers demand is the win–win situation. Text mining is a powerful tool to make this happen. The easiest way is the already mentioned binary classification and analysis of comments under the post for example "Would like to have a feature X in our product?" But consider post title "What would

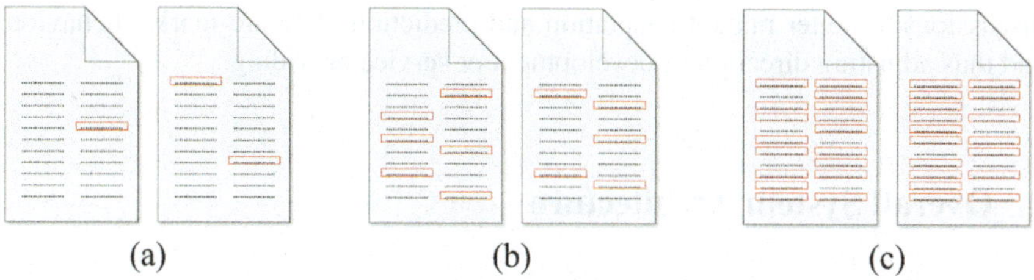

Fig. 3 Articles similarity matching and detection, where **a** shows different texts, **b** has some matches and **c** are almost identical or very similar articles

you improve in our product?" Classification cannot be used because of diversity of comments and there is no way to train a classifier. Again, this feedback evaluation is so difficult, maybe impossible for humans to read and analyse each comment and prepare some statistics and outcomes.

Clusterization methods [36] create a defined number of groups containing the most similar comments. These groups greatly simplify an analysis process of all comments because comments in one group should deal about the same or very similar content. Furthermore, a histogram created from all groups underlines the most important groups. Clusterization can be improved by similarity measuring and the whole process should be able to determine the meaning of the groups or divide the groups semantically.

Sometimes there is a need to test new features or a new product (service) conditions, because no one knows if customers appreciate it or it is inconvenient. Early feedback analysis is critical and using this approach it is possible to get the most important issues almost automatically (see example in Fig. 4). This process can be described in general as obtaining indirect feedback from social media and public

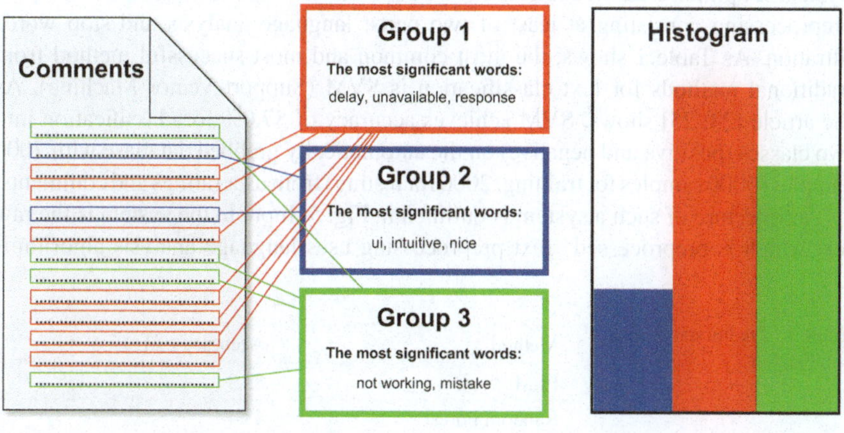

Fig. 4 Example of comments clusterization and histogram of importance

discussions to better market adaptation and prediction of future market behaviour and thus adjusting direction of development or service providing.

4 Overall System Architecture

Accuracy of knowledge extraction from text can be assessed on plenty of public benchmarks. One of the well-known problems and benchmarks is opinion mining (or sentiment classification). Author's opinion can often be understood from the whole context of his/her message. The automatic opinion extraction is complicated and it is not possible to be clearly determined from author's selected few key words, especially in the case of sarcasm or irony. This problem would be used to demonstrate the knowledge extraction system design, which can be applied in different use cases mentioned.

To meet the objectives—i.e. to classify meaning of the input text for operations management—several different approaches can be used: (a) a method based on traditional approaches, which would be strongly language dependent, however, does not need big amount of training data to achieve satisfying accuracy; (b) a method based on traditional approaches without language dependent preprocessing, but trained with big amount of data to achieve similar accuracy; (c) using an embedding model followed by deep neural network; (d) a novel approach based on deep learning which can be completely language independent (CNN based solutions).

4.1 Traditional Methods

System of opinion classification based on traditional methods should contain complex preprocessing consisting at least of two parts: language analysis and stop words filtration. As Table 1 shows, the most common and most successful method from traditional methods for text classification is SVM (Support Vector Machine). As the article [24, 25] shows, SVM achieves accuracy of 87.0% for classification into two classes (positive and negative) on the automatically grabbed database with 7000 samples (5000 samples for training, 2000 for testing) from user ratings and comments.

Architecture of such a system is outlined in Fig. 5. Input to the system is the raw text which is preprocessed. Text preprocessing uses language analysis algorithms

Table 1 Traditional methods comparison

Method	Accuracy (%)
FLM	83.3
Random Forest	48.4
k-NN	50.9
SVM	87.0

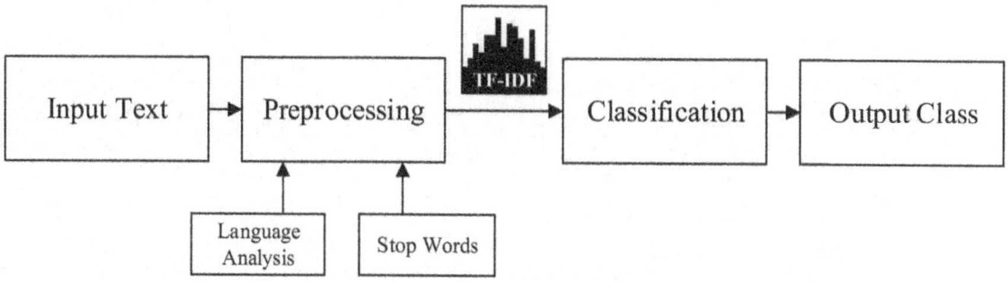

Fig. 5 Architecture of text classification system based on traditional approach

and a stop words database for a given language. At the end of preprocessing, data are encoded using the TF-IDF method introduced in [30]. Data encoded in this way is then fed into single or multiple SVM-based classifiers to predict the class the text belongs to. However, not each parameter of TF-IDF may carry an information. These parameters can be omitted.

Text data encoded in a n-dimensional feature vector consist of numerical values, where not every parameter has to have influence on the classification. Removing irrelevant parameters is an important task in classification model design because high dimensionality of the feature vector can lead to overfitting and worse results. Removing these features will reduce the level of noise in the training data, decrease the probability of model overfitting and confusion and increase prediction accuracy. Trying all possible combinations is not computationally feasible. Therefore, this is a crucial disadvantage of the traditional approach.

A common approach to reduce the dimensionality of the feature vector is the forward selection algorithm [18]. The method adds a single parameter at each step, so each used parameter has to have a positive effect on the resulting accuracy of the trained model. In each step, the parameter with the most significant increase in model accuracy will be used. This process is repeated until no parameter has a positive effect or the maximum number of parameters is achieved. Opposite to forward selection is the backward elimination algorithm [12] that starts with a full length of the feature vector and starts to remove parameters one by one while those removals have positive impact on the resulting accuracy. When compared to forward selection, the resulting vector length is commonly longer. For multi-model cases, the sequential elimination was introduced in Povoda et al. [24, 25] which emphasizes the whole system accuracy, not just a single model.

4.2 Abstraction Using Big Data

Language dependency of text classification systems can be minimized by removing the language dependent parts from the preprocessing step which leads to a radical reduction of accuracy. However, if the proposed solution has a large amount of training data (often referred to as "Big Data"), it is possible to achieve similar

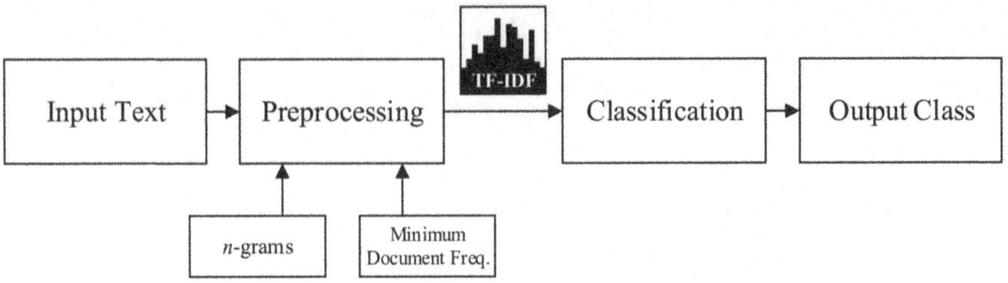

Fig. 6 Architecture of big data abstracted classification system

results without the need for complex preprocessing. So, the basic features of the language can be learned by the classifier—some frequently occurring combinations of words, phrases, different word forms, understand negation, etc. Much more complex analysis can be done by using the combination of distributed computing and Big Data approach. Parallel computing by distributing the computation with several independent tasks can accelerate computing and significantly reduce the training time. Figure 6 shows the language dependent parts replaced with different approaches—n-grams and minimum document frequency.

The experiments described in Povoda et al. [24, 25] compares various configurations of n-grams and minimum document frequency and proves that using bi-grams performs well with Big Data approach. Bi-grams can actually handle the problem with negation and different word forms in different situations (e.g. in some languages preposition changes form of the word), as shown in Table 2. The experiments proves suitability for different language families: Anglo-Frisian, Germanic, Slavic and Romanesque. Different n-gram levels increase input dimensionality (words are connected in sentence context and new input parameters are created), while minimum document frequency reduces dimensionality by filtering out the parameters that do not occur in the training dataset in a sufficient quantity.

A better performance can be achieved by using different values of minimum document frequency; however, the accuracy can be even higher using parameter optimization. In Big Data approach, the simple selection described in previous chapter cannot be used. Therefore, utilizing an optimization method based on a genetic algorithm may achieve further accuracy improvement. The method aims to find the optimal configuration of n-gram level, the value of minimum document frequency, and the subset of allowed input parameters (words and phrases), which will achieve the

Table 2 Comparison of different n-grams and their accuracies

Language	1-g (%)	2-g (%)	3-g (%)	4-g (%)
English	93.8	95.3	94.2	88.4
German	87.1	88.3	87.9	87.7
Spanish	91.4	93.2	**93.2**	93.1
Czech	88.7	89.1	88.7	88.4

highest possible classification accuracy on the validation set. The results of this optimization method from article [26] have shown that using 99.5% of the number of parameters may result in 1% accuracy increase.

4.3 Universal Language-Independent Method

Removing all language dependent parts brings demands on the implementation language and grammar analysis in different parts of the system. As Fig. 7 shows, if the preprocessing part is completely missing, its work has to be done in another part of the system, which can see the original raw data from the input. Prior knowledge of the language and its grammar is the main precondition to text understanding and is necessary for knowledge extraction. Current state-of-the-art methods use embedding models mentioned before for this purpose. They have been used as the input of the classification block or as the first layer of the deep neural network. Here each word is converted to its embedded representation—i.e. expressing relations to other words and its meaning even when the word has more than one possible meaning. In this case, the language analysis and grammar processing are done by embedding. The embedding model is commonly trained on a large corpus of different sentences (or documents) in a given language in the unsupervised way. The disadvantage is, that the model is language dependent on a particular language and the embedding model must be trained for each model separately.

Using the raw text input on the input layer of a deep neural network (without an embedding model) is the alternative approach to removing the preprocessing part. This approach is used by Facebook AI Research in articles [4, 2] for example in their very deep convolutional network. Similar approach was used by Google DeepMind in [34] for their recurrent neural network. In this chapter, we introduce a novel method based on deep learning methods utilizing big data training with similar input handling as in previously mentioned works. The method is compared to the current state-of-the-art methods using two databases—Yelp Review dataset (from 10th round of Yelp challenge) with 4.7 million of examples, and Amazon Review dataset published by Stanford Network Analysis Project with 82.4 million of examples. Both databases contain numeric representation of reviews within the interval from 1 to 5 stars. Only values 1 (negative) and 5 (positive) are used in the binary classification, which was the method used for comparison.

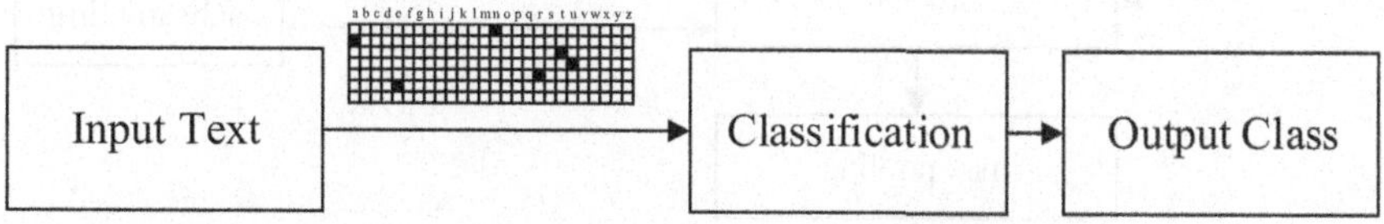

Fig. 7 Architecture of universal language independent classification system

The proposal of the novel deep neural network architecture was based on structures used in image processing—VGG network presented in [31], Inception kernel proposed by Google in [32], multicolumn architecture from technical report [3] and residual learning described in [6]. The method is trying to approach the process of human text data understanding. Firstly, the representation of typography of the language is created and beginnings and endings of words are detected (if it is possible in a given language). Letters and typographic marks are then formed into syllables and words, words to phrases, finally to higher abstraction representing resulting knowledge. The proposed method simulates the process of reading by humans. This has been done by utilizing several convolution layers and complex structures.

In the proposed structure of deep neural networks, repeating substructures were identified and extracted. This repeated substructure of network was named as "Recurring Kernel" (RCK, shown in Fig. 8) based on how these structures actually work. RCK kernels are carrying the same information to different branches of the network in the deeper layers of the network (partially subsampled and partially with extracted information), which allows deep neural networks to see the input text through different degrees of abstraction. This kernel is designed to get as much information as possible at a given network depth. By integrating several cores behind one another, it is possible to create a very deep architectures with high information throughput across the whole network structure. Even the deepest layers can partially access data

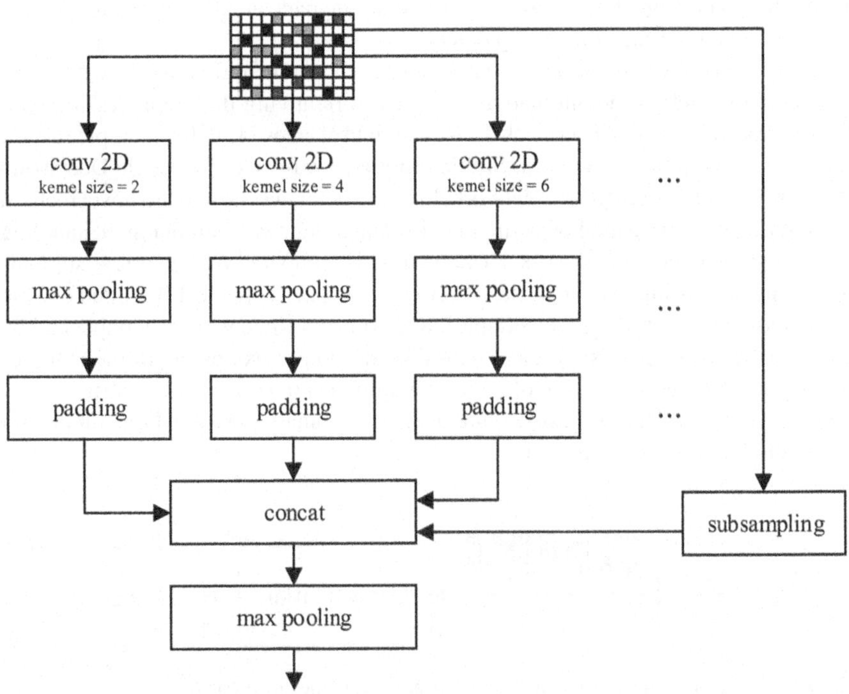

Fig. 8 Recurring kernel structure

from the first layers. All data and partially extracted information are carried to the lowest layers, where classification is done using dense layers. Convolution layers create a "fingerprint" (FP) at each level of the network in each column of the RCK kernel. These FPs are collected and extended with the residual part – with the input of the given kernel. Since it is a memory demanding operation, data are subsampled before outputting the kernel. This can be done due to the fact that some information is already duplicated in the FPs.

Two experiments were performed: (1) to demonstrate language independence, and (2) to compare performance with current state-of-the-art methods. Previous solution based on the Big Data abstracted statistical method was not fully independent because the proposed classifier counts on the existence of separate words in the input text. Such a solution is not applicable for languages such as Chinese where words are not separated by a space or another white character. Tested languages were: English—as an Anglo-Frisian language, a set of test texts targeted everyday communication; German—as a Germanic language, it tests the ability to process longer words and compounds, and complex grammar; Czech—as a representative of Slavic languages, this set focuses on words with and without accents, as well as the common mistakes in spelling and typos; Spanish—a Romanesque language, where this set of texts is mainly focused on short texts; and Chinese—to prove that the method is able to work with different alphabets and that the method does not rely on separated words. The second experiment compares performance of the proposed method with current methods on public text databases—the Yelp Review dataset and the Amazon Review dataset. These databases provide big amount of data to be used to train and test deep neural networks. Both experiments divide the text database into 3 parts: a training set, a validation set, and a test set. The accuracy shown in the tables are measured on the test set. The validation set was used only to control the training process. Table 3 shows selected RCK-based deep neural networks with parameters of the structure and their number of trainable parameters.

Table 4 shows the achieved accuracies on different languages. The variance of 12.7% and the average accuracy of 87.7% show that the method is applicable to a variety of languages. As for any machine learning method, the quality of the text database directly influences the achieved accuracy. Results from public text datasets

Table 3 Summary of proposed RCK-based deep neural networks

Network	Depth	RCK sizes	Dense	Size
1	4	64/64/64/32	128/128	3.8 M
2	4	64/64/64/32	256/128	6.8 M
3	5	64/64/64/32/32	128/128	2.8 M
4	5	64/64/128/256/256	512/256	27.7 M
5	6	64/64/64/32/32/32	256/128	3.3 M
6	7	64/64/64/64/64/32/32	128/128	3.1 M
7	7	64/64/64/64/64/32/32	256/128	3.8 M

Table 4 Achieved accuracies for binary classification on different languages and databases

Database/Language	Network	Accuracy (%)
English	6	84.8
German	4	83.4
Spanish	6	92.5
Czech	5	86.8
Chinese	7	91.2
Yelp Reviews dataset	6	**96.2**
	2	96.1
	5	96.0
Amazon Reviews dataset	3	**94.6**
	5	**94.6**
	2	94.0

are summarized in Table 4, where network no. 6 achieves on Yelp Review dataset a state-of-the-art accuracy. This network consists of 7 RCKs where the number of convolution kernels does not exceed 64. The trained model has 3.3 million training parameters that can be represented in memory by using 24 MB of data (85% lower than the simplest GloVe model from the article [22]). Compared to solutions based on word embedding, this method is much less memory demanding. Embedding models require hundreds of MBs, for example the simplest pretrained GloVe model has 171 MB, more complex models occupy more than 1 GB of memory, the Word2Vec [20] occupies 1.5 GB, BERT [5] model has 1.3 GB, even the fastText model [2] for English has 280 MB which is more than 10 times higher than model with 7 RCK kernels.

5 Conclusion and Direction for Future Research

This chapter presented the ideas of using artificial intelligence for extraction of knowledge from text data and how to use this knowledge in operations management. Thanks to current trends, increased interest in processing and analysing these data is anticipated. The development of new computing hardware with a focus on massive parallelism makes this analysis considerably faster. However, understanding of the content of natural languages as well as machine generated texts, all in fast and scalable manner, can be crucial to be able to utilize the information in knowledge-driven service operations management.

In particular, we discuss machine text classification and compare different approaches for this purpose. It was proven that traditional methods need additional optimization techniques to achieve higher accuracies. However, such solution uses complex preprocessing with language analysis which brings a strong language dependency. Abstracting this method by removing language-dependent parts and using Big Data approach, a more generalized system was described. In this case, optimization techniques have to be computationally efficient, so traditional techniques like forward attribute selection and backward elimination cannot be used. Therefore, possible optimization using a genetic algorithm was described. However, it was not possible to achieve a higher degree of linguistic independence using traditional methods. The solution was not suitable e.g. for Chinese language whose syntax does not use words. Therefore, methods based on deep learning were emphasised. In this field, methods utilizing embedding models are currently representing the state of the art. However, embedding models bring high memory demands which can be in some cases crucial. Therefore, we decided to emphasize deep neural networks without embedding models on their input. The input data conversion and the structure of deep neural network was described. The design of proposed solution was based on existing structures often used in image processing. Repeated patterns were identified and extracted in the repeating kernel.

The main output of this chapter is a novel method for text data-mining, which was experimentally verified on a selected case of text classification and compared with current state-of-the-art methods. Here, similar accuracies were achieved as current methods (in one specific case even better by 0.5%) while memory demands were significantly reduced (85% lower than the simplest GloVe model) when compared with the current methods based on embedding models. The proposed method is universal for all languages (natural or machine), it does not require any preliminary language knowledge or grammar preprocessing. The functionality was presented on 5 private datasets (5 different languages) and on the binary classification of publicly available datasets—Yelp Reviews dataset (with accuracy 96.2%) and Amazon Reviews dataset (94.6%).

For future research, this method should be subjected to experiments with transfer learning to identify the limits of this method. However, it is expected that pretrained language models would be narrowly focused for a given use case (like for example sentiment analysis described in this chapter). The method based on RCK kernels has significantly lower demands on hardware and is suitable for embedded devices. Even that in future we can expect more widespread use of TPUs, GPUs with bigger memories and probably even new hardware focused on artificial intelligence, there will still be use cases for this method to run on small limited hardware.

Acknowledgements Research described in this chapter was financed by the National Sustainability Program under grant LO1401 and MPO-CR FV20044. For the research, infrastructure of the SIX Center was used.

References

1. Bai, S., Kolter, J. Z., & Koltun, V. (2018). An empirical evaluation of generic convolutional and recurrent networks for sequence modeling. arXiv preprint arXiv:1803.01271.
2. Bojanowski, P., Grave, E., Jouilin, A., et al. (2017). (2017) Enriching word vectors with subword information. *Transactions of the Association for Computational Linguistics, 5*, 135–146.
3. Cireşan, D., Meier, U., & Schmidhuber, J. (2012). Multi-column deep neural networks for image classification. In: *2012 IEEE Conference on Computer Vision and Pattern Recognition*. https://doi.org/10.1109/cvpr.2012.6248110.
4. Conneau, A., Schwenk, H., Barrault, L. et al. (2016) Very deep convolutional networks for text classification. In: *Proceedings of the 15th Conference of the European Chapter of the Association for Computational Linguistics*. https://doi.org/10.18653/v1/e17-1104.
5. Devlin, J., Chang, M. W., Lee, K. et al. (2018). Bert: Pre-training of deep bidirectional transformers for language understanding. arXiv preprint arXiv:1810.04805.
6. He, K., Zhang, X., Ren, S. et al. (2016). Deep residual learning for image recognition. In: *2016 IEEE Conference on Computer Vision and Pattern Recognition (CVPR)*. https://doi.org/10.1109/cvpr.2016.90.
7. Howard, J., Ruder, S. (2018). Universal language model fine-tuning for text classification. arXiv preprint arXiv:1801.06146.
8. Huang, X., Alleva, F., Hon, H. W., et al. (1993). The SPHINX-II speech recognition system: An overview. *Computer, Speech and Language, 7*(2), 137–148. https://doi.org/10.1006/csla.1993.1007
9. Chakraborty, G., & Krishna, M. (2014). Analysis of unstructured data: Applications of text analytics and sentiment mining. *SAS Global Forum, 2014*, 1288–2014.
10. Ishikiriyama, C. S., Gomes, C. F. S. (2019). Big data: A global overview. In: *Big Data for the Greater Good* (pp. 35–50).https://doi.org/10.1007/978-3-319-93061-9_3.
11. Joachims, T. (1998). Text categorization with support vector machines: Learning with many relevant features. In: *European Conference on Machine Learning*. Springer, Berlin, Heidelberg. https://doi.org/10.1007/bfb0026683
12. Koller, D., & Sahami, M. (1996). *Toward optimal feature selection*. Stanford InfoLab.
13. Krizhevsky, A., Sutskever, I., & Hinton, G. E. (2012). Imagenet classification with deep convolutional neural networks. *Advances in Neural Information Processing Systems*:1097–1105.
14. Lai, S., Liu, K., He, S., et al. (2016). How to generate a good word embedding. *IEEE Intelligent Systems, 31*(6), 5–14. https://doi.org/10.1109/mis.2017.2581325
15. Lauren, P., Qu, G., Yang, J., et al. (2018). Generating word embeddings from an extreme learning machine for sentiment analysis and sequence labeling tasks. *Cognitive Computation, 2018*, 1–14. https://doi.org/10.1007/s12559-018-9548-y
16. Lewis, M. P. (2009). *Ethnologue: Languages of the world*. SIL International.
17. Liu, B., Zhang, T., Niu, D. et al. (2018). Matching long text documents via graph convolutional networks. arXiv preprint arXiv:1802.07459.
18. Luo, S., & Ghosal, S. (2016). Forward selection and estimation in high dimensional single index models. *Statistical Methodology, 33*, 172–179. https://doi.org/10.1016/j.stamet.2016.09.002
19. McAuley, J., Yang, A. (2016). Addressing complex and subjective product-related queries with customer reviews. In: *Proceedings of the 25th International Conference on World Wide Web. International World Wide Web Conferences Steering Committee* (pp. 625–635). https://doi.org/10.1145/2872427.2883044.
20. Mikolov, T., Chen, K., Corrado, G. et al. (2013). Efficient estimation of word representations in vector space. arXiv preprint arXiv:1301.3781.
21. Paolillo, J. C., & Das, A. (2006). *Evaluating language statistics: The Ethnologue and beyond* (pp. 3–5). UNESCO Institute of Statistics. Retrieved 17 Nov 2018.
22. Pennington, J., Socher, R., & Manning, C. (2014). Glove: Global vectors for word representation. In: *Proceedings of the 2014 Conference on Empirical Methods in Natural Language Processing (EMNLP)*. https://doi.org/10.3115/v1/d14-1162

23. Peters, M. E., Neumann, M., & Iyyer, M. et al. (2018). Deep contextualized word representations. In: *Proceedings of the 2018 Conference of the North American Chapter of the Association for Computational Linguistics: Human Language Technologies*. https://doi.org/10.18653/v1/n18-1202.
24. Povoda, L., Burget, R., & Dutta, M. K. (2016a). Sentiment analysis based on support vector machine and big data. In: *2016 39th International Conference on Telecommunications and Signal Processing (TSP)* (pp. 543–545). https://doi.org/10.1109/tsp.2016.7760939.
25. Povoda, L., Burget, R., Masek, J., et al. (2016b). Optimization methods in emotion recognition system. *Radioengineering, 25*(3), 565–572. https://doi.org/10.13164/re.2016.0565
26. Povoda, L., Burget, R., Dutta, M. K. et al. (2017). Genetic optimization of big data sentiment analysis. In: *2017 4th International Conference on Signal Processing and Integrated Networks (SPIN)*. https://doi.org/10.1109/spin.2017.8049932
27. Power, D. J. (2014). Using 'Big Data' for analytics and decision support. *Journal of Decision Systems, 23*(2), 222–228. https://doi.org/10.1080/12460125.2014.888848
28. Radford, A., Wu, J., Child, R., et al. (2019). Language models are unsupervised multitask learners. *OpenAI Blog, 1*, 8.
29. Ramakrishnan, K., Scholte, S., Lamme, V., et al. (2015). Convolutional neural networks in the brain: An fMRI study. *Journal of Vision, 15*(12), 371–371. https://doi.org/10.1167/15.12.371
30. Salton, G., & Buckley, C. (1988). Term-weighting approaches in automatic text retrieval. *Information Processing and Management, 24*(5), 513–523. https://doi.org/10.1016/0306-4573(88)90021-0
31. Simonyan, K., & Zisserman, A. (2014). Very deep convolutional networks for large-scale image recognition. arXiv preprint arXiv:1409.1556.
32. Szegedy, C., Vanhoucke, V., Ioffe, S., et al. (2016). Rethinking the inception architecture for computer vision. In: *Proceedings of the IEEE Conference on Computer Vision and Pattern Recognition*. https://doi.org/10.1109/cvpr.2016.308
33. Yang, Z., Dai, Z., Yang, Y. et al. (2019). XLNet: Generalized autoregressive pretraining for language understanding. arXiv preprint arXiv:1906.08237.
34. Yogatama, D., Dyer, C., Ling, W., et al. (2017). Generative and discriminative text classification with recurrent neural networks. *Transactions of the Association for Computational Linguistics, 5*, 135–146. https://doi.org/10.1162/tacl_a_00051
35. Zhang, X., Zhao, J., & LeCun, Y. (2015). Character-level convolutional networks for text classification. *Advances in Neural Information Processing Systems*, 649–657.
36. Zhukov, D., Andrianova, E., Otradnov, K., et al. (2018). Soft clustering method for text mining, with an opportunity to attribute them to different semantic groups. In: *ITM Web of Conferences* (Vol. 18, p. 03004).
37. Zicari, R. V., Rosselli, M., Ivanov, T., et al. (2016). Setting up a big data project: Challenges, opportunities, technologies and optimization. *Studies in Big Data*, 17–47. https://doi.org/10.1007/978-3-319-30265-2_2

Lukas Povoda was born in 1990. He received his Ph.D. from the Faculty of Electrical Engineering and Communication, Brno University of Technology, Czech Republic in 2018. His research interests include data-mining (especially text-mining), machine learning and optimization techniques in this field. His primary objective is to determine the meaning of texts with deep learning methods and deep text understanding methods.

Radim Burget is an Assoc. Professor at Brno University of Technology and is heading the Signal processing program at SIX Research Centre. He has been involved in research of artificial intelligence for many years and in plenty of research projects which include projects funded on European level, national level or privately funded projects. Companies he is cooperating with include Honeywell, Mitsubishi Electric, Rapidminer, Konica-Minolta and others.

Martin Rajnoha was born in 1991. He is Ph.D. candidate at the Faculty of Electrical Engineering and Communication, Brno University of Technology, Czech Republic. His research area is in general machine learning but main focus is on image processing, especially image-to-image translations and face recognition. His primary objective is reconstruction or superresolution of face from a sequence of low quality input images. Currently, he is working as a software developer dealing with creating and integration of scientific methods and models including NLP and image processing into production solutions. His interests except artificial intelligence are also data analysis, latest technologies, continuous integration and continious delivery and algorithms.

Peter Brezany serves as a Professor of Computer Science at the University of Vienna, Austria and the Technical University of Brno, Czech Republic. He completed his doctorate in computer science at the Slovak Technical University Bratislava. His focus was first on the programming language and compiler design, automatic parallelization of sequential programs and later high-performance input/output, parallel and distributed data analytics, and other aspects of data-intensive computing. Since 2001 he has led national and international projects developing Grid- and Cloud-based infrastructures for high-productivity data analytics. Dr. Brezany has published one book, several book chapters and over 130 papers. He co-authored a book addressing big data challenges and advanced e-Science data analytics; it was published in Wiley in 2013.

Toward a Comprehensive Framework of Social Media Analytics

Vala Ali Rohani and Shahid Shayaa

Abstract Social media is rapidly becoming an essential platform for social connectivity and information diffusion leading to more efficient service operations. Despite the emergent interest in studying the online behavior of social media users from different viewpoints, such as Sentiment Analysis, Topic Modeling, Emotion Mining, and Social Network Analysis, a gap remains in utilizing all such methods concurrently toward building a comprehensive analytical model. In this chapter, the mentioned aspects are combined to propose a practical analytics framework for gaining more actionable insights from social media content. In doing so, the framework is developed based on a series of machine learning and data analysis algorithms along with the required ETL modules. To evaluate the performance, our framework is exposed to a real social media dataset comprising around 600,000 records. The high level of compatibility between the actionable insights generated by designated machine learning algorithms and actual events that occurred in the same period reveals the merit of the proposed approach in processing the social media datasets. Having the ability to get embedded in big data clusters, the proposed analytics engine can be utilized in analyzing the large social media datasets through big data analytics solutions.

Keyword Social Media Analytics · Sentiment Analysis · Emotion Mining · Topic Modeling · Social Network Analysis

---

V. A. Rohani (✉)
Department of Research and Development, Data Corner, Lisbon, Portugal
e-mail: vala@Data-Corner.com

S. Shayaa
Department of Data Analytics, Berkshire Media, Kuala Lumpur, Malaysia
e-mail: shahid@BerkshireMedia.com.my

A. Emrouznejad and V. Charles (eds.), *Big Data and Blockchain for Service Operations Management*, Studies in Big Data 98,
https://doi.org/10.1007/978-3-030-87304-2_7

List of acronyms

SMA	Social Media Analytics
SNA	Social Network Analysis
PLSI	Probabilistic Latent Semantic Indexing
LDA	Latent Dirichlet Allocation
TME	Topic-level Maximum Entropy
MAP	Maximum a Posteriori

1 Introduction

The concept of Social Media Analytics (SMA) has attracted significant attention in recent years. SMA entails continuously collecting, monitoring, analyzing, summarizing, and visualizing information from social media [1–3]. These activities require considerable effort due to the wide variety of social media channels as well as huge volumes of unstructured data on online platforms. The core proposition is that studying the interactions among social media users and analyzing their online behavior provides an opportunity to generate actionable insights toward making more effective decisions [4].

The social media domain and respective applications are recently being studied from various perspectives. Some researchers focus on Sentiment Analysis to predict public perceptions from online communications among social media users [5–9]. Other researchers perform trend analysis on the emotions of social media users by applying emotion mining techniques [10, 11]. Topic modeling [12, 13] and text mining [14, 15] are also currently employed as data science approaches to capture topics from social media content. From another standpoint, analytical advances have enabled researchers and data scientists to detect influencers in different domains. In this context, Social Network Analysis (SNA) techniques have been employed effectively to detect such important users [16–18]. Although the above mentioned techniques are helpful for gaining information from social media, we believe that combining multiple approaches would give us more coherent insights with a wider perspective. Combining these methods makes us enabled to have a more clear root cause analysis toward finding the main reasons behind the insights obtained in different dimensions separately. Each of the existing social media analytics methods provides some insights from their single point of view, while end-users need more reliable actionable insights supported by analysis results in different dimensions to utilize them in making data-driven decisions [17, 18].

In this chapter, a multi-dimensional SMA framework is proposed to elicit actionable insights from social media content. Several analytical techniques are combined to build this framework, including Sentiment Analysis, Emotion Mining, Topic Modeling, Text Mining, and Social Network Analysis (SNA). The chapter is structured as follows. The Related Works section presents a summary of recent works on the SMA context. The Proposed Framework section introduces the proposed comprehensive model and the theories behind it. The proposed model results and output are discussed in the Actionable Insights section. The Conclusion and Future Work section concludes with future directions.

2 Related Works

Sentiment analysis, or opinion mining, is one of the most active research areas for analyzing people's opinions, sentiments, attitudes, evaluations, and preferences from text documents [19]. It encompasses several activities, such as sentiment extraction, subjectivity classification, opinion summarization, and spam detection, among others [20]. Earlier studies have indicated two main categories of sentence-level sentiment classification: term-counting methods (lexicon-based) and machine learning approaches (corpus-based). A term-counting approach involves deriving a sentiment measure by calculating the total number of negative and positive terms or keywords. This approach has been widely used as the most basic means of analyzing simple text [21, 22]. More recently, however, new machine learning approaches have emerged based on computing technology advancements that offer superior performance in sentiment prediction from complex documents and sentences. In 2015, Olga Kolchyna et al. covered two lexicon-based and machine learning methods and showed how combining sentiment lexicons with emoticons, abbreviations, and social media slang expressions can increase the accuracy of lexicon-based classification for Twitter [23]. In another research, Vala and Shahid introduced SentiRobo, a new supervised machine learning approach for calculating sentiment scores. They developed an enhanced version of the Naïve Bayes algorithm to predict the sentiment polarity of large social media datasets [8].

Nowadays, social media have been increasingly adopted as a useful environment to improve operational efficiency in organizations by facilitating the information flow and knowledge and strengthening the firm-customer interactions [24]. Wenjing Duan used the sentiment analysis technique to decompose user reviews into different dimensions to measure hotel service quality and performance. They found that different dimensions of user reviews have significantly different effects in forming user evaluation and driving content generation [25]. In another recent study, the researchers investigated how the social media marketing activities of airlines influence brand equity and customer response. According to their results, trendiness was found as the most significant social media marketing activity component affecting the airline's brand and customer satisfaction [26]. Also, big data technologies are essential to address the issues related to the huge size of data in the social media analysis domain [27]. Big data solutions enable managers and decisions makers to use large-scale monitoring systems towards making data-driven decisions [28]. In this context, social media analysis can be considered as one of the real-world monitoring systems that provide useful actionable insights for strategists and decision makers.

Social network analysis (SNA) has theoretical roots in early research by twentieth century sociologists, such as Georg Simmel and Émile Durkheim, who studied the patterns of interaction among social actors [29]. Since then, social scientists have used the term "social networks" to denote complex sets of relationships among members of social systems [18, 30]. More recently, researchers like Marc Smith [31] and Mathieu Bastian et al. [32] have introduced the NodeXl and Gephi tools to visualize the relationships between nodes in SNA studies.

Recent years have witnessed a growing interest in adopting SNA techniques with an emphasis on analyzing the interactions between users of online social media websites such as Twitter and Facebook [33]. In 2015, Hywel T.P. Williams et al. found that most Twitter users interact only with like-minded persons in communities dominated by a single view [16]. In another work, SNA techniques were adopted to detect influencers in social media and categorize them into three different typologies: disseminators, engagers, and leaders. It was claimed this approach provides high resource optimization levels to create effective online communication policies [34]. Also in 2015, Helmi Norman et al. explored the role of social participation in mobile social media learning through a social network analysis study [35].

Topic modeling provides an effective means of projecting text documents into a topic space. Current studies indicate a growing interest in topic modeling. Early research on topic modeling was conducted by Deerwester et al. in 1990. They proposed a semantic structure for enhancing the detection of relevant documents based on terms found in queries [36]. In 1999, Hofmann proposed a new Probabilistic Latent Semantic Indexing (PLSI) model for handling domain-specific synonymy as well as polysemous words [37]. In 2003, David Blei et al. developed the latent Dirichlet allocation (LDA) to allow documents to contain a mixture of topics [38], after which LDA became a standard tool in topic modeling. Numerous authors are utilizing LDA for social networks and social media. In 2007, Zhang et al. worked on a hierarchical Bayesian model inferred from the widely-received LDA model to find probabilistic community profiles on social networks. In this model, communities are considered latent variables and defined as distributions over the social actor space [39]. In 2008, Nallapati et al. introduced a model that combines LDA with a community detection process [40]. Most recently, in 2016, the authors presented their research findings on applying the unsupervised LDA algorithm to social media datasets with 90,527 records [13].

Emotion mining refers to the extraction of emotions from textual data [41]. With the rapid proliferation of Web 2.0, eliciting emotions embedded in social media content has become essential. In 2011, Johan Bollen et al. found that social, political, cultural, and economic events have a considerable effect on the public mood in various dimensions [42]. Yanghui Rao et al. also studied social emotion mining in the news domain. They proposed a framework for modeling topics and emotions jointly [43]. In another research, social media was investigated as an important channel for public health to predict the forthcoming influence of childbirth on the mood and behavior of new mothers. The authors used social media datasets to detect mothers at risk of postpartum depression and consequently to design low-cost, early warning systems and intervention programs aimed at promoting postpartum wellness [44].

Subsequently, in 2016, the researchers proposed a topic-level maximum entropy (TME) model for social emotion classification of short text [45].

In this section, several existing research works were discussed in the SMA domain. From Sentiment Analysis techniques that detect polarity in social media, to emotion mining algorithms that go deeper with exploring the emotion trends in studied topics. Along with the SNA methods that were discussed aiming to detect the social media influencers, topic modeling approaches also explored which enable social media researchers to automatically detect the discussed topics in online content. In the next section, a novel framework is proposed aiming to provide more reliable actionable insights through merging the four discussed SMA together in one picture.

3 Proposed Framework

In the present article, an attempt is made to adopt all mentioned techniques to build a comprehensive framework for social media analytics. Referring to the authors´ previous research works in analyzing millions of social media data [13, 46, 47], It is believed that integrating the Sentiment Analysis, Topic Modeling, Emotion Mining, and Social Network Analysis techniques into a single framework would enable providing more responsive insight to help decision-makers make strategic decisions in the domain. Figure 1 illustrates the different layers of the proposed social media analytics framework.

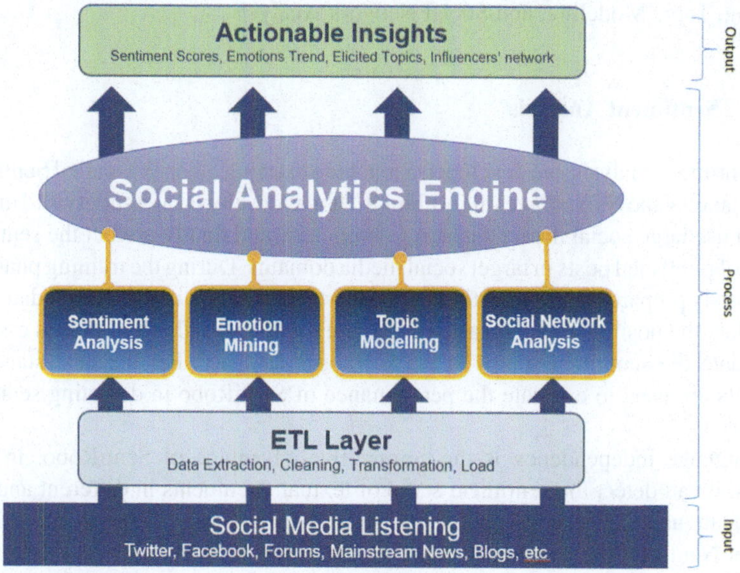

Fig. 1 The proposed framework of social media analytics

3.1 Input

The input data for the proposed model was extracted using commercial social media listening tools, such as Radian 6, GeoFedia, and Meltwater. Social media listening is the process of monitoring social media by defining specific topics based on related keywords. The tools built in this domain enable searching, filtering, and refining the data to extract information from a variety of social media content providers like Twitter, Facebook, Instagram, and forums, mainstream news, and other channels. In this research, Radian 6 is employed to collect the data for the present research, including 599,739 social media records from the education domain from the last three months of 2016. The data includes social media posts published in English, Bahasa Malaysia, and combinations of both languages.

3.2 Process

The raw data extracted using social media listening tools need to be cleaned and prepared for usage in higher layers. To this end, we developed an ETL layer for the extraction, transformation, and loading of social media data collected. Several data cleaning modules were developed in different programming languages including R and VBA to find the best alternative for null authors, identify unrelated posts, detect anomalies, and so on. The ETL layer output, which is the cleaned data, was passed to the social analytics engine for processing including Sentiment Analysis, Emotion Mining, Topic Modeling, and Social Network Analysis.

3.2.1 Sentiment Analysis

For sentiment analysis, we developed a machine learning algorithm called SentiRobo by enhancing the Naïve Bayes algorithm [8]. It was developed as a supervised method which use large social media datasets to train, test, and finally predict the sentiment value of published posts in target social media domains. During the training phase, the previously prepared list of negative and positive keywords is used for each dataset, to calculate the positivity and negativity score of each keyword. This stage is essential to update the sentiment scores in the keywords database. For the next stage, test datasets are used to evaluate the performance of SentiRobo in detecting sentiment polarity.

Language independency is the competitive advantage of SentiRobo. In other words, it can detect the sentiment score of textual documents in different and even mixed languages. This approach assigns a given document d class c (Positive, Negative, or Neutral). The formula for detecting the Maximum a Posteriori (MAP) class that is the best sentiment class sought to assign document d is as follows:

$$C_{\text{MAP}} = \arg\max_{c_j \in c} p(d|c_j)p(c_j)$$
$$= \arg\max_{c_j \in c} p(x_1, x_2, \ldots, x_n) \cdot p(c_j) \tag{1}$$

Where document d is represented by features $x_1, x_2, \ldots, x_n$. The Bag-of-Words assumption is taken, whereby the position of words in documents does not matter. We also assume that the feature probabilities $P(x_i|c)$ are independent given class c. Based on this assumption, formula (1) can be simplified as:

$$C_{\text{MAP}} = \arg\max_{c_j \in c} p(x_1|c_j) \cdot p(x_2|c_j) \cdots p(x_n|c_j) \cdot p(c_j) \tag{2}$$

In other words,

$$C_{\text{MAP}} = \arg\max_{c_j \in c} p(c_j) \cdot \prod_{x_i \in X} p(x_i|c_j) \tag{3}$$

In this algorithm, the sentiment score of each keyword in a sentence is considered as an indicator of the overall sentiment polarity of the whole sentence. Thus, the positivity and negativity of each keyword in the list of already detected positive and negative keywords (k) should initially be calculated using the following formulas:

$$\text{Positivity}_k = \frac{\text{Positivecount}_k}{\text{Positivecount}_k + \text{Negativecount}_k} \tag{4}$$

$$\text{Negativaty}_k = \frac{\text{Negativecount}_k}{\text{Positivecount}_k + \text{Negativecount}_k} \tag{5}$$

Where Positivity_k and Negativaty_k are the numbers of occurrences of the keyword K in positive and negative contents respectively, in the training dataset.

After calculating the positivity and negativity of all keywords in the sentiment analysis database, the sentiment score of each social media content can be calculated as:

$$\text{Sent\_ Score} = \sum\nolimits_{\forall t \in term} \text{Positivity}_t \text{ - Negativity}_t \tag{6}$$

where t is a negative or positive keyword that exists in the content. SentiRobo succeeded in predicting the sentiment value of mixed English-Malay tweets with an average accuracy rate of 75% [8] which outperforms other existing commercial tools on sentiment analysis with an average accuracy rate of 54% [48]. Figure 2 depicts the sentiment trend in the studied datasets detected by SentiRobo. As illustrated in Fig. 2, the overall sentiment for the last three months of 2016 in the studied domain is mainly negative with two peaks at the end of October and mid-November. However, there is a sign of positive talks in November 2016. Additional details on actionable insights elicited are discussed in Sect. 4.

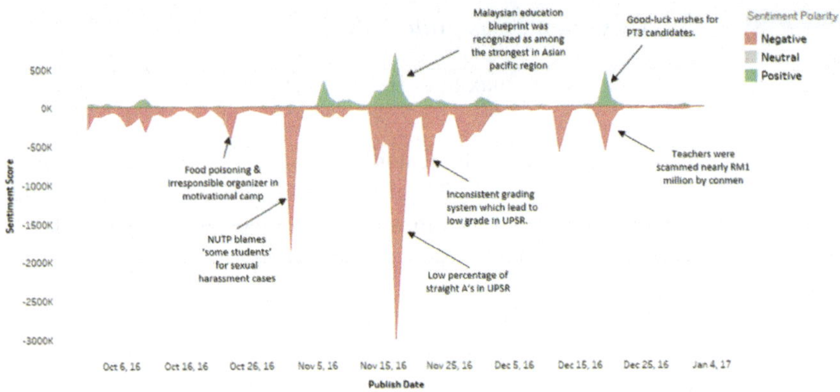

Fig. 2 Sentiment trend over time

3.2.2 Emotion Mining

Emotion mining can be done at two levels: binary and discrete. The binary level defines a particular text with positive or negative polarity. But previous research has indicated that the binary sentiment analysis level is not sufficient to analyze all emotions related to a particular comment [46].

Referring to our proposed framework in Fig. 1, although SentiRobo could detect the polarity of user discussions on social media, we attempted to achieve more detailed information enriched by emotion mining methods. To this end, we built a database of emotion keywords in both English and Bahasa Malaysia and grouped them into six main categories: anger, fear, joy, love, sadness, and surprise that corresponds to Ekman's six basic emotions [49]. Then we used text mining to look up the categorized emotion keywords in every post published throughout the study duration. Upon analyzing the detected emotions, the overall trend was generated (Fig. 3). In this figure, each color represents a specific emotion. The majority of emotional content was published during November 2016, although some peaks are also shown in early October and mid of December. In the middle of November as illustrated in Fig. 3, social media users got angry about the low percentage of A marks, while we see a peak of joy feeling during the same days after announcing that the Malaysian education blueprint was recognized as among the strongest in Asian pacific region.

3.2.3 Topic Modeling

For the Topic Modeling element in the proposed SMA engine, we adopted an unsupervised topic modeling approach that incorporates the LDA (latent Dirichlet allocation) algorithm toward discovering topics from the imported datasets. LDA is a technique for automatically finding topics that contain certain words. It represents a combination of topics that discharge words with certain probabilities. LDA can

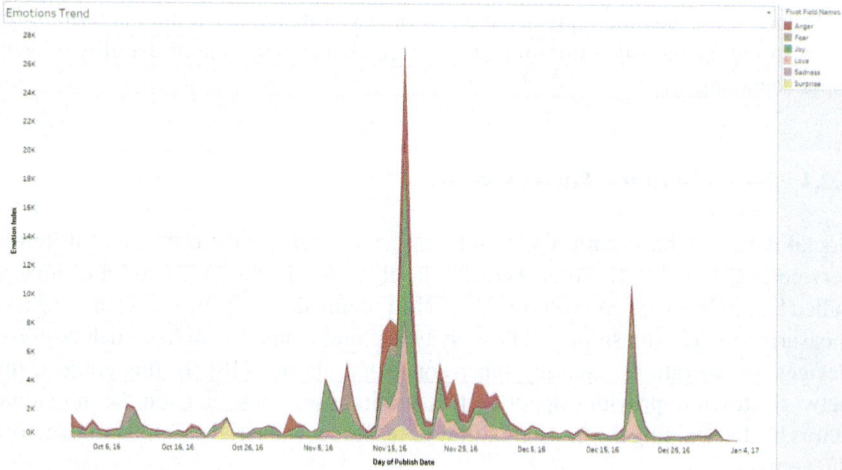

Fig. 3 Emotion trend over time

discover topics in a collection of documents automatically. The goal of LDA is to derive the latent topic from words and documents. LDA first recreates the documents in the corpus and functions based on the importance of related topics from words and documents [13]. The following is the LDA algorithm pseudo-code for the generative process of each document w in a corpus D:

1. Choose N ~Poisson(ε).
2. Choose θ~Dir(α).
3. For each of the N words w_n:
   a. Choose a topic z_n ~Multinomial(θ).
   b. Choose a word w_n from $p(w_n \mid z_n, \beta)$, a multinomial probability conditioned on topic z_n.

The following are further explanations:

1. The dimensionality k of the Dirichlet distribution and the dimensionality of topic variable z is known and fixed.
2. The word probabilities are defined as k × V (matrix β) where $\beta_{ij} = p(w^j = 1|z^i = 1)$, which is considered a fixed quantity for estimation.
3. The Poisson assumption is not important to anything subsequent, and documents with distributions of more realistic lengths can be used if necessary.
4. N is independent of all data-generating variables like θ and z. It is also a secondary variable and its randomness is ignored in subsequent development.

θ is a k-dimensional Dirichlet random variable that can take values in the (k−1)-simplex (a k-vector θ lies in the $(k-1)$-simplex if $\theta\, i \geq 0$, $\sum_{i=1}^{k} \theta_i = 1$).

Figure 3 presents the topic detected by the LDA machine learning algorithm for every month of the study duration. The figure also presents the probability of each topic occurring.

3.2.4 Social Network Analysis (SNA)

Social networks have both direct and indirect impacts on the efficacy of different service operations [50]. To analyze the level of this impact, we need a technique called Social Network Analysis (SNA). SNA is defined as the process of mapping and measuring the relationships and flows between interconnected nodes, such as users, devices, organizations, and any other connected entities [18]. In this context, the network structure provides opportunities and imposes constraints on the individual actors by determining the transfer or flow of resources (material or immaterial) across the network.

Social network visualization enables making sense of network data, detecting the communities, and identifying the influencers in the domains being studied [51]. Accordingly, many SNA tools have been developed by research groups and startup companies. NodeXL [17], Gephi [32], UCINET [52], and Pajek [53] are some examples of such tools.

In this research, we used Gephi to visualize the SNA graph in the studied duration. Before that, we developed some data cleansing modules in R and VBA to prepare the clean data for processing. As depicted in Fig.5, one of the novelties of the current research work is grouping social media content into two categories of positive and negative posts and then visualizing the related SNA graphs separately. In this way, it would be easier to detect top influencers in both categories of negative and positive online discussions. Moreover, considering six different types of social media networks [51], it is possible to generate some amazing insights based on the shape of SNA graphs which will be discussed further in Sect. 4.

According to Marc Smith and Michael Lieberman, there are six types of SNA maps called Polarized, In-group, Brand, Bazaar, Broadcast, and Support. Polarized networks are usually seen in political discussions on social media. This kind of pattern emerges when two main groups are shaped based on their opinions. Figure 5.e displays this pattern for a network of positive talking influencers in December 2016. The In-Group type is mostly seen among tightly knit groups of people, for example at conferences. From a marketing viewpoint, this kind of SNA community is most useful for technical products where users converse over social media. The Brand type is recognized as small disconnected groups. In this specific topology of SNA graphs, it is much easier to detect the major influencers and determine what they are talking about. These influencers are customers that brand owners might be interested to contact directly to improve their sales and marketing strategies. Bazaar means 'marketplace'. Therefore, Bazar type SNA graphs are known for their diverse SNA clusters. They are usually seen for medium-sized companies or political issues with various community involvement levels. Another useful SNA map is the Broadcast type, where an individual account dominates the map. Examples include celebrities like Adele, a news channel, or a sports club like Chicago Bulls. Support networks are known for being used for customer service. Several organizations usually provide support to various companies. These maps are the least commonly found in social network analysis.

3.3 Output

After applying all above mentioned analytical techniques embedded in the social analytics engine (Fig. 1), the generated outputs need to communicate to clients to help them in making data-driven decisions. A communicated insight creates an effective signal that prepares a pathway for better actions to take.

Each element of the proposed social analytics engine provides some useful information from different viewpoints. SentiRobo utilizes machine learning techniques to assign to every published post a sentiment score, identifying that it is positive, negative, or neutral (Fig. 2). Applied emotion mining techniques provide more detailed information for every post, which is called emotions (Fig. 3). Taking advantage of topic modeling techniques, we could detect the main topics discussed in every month of study duration (Fig. 4). SNA graphs developed by social network analysis techniques recognize the influencers and mayors in the studied domain. It also clarified the structure of communities around them (Fig. 5). These graphs help to detect the key influencers in both positive and negative talking atmospheres. By detecting them, marketing strategists can easily focus on their online behaviors in social media to change the overall sentiment from negative to positive with much less marketing budgets and efforts.

But whatever is generated by social analytics engine, from sentiment analysis and emotion mining to elicited topics and SNA graphs, all are just information but not insights. Insights are generated by analyzing information and drawing conclusions. Both data and information set the stage for providing insights that can help decision makers to make real data-driven decisions. To provide actionable insights, domain experts are needed to explore all generated information and graphs. The next section represents a list of actionable insights detected in this study by experts with sophisticated knowledge and experience in social media analytics in the field of education. As a social media analysis team of Berkshire Media company, they have more than five years of field experience analyzing more than three million social media content in the education domain.

4 Actionable Insights

A significant gap in analytics projects is with providing actionable insights to help management gain a clear view of the data studied and support data-driven decisions.

	Oct 2016		Nov 2016		Dec 2016
T1	Sexual Harassment	T1	The low percentage of straight As	T1	Teachers defrauded by rich quick scheme
T2	Food poisoning at schools	T2	The alleged leak of a national examination paper	T2	Comparison among student batch on lifestyle
T3	Hot PT3 Question	T3	Malaysia has among strongest vision in Education Planning in the Asia Pacific	T3	Education Ministry will defend teachers sued by parents
T4	Helicopter crashes into school	T4	Educational performance is not satisfactory	T4	Good luck wishes to candidates on PT3 exam results
T5	Drug abuse cases among students	T5	Transferring mole teachers	T5	Candidates showing excitement after finishing SPM Exam

	Oct 2016					Nov 2016					Dec 2016				
Day	T1	T2	T3	T4	T5	T1	T2	T3	T4	T5	T1	T2	T3	T4	T5
1	0	0	0	0	0.99	0.00	0.07	0.00	0.00	0.92	0.00	0.99	0.00	0.00	0.00
2	0.01	0	0	0.01	0.98	0.00	0.10	0.00	0.00	0.90	0.01	0.99	0.00	0.00	0.00
3	0.03	0.01	0.01	0.06	0.88	0.00	0.14	0.00	0.00	0.85	0.01	0.98	0.01	0.00	0.00
4	0	0	0	0.73	0.26	0.01	0.51	0.00	0.01	0.47	0.00	0.97	0.01	0.01	0.00
5	0.01	0.01	0	0.84	0.14	0.00	0.93	0.00	0.01	0.06	0.00	0.33	0.67	0.00	0.00
6	0.02	0	0.01	0.9	0.08	0.00	0.98	0.00	0.00	0.01	0.00	0.15	0.84	0.00	0.00
7	0	0	0	0.99	0	0.01	0.96	0.00	0.01	0.01	0.01	0.10	0.88	0.01	0.00
8	0	0	0	0.99	0	0.02	0.94	0.01	0.02	0.01	0.00	0.04	0.95	0.00	0.00
9	0.8	0	0	0.2	0	0.00	1.00	0.00	0.00	0.00	0.01	0.03	0.95	0.00	0.00
10	0.15	0	0.01	0.83	0	0.00	0.99	0.00	0.00	0.00	0.75	0.01	0.24	0.00	0.00
11	0.06	0	0.01	0.93	0.01	0.00	0.99	0.00	0.00	0.01	0.73	0.02	0.23	0.00	0.02
12	0.04	0.01	0.02	0.93	0.01	0.01	0.95	0.01	0.01	0.02	0.00	0.00	0.00	0.00	0.99
13	0	0	0.96	0.04	0	0.00	1.00	0.00	0.00	0.00	0.01	0.05	0.02	0.01	0.91
14	0	0	1	0	0	0.00	0.01	0.99	0.00	0.00	0.01	0.00	0.79	0.00	0.19
15	0.05	0.01	0.9	0.03	0.03	0.00	0.00	0.99	0.00	0.00	0.02	0.02	0.89	0.01	0.05
16	0.04	0	0.29	0.01	0.65	0.75	0.00	0.25	0.00	0.00	0.01	0.01	0.92	0.03	0.03
17	0.09	0.01	0.11	0.04	0.74	0.99	0.00	0.01	0.00	0.00	0.04	0.01	0.88	0.05	0.02
18	0.02	0	0.04	0	0.94	0.30	0.00	0.01	0.00	0.69	0.77	0.00	0.11	0.10	0.01
19	0.02	0	0.01	0	0.97	0.13	0.00	0.02	0.00	0.84	0.00	0.00	0.00	0.99	0.00
20	0.05	0.01	0.02	0.02	0.9	0.07	0.00	0.03	0.01	0.88	0.01	0.00	0.00	0.98	0.00
21	0.01	0.94	0.01	0.01	0.04	0.04	0.00	0.01	0.04	0.91	0.37	0.01	0.01	0.60	0.01
22	0.05	0.92	0.01	0	0.02	0.02	0.00	0.01	0.93	0.04	0.49	0.06	0.06	0.39	0.01
23	0	0.99	0	0	0	0.02	0.01	0.02	0.92	0.03	0.78	0.04	0.03	0.14	0.02
24	0.01	0.97	0.01	0.01	0.01	0.01	0.00	0.02	0.97	0.00	0.82	0.03	0.04	0.08	0.02
25	0.01	0.98	0.01	0	0	0.02	0.01	0.02	0.94	0.01	0.80	0.06	0.03	0.09	0.02
26	0.46	0.5	0.01	0.01	0.02	0.01	0.06	0.01	0.90	0.02	0.84	0.07	0.02	0.06	0.01
27	0.91	0.08	0	0	0	0.00	0.01	0.00	0.99	0.00	0.87	0.05	0.02	0.04	0.02
28	0.91	0.06	0.02	0.01	0.01	0.00	0.01	0.00	0.98	0.01	0.81	0.07	0.03	0.07	0.02
29	0.97	0.02	0	0	0	0.00	0.01	0.01	0.98	0.00	0.69	0.19	0.02	0.06	0.04
30	0.98	0.01	0	0	0	0.00	0.04	0.01	0.94	0.01	0.92	0.04	0.01	0.02	0.02
31	0.91	0.03	0.01	0.04	0.01						0.99	0.00	0.00	0.00	0.00

Fig. 4 Probability of topics detected by the LDA topic mining algorithm

This section presents the actionable insights elicited by the team of domain experts after analyzing the generated graphs from different perspectives.

For October 21st–24th 2016, the volumetric sentiment score trend line displays high negative spikes. The negative spikes were caused by two main accidents, which were food poisoning cases in a school and an irresponsible organizer in a student motivational camp. The spikes are correlated with the Emotion trend line graph, whereby on the mentioned dates, 'Anger' increased. This might have been caused by posts of parents who were worried and concerned about their children's wellbeing, as

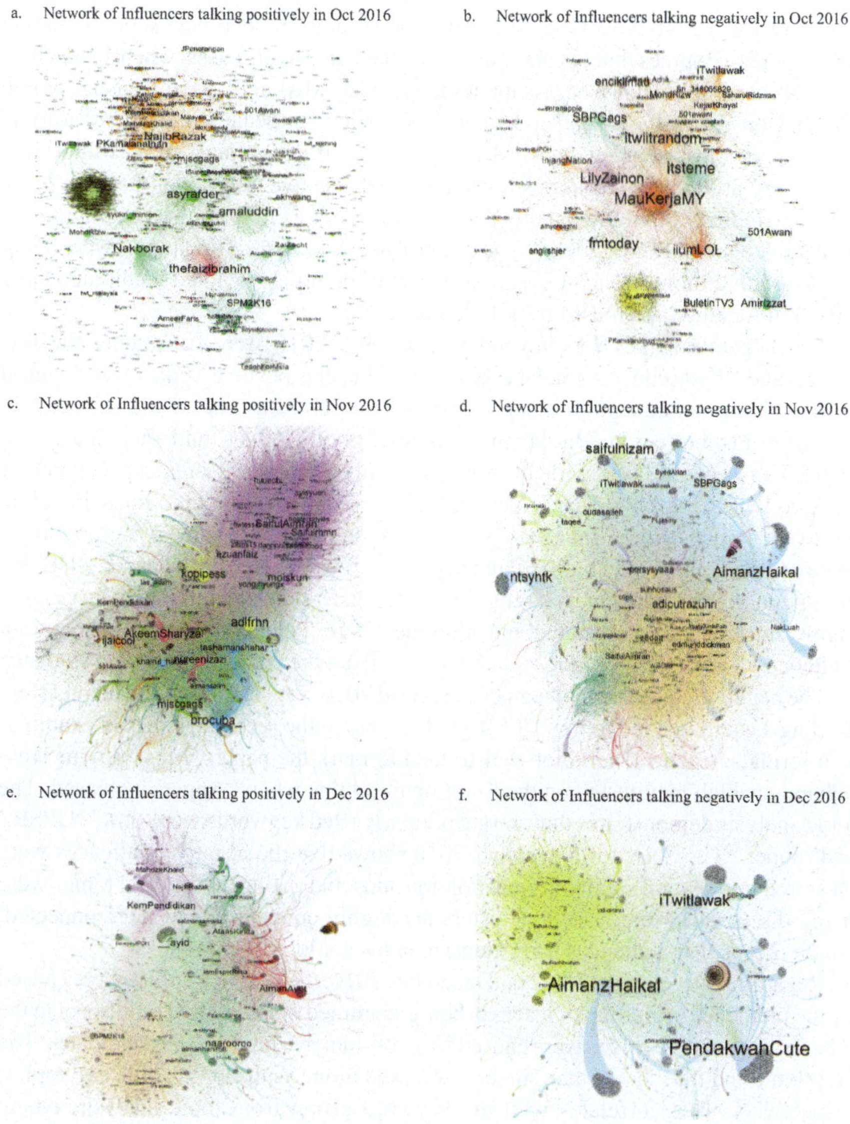

Fig. 5 SNA graphs illustrating top social media influencers in the domain under study

well as parents blaming the school organization. Besides, during the same period, the topic analysis shows that most posts related to these topics contained keywords such as 'keracunan,' 'cikgu,' or 'student.' The Social Network Analysis (SNA) indicates that in October 2016, the main negative influencers were from mainstream media and students. The connection between them is evident from the items clustered and linked by a dotted line.

For October 31st–November 3rd, 2016, the volumetric sentiment score trend line shows negative spikes that occurred due to sexual harassment cases, when 15 teachers were accused of sexually harassing students from 2010 until 2016, causing public uproar. Consequently, 'Anger' spiked on the Emotion trend line. Referring to the topic analysis, the most frequent keywords were 'cabul,' 'transfer,' and 'guru,' which were commonly used in conversations related to the respective issues. For October 2016, the SNA indicates that the main negative influencers were from the general public, mainstream media, and students. The graph also shows that several topics were being discussed that month. The connection between the topics is noted as the items are clustered together and linked by a dotted line.

The negative spikes for November 16th-19th 2016 were caused by the low percentage of straight As for UPSR. Meanwhile, the positive spikes were caused by the Malaysian education blueprint being recognized as one of the strongest in the Asian Pacific region. On the emotion graph, both 'Anger' and 'Joy' increased, which can be correlated with both issues mentioned. The decreasing number of students who achieved good results for UPSR caused the 'Anger' emotion to increase, while the praise of the Malaysian education system leads to the 'joy' emotion. In topic analysis, among the more frequently used keywords were 'rancangan,' 'UPSR,' and 'pelajar.' The SNA analysis for November 2016 shows that the audiences were highly condensed and clustered, meaning there were intense conversations between audiences. The majority of the audiences were from the general public and students.

The negative spikes for November 21st-23rd 2016 were caused by the inconsistent grading system that led to low UPSR grades. Due to the recent changes in examination formats, teachers were not well-trained to mark the papers. Most parents were unhappy, which manifested on the Emotion trend line where 'Anger' increased. The topic analysis demonstrates that among regularly used keywords were 'low,' 'UPSR,' and 'paper.' The SNA for November 2016 shows that the negative audiences were clustered into several small different color groups, indicating that several topics were being discussed. Moreover, the clusters are highly condensed and interconnected, which means that audiences had extensive conversations.

Regarding December 17th–21st December 2016, the negative spikes were caused by the incidence of a group of teachers being scammed of nearly RM1 million. On the other hand, positive spikes were caused by good-luck wishes for PT3 candidates. The emotion trend line shows that 'anger' increased along with 'joy' for the two topics, respectively. This correlates with the keyword group from topic analysis, where 'pt3,' 'guru,' and 'keputusan' were among the most regularly used keywords. The SNA of positive influencers indicates several small clusters that are mildly dense. This denotes that audiences were discussing several positive topics, for instance, good luck wishes for PT3 candidates. However, negative audiences in the SNA for December 2016 were clustered into two large different color groups, meaning that two topics were being discussed intensely. The majority of the audiences were from mainstream media, the general public, and students.

5 Conclusion And Future Works

With the huge volumes of conversations occurring on social media every day, the science of extracting and analyzing unstructured datasets from such live information channels has become a rapidly evolving discipline. In this context, there are significant advantages to making data-driven decisions, whereby companies, organizations, and government agencies can apply analytics techniques to elicit actionable insights from such noisy and unstructured data.

This chapter presented practical research efforts in building a comprehensive framework of social media analytics. The contribution of this chapter is three-fold: (i) providing language-independent analytical solutions capable of analyzing textual documents in different languages and even mixed languages, (ii) combining several related techniques like Sentiment Analysis, Emotion Mining, Topic Modeling, and Social Network Analysis in a single integrated framework, and (iii) generating actionable insights by exposing the proposed model to a real dataset with around 600,000 social media records from the education domain. The language independency feature considered in the proposed solution effectively addresses the issue of analyzing the multi-language contents in social media. Furthermore, the novel idea of combining several SMA techniques provides insights that are more reliable for decision makers and business owners interested in hearing the real voice of customers. The proposed framework along with all discussed machine-learning algorithms can be developed in big data clusters to analyze even larger datasets collected from social media in any targeted topic..In this research, we had the limitation of accessing just public content in social media. The existing social listening tools that were used in this study, can collect just the public content according to their agreement with social media content providers.

By analyzing large volumes of social media content in target domains, business owners can gain useful information about public perception and customer preferences to develop marketing policies and enhance sales. For future work, it is suggested to add some new components to the proposed framework to broaden its functionality. Predictive modeling, stochastic analysis, and exploring information diffusion patterns may be good addition options to the proposed analytical framework in the future. Besides, it is suggested to repeat analyzing the sentiments using some more recent methods such as deep learning, to compare its accuracy with the methods already used in this chapter. Maybe we find in the future that combining the existing and new sentiment analysis algorithms can come up with a more sophisticated hybrid framework with a higher accuracy level.

References

1. Stieglitz, S., & Dang-Xuan, L. (2013). Social media and political communication: A social media analytics framework. *Social Network Analysis and Mining, 3*(4), 1277–1291.
2. Fan, W., & Gordon, M. D. (2014). The power of social media analytics. *Communications of the ACM, 57*(6), 74–81.
3. Ghani, N. A., et al. (2019). Social media big data analytics: A survey. *Computers in Human Behavior, 101*, 417–428.
4. Tang, J., et al. (2016). A survey of signed network mining in social media. *ACM Computing Surveys (CSUR), 49*(3), 42.
5. Hu, X., et al. (2013) Exploiting social relations for sentiment analysis in microblogging. In: *Proceedings of the Sixth ACM International Conference on Web Search and Data Mining*. ACM.
6. Nakov, P., et al. (2016). Developing a successful SemEval task in sentiment analysis of Twitter and other social media texts. *Language Resources and Evaluation, 50*(1), 35–65.
7. Nguyen, T. H., Shirai, K., & Velcin, J. (2015). Sentiment analysis on social media for stock movement prediction. *Expert Systems with Applications, 42*(24), 9603–9611.
8. Rohani, V. A., & Shayaa, S. (2015) Utilizing machine learning in sentiment analysis: SentiRobo approach. In: *2nd International Symposium on Technology Management and Emerging Technologies*, Langkawi.
9. Proksch, S. O., et al. (2019). Multilingual sentiment analysis: A new approach to measuring conflict in legislative speeches. *Legislative Studies Quarterly, 44*(1), 97–131.
10. Dai, W., et al. (2015). Emotion recognition and affective computing on vocal social media. *Information and Management, 52*(7), 777–788.
11. Zhao, J., et al. (2014) PEARL: An interactive visual analytic tool for understanding personal emotion style derived from social media. In: *2014 IEEE Conference on Visual Analytics Science and Technology (VAST)*. IEEE.
12. Nguyen, T.H., & Shirai, K. (2015) Topic modeling based sentiment analysis on social media for stock market prediction. In: *Proceedings of the 53rd Annural Meeting of the Association for Computational Linguistics*.
13. Vala Ali Rohani, S.S., & Ghazaleh Babanejaddehaki (2016) Topic modeling for social media content: A practical approach. In: *3rd International Conference on Computer and Information Sciences*. Kuala Lumpur, Malaysia.
14. Ampofo, L., et al. (2015) Text mining and social media: When quantitative meets qualitative and software meets people. Innovations Digital Res Meth:161–91
15. Gémar, G., & Jiménez-Quintero, J. A. (2015). Text mining social media for competitive analysis. *Tourism & Management Studies, 11*(1), 84–90.
16. Williams, H. T., et al. (2015). Network analysis reveals open forums and echo chambers in social media discussions of climate change. *Global Environmental Change, 32*, 126–138.
17. Smith, M. A. (2014). NodeXL: Simple network analysis for social media. *Encyclopedia of Social Network Analysis and Mining* (pp. 1153–1170). Springer.
18. Rohani, V. A., & Hock, O. S. (2009). On social network web sites: Definition, features, architectures and analysis tools. *Journal of Computer Engineering, 1*, 3–11.
19. Liu, B. (2012). Sentiment analysis and opinion mining. *Synthesis Lectures on Human Language Technologies, 5*(1), 1–167.
20. Balahur, A., & Turchi, M. (2012) Comparative experiments for multilingual sentiment analysis using machine translation. In: *SDAD 2012 The 1st International Workshop on Sentiment Discovery from Affective Data*.
21. Ohana, B., & Tierney, B. (2009) Sentiment classification of reviews using SentiWordNet. In: *9th It & T Conference*.
22. Hamouda, A., & Rohaim, M. (2011) Reviews classification using sentiwordnet lexicon. In: *World Congress on Computer Science and Information Technology*. IAENG.
23. Kolchyna, O., et al. (2015) *Twitter sentiment analysis: Lexicon method, machine learning method and their combination*. arXiv preprint arXiv:1507.00955.

24. Lam, H. K., Yeung, A. C., & Cheng, T. E. (2016). The impact of firms' social media initiatives on operational efficiency and innovativeness. *Journal of Operations Management, 47*, 28–43.
25. Duan, W., et al. (2016). Exploring the impact of social media on hotel service performance: A sentimental analysis approach. *Cornell Hospitality Quarterly, 57*(3), 282–296.
26. Seo, E.-J., & Park, J.-W. (2018). A study on the effects of social media marketing activities on brand equity and customer response in the airline industry. *Journal of Air Transport Management, 66*, 36–41.
27. Charles, V., & Emrouznejad, A. (2019). Big data for the greater good: An introduction. *big data for the greater good* (pp. 1–18). Springer.
28. Emrouznejad, A. (2016) *Big data optimization: Recent developments and challenges*, vol. 18. Springer.
29. Simmel, G., & Wolff, K. H. (1950) *The sociology of georg simmel*, vol. 92892. Simon and Schuster.
30. Marin, A., & Wellman, B. (2011) *Social network analysis: An introduction.* The SAGE handbook of social network analysis, p 11.
31. Smith, M.A., et al. (2009) Analyzing (social media) networks with NodeXL. In: *Proceedings of the Fourth International Conference on Communities and technologies.* ACM.
32. Bastian, M., Heymann, S., & Jacomy, M. (2009). Gephi: An open source software for exploring and manipulating networks. *ICWSM, 8*, 361–362.
33. Scott, J. (2012) *Social network analysis.* Sage.
34. del Fresno García, M., Daly, A. J., & Segado Sánchez-Cabezudo, S. (2016) Identifying the new Influences in the Internet Era: Social media and social network analysis. Revista Española de Investigaciones Sociológicas (153).
35. Norman, H., et al. (2015) Exploring the roles of social participation in mobile social media learning: A social network analysis. Int Rev Res Open Distribut Learn 16(4).
36. Deerwester, S. C., Dumais, S. T., & Harshman, R. A. (1990) *Indexing by latent semantic analysi.*
37. Hofmann, T. (1999) Probabilistic latent semantic indexing. In: *Proceedings of the 22nd Annual International ACM SIGIR Conference on Research and Development in Information Retrieval.* ACM.
38. Blei, D.M., Ng, A.Y., & Jordan, M.I. (2003) Latent dirichlet allocation. J Mach Learn Res 3:993–1022.
39. Zhang, H., et al. (2007) Probabilistic community discovery using hierarchical latent gaussian mixture model. In: *AAAI*.
40. Nallapati, R.M., et al. (2008) Joint latent topic models for text and citations. In: *Proceedings of the 14th ACM SIGKDD International Conference on Knowledge Discovery and Data Mining.* ACM.
41. Saini, A., et al. (2014) EmoXract: Domain independent emotion mining model for unstructured data. In: *2014 Seventh International Conference on Contemporary Computing (IC3).* IEEE.
42. Bollen, J., Mao, H., & Pepe, A. (2011). Modeling public mood and emotion: Twitter sentiment and socio-economic phenomena. *ICWSM, 11*, 450–453.
43. Rao, Y., et al. (2014). Sentiment topic models for social emotion mining. *Information Sciences, 266*, 90–100.
44. De Choudhury, M., Counts, S., Horvitz, E. (2013) Predicting postpartum changes in emotion and behavior via social media. In: *Proceedings of the SIGCHI Conference on Human Factors in Computing Systems.* ACM.
45. Rao, Y., et al. (2016). Social emotion classification of short text via topic-level maximum entropy model. *Information & Management, 53*(8), 978–986.
46. Rohani, V. A., & Shayaa, S. (2015) Utilizing machine learning in Sentiment Analysis: SentiRobo approach. In: *2015 International Symposium on Technology Management and Emerging Technologies (ISTMET).* IEEE.
47. Rohani, V. A., Shayaa, S., & Babanejaddehaki, G. (2017). How social media influencers govern sentiment territory. *International Journal of Applied Evolutionary Computation (IJAEC), 8*(1), 49–60.

48. Cieliebak, M., Dürr, O., & Uzdilli, F. (2013) Potential and limitations of commercial sentiment detection tools. In: *ESSEM@ AI* IA*. Citeseer.
49. Ekman, P. (1999). Basic emotions. *Handbook of cognition and emotion, 98*(45–60), 16.
50. Marra, M., Emrouznejad, A., Ho, W., & Edwards, J. S. (2015). The value of indirect ties in citation networks: SNA analysis with OWA operator weights. *Information Sciences, 314*, 135–151.
51. Lieberman, M. (2014) Visualizing big data: Social network analysis. In: *Digital Research Conference.*
52. Borgatti, S. P., Everett, M. G., & Freeman, L. C. (2014). Ucinet. *Encyclopedia of Social Network Analysis and Mining* (pp. 2261–2267). Springer.
53. De Nooy, W., Mrvar, A., & Batagelj, V. (2011) *Exploratory social network analysis with Pajek*, vol. 27. Cambridge University Press.

Dr. Vala Ali Rohani received his Ph.D. in software engineering from University of Malaya (UM; 59th at QS world ranking) in 2014, with several awarded research grants in his records. He also accomplished a post-doc at UM as a data scientist, which resulted in several publications in top international journals. Meanwhile, as a lead data scientist, Vala has managed several national level analytics projects to analyze millions of dataset records and generate actionable insights in different area such as recommender systems, upskilling, sentiment analysis algorithms, logistics optimization, and credit scoring in financial institutions.Having more than 20 professional certificates in data science, Vala currently is based in Portugal as an invited associate professor in IPS, and the founder and chief data scientist of Data Corner with two branches in Malaysia and Portugal, which recently brought him invitation to join the Enterprise European Network (EEN) as a flagship startup in Portugal.

Shahid Shayaa founded Berkshire Media after spending more than a decade in business strategy and transformation. He is instrumental in pushing the boundary of social analytics through data science and data-driven approaches in order to obtain actionable insights for strategic decision makers. He now advises top leaderships and senior executives on strategic communications for publicly traded companies and the public sector. Shahid holds a First-Class Honours in Mechanical Engineering and Masters of Philosophy (M.Phil) from the University of Sheffield (UK). He was the recipient of Asia Award 2018 by the CEO Magazine in United Kingdom, Top 40 Under 40 by The Prestige, The Entrepreneur, Astro Awani and other many other media.

Data Mining Approach in Repair and Service Systems of Electronic Products Under Warranty

Filiz Ersöz and Deniz Merdin

Abstract The use and management of technology have an important role in increasing efficiency in the service sector. Especially in the electronics sector, which has a dynamic rate of change, technological support of service activities and big data have shown to enhance operational and tactical decision-making processes. In this way, efficiency can be increased by making faster and more accurate decisions in the organization. In this study, the researches covered technical service for the TV repairs in the electronics sector of Turkey by analyzing incoming data to create models and data mining techniques. This work is the only study in the literature aimed at contributing to operational processes by evaluating the only type of product in the electronic repair service sector using data mining methods relative to the type of repair. The results of the research are suggested to increase service efficiency and customer satisfaction from the feedback on sales and production processes with the proposed changes based on repair service information.

Keywords Repair service systems · Data mining · Clustering · Electronic product

Abbreviations

CART	Classification and Regression Trees
SOM	Self Organizing Maps
IH	In Home
II	Initial Installation
CI	Carry In
RS	Inventory Device Repair

F. Ersöz (✉)
Department of Industrial Engineering, Karabük University, Karabük, Turkey
e-mail: fersoz@karabuk.edu.tr

D. Merdin
Department of Transport Services, Tokat Gaziosmanpaşa University, Tokat, Turkey
e-mail: deniz.merdin@gop.edu.tr

A. Emrouznejad and V. Charles (eds.), *Big Data and Blockchain for Service Operations Management*, Studies in Big Data 98,
https://doi.org/10.1007/978-3-030-87304-2_8

RH	Return Handling
SH	Service Handling
RC1	Software Reinstallation
RC2	Electrical Parts Replacement
RC3	LCD Replacement
RC4	PCB Replacement
RC5	Accessory Replacement
RC6	Product Replacement as Requires DNA Part
RC7	Software Upgrade
RC8	First Assembly
RC9	Product Change
RC10	No Error Found
RC11	Cleaning
RC12	Mechanical Parts Replacement
RC13	Functional Control
RC14	Refund
FC1	Open Circuit
FC2	Dirty
FC3	Software Error
FC4	Short Circuit
GDP	Gross Domestic Product
KDD	Knowledge Discovery in Databases
SVM	Support Vector Machine
ANN	Artificial Neural Network

1 Introduction

The rapid development of technology and the use of big data and information technologies continue to offer great advantages in manufacturing and service businesses. These businesses can now manage big data, obtain information from the data, and make more effective and accurate decisions. According to findings from the forum of IDC (International Data Corporation), the total amount of data to be generated worldwide in 2025 is expected to reach 163 zettabytes. That is ten times the 16.1 ZB of data generated in 2016. In 2017, in the studies of Reinsel et al., the strength of the data, the value of the systems that make up this data, and the value of data mining, which is considered as data analytics, began to be understood [50]. As big data usage and its potential benefits get more and more widespread, public and private institutions around the world have come to realize the need for incorporating comprehensive and robust technologies into their business processes. Undoubtedly, data analytics and data-based scientific management concept have come to stay, especially in service-based sectors.

The proliferation of the production processes and increase in product diversity have made it extremely difficult for some companies to maintain a decent market share. Successful firms know that responding quickly to customer demands through market flexibility and logistics activities can accrue several benefits. Such firms make it a point to factor in the demands and complaints that come with after-sales [4]. These companies have started to give importance to reverse logistics activities and after-sales services to meet customer demands fast, offer high customer satisfaction, fulfill their legal and environmental responsibilities, provide a competitive advantage, and reduce the total production cost.

As firms introduce more and more intelligent systems into their business processes, they will be able to better optimize their resources, develop new business models, and develop new social infrastructure that can better accommodate individual differences at the workplace. Employees, more than ever before, have a high need for flexible work structure and sustainable work-life balance.

With regards to a firm's customers, there is also an equal need for the development of intelligent software and technologies that will handle an immediate response to the consumer's requests, as well as, offer instant response to their problems.

Taking into account the above needs, this study mentioned the exploration process of data mining in the input section, industry 4.0 inservice systems, and the use of big data and data mining. Thereafter, the importance of data mining in the electronic service system was emphasized and inventory management of the products under warranty was mentioned. Additionally, a part is presented on the importance and purpose of the work carried out in other related studies. After that, data mining and clustering methods are explained in the material and method section. Result from the analysis which were based on data mining stages are given in the findings section. In the discussion section, the results of the study are given, and the future contributions of the study and the works that can be suggested in the future are mentioned.

1.1 Data Mining Discovery Process

Data mining is basically a technique that arises when numerous disciplines such as statistics, computer science, machine learning are used together. Figure 1 includes the exploration process of data mining [1, 7, 11, 12, 30, 32, 35, 41, 57, 58, 63, 70].

The history of data mining actually dates back a long way. After the Bayes theorem [7], which showed the relationship between probabilities by Thomas Bayes in the 1700s, Regression analysis was introduced, showing the relationship between the variables in the 1800s by Galton [61]. Data mining, which has formed the basic infrastructure in statistics, has entered the computer age. In an article published by Allen Turing in 1936 (On computable numbers, with an application to the Entscheidung problem), it states that any mathematical account can be done with the virtual machine recommended in the article [63]. This idea of a virtual machine is seen as one of the key stones of Eniac's discovery, which was fully operated in 1947 and is considered the first computer [3]. After the invention of the first computer, the process

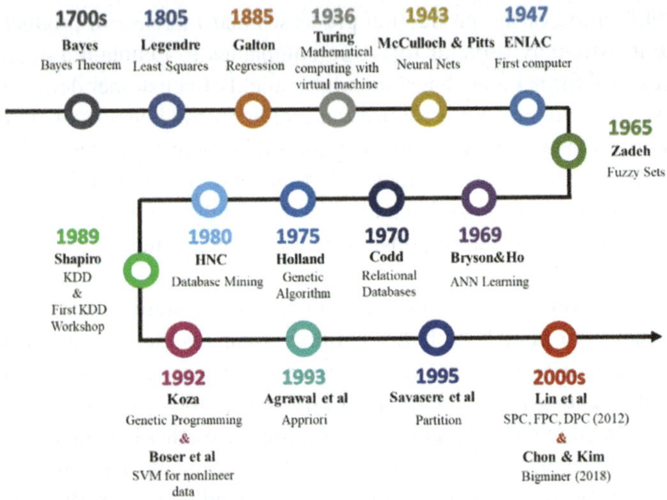

Fig. 1 Data mining discovery process

continued with the discovery of algorithms and databases to be used. The artificial neural networks technique [42], first created by Warren McCulloch and Walter Pitts in 1943, continues to develop today with the contributions of different scientists and a mathematical system inspired by the information processing of the human brain. Later, Evolutionary computing techniques such as Ant colony optimization, Evolutionary algorithms, learning classifier systems were developed. Using such algorithms, the relational model proposed by Edgar F. Codd in the 1970s allowed applications to search according to content [13]. In this way, databases, which are the areas where databases and related information are stored, started to be used. Genetic Algorithm basics were first identified by John Holland in 1975 for optimization problems in complex multidimensional search space due to increasing numbers of data and difficult analysis [22], Genetic algorithm is are an algorithm that search for the best by searching on many sets of solutions. The discovery of the genetic algorithm can be considered an important factor in the development of data mining methods in terms of evaluating all data in a certain order. With the increasing use of the computer, the concept of data mining has been developed and various algorithms have been developed by analyzing the increasing data and making meaningful results. In the 1980s, the name "Database Mining" became a trademark by HNC, when the discussions were made to prove which one amongst data mining, knowledge mining and knowledge extraction was more meaningful. Meanwhile, Shapiro, who worked on optimizing machine learning for his doctorate, developed technologies by going to various conferences and developed the term "KDD- Knowledge Discovery in Databases". In 1989, the first KDD workshop (KDD-89) was held in Detroit. The main topics in this workshop, in which there were participation from 12 countries, are, expert database systems, scientific discovery, fuzzy rules, dealing with text and other complex data etc. [58]. After these developments, in 1992, Boser et al. developed the support vector machine, which analyzed data by classification and regression, to allow the creation of nonlinear classifiers [9]. Data mining, which began to take place in the concept of data science, was then widely used in academic studies. Some of the first studies with data mining are as follows:

Fayyad and Stolorz [18] defined data mining and knowledge discovery (KDD) in databases and stated that interesting models and patterns should be removed from large databases. He also gave information about the techniques used in data mining [18].

Ling and Li [36] focused on several problems that arise in the use of data mining in direct marketing. Examples such as the inability to use predictive accuracy as an appropriate evaluation criterion for data mining, even if there are excessive imbalances in s-class distributions or logical models. He commented on how to eliminate these problems [36].

Goebel and Gruenwald [20] discussed software tools used in data mining from a general perspective. He mentioned the problems that need to be solved in order to implement a successful data mining. These are; using a single technique or a limited technical set to perform data analysis with existing tools, different techniques perform better than each other in terms of different problems, data collection tool's inability to process data and lack of data processing, the lack of tools that are easy to use in order to get clear, fast answers, lack of standard data and inability to process it with standard analysis methods [20].

In his study, Domingos and Hulten [16] defined the concept of VFDT (Very Fast Decision Tree Learner). He used Hoeffding trees, an online learning method from the increasingly large volume of data streams [16].

Rygielski et al. [53] examined the concepts of data mining, one of CRM and its components. He compared Chi-Square Automated Interaction Detection (CHAID) and Artificial Neural Network applications based on their strengths and weaknesses [53].

In their studies, Keogh et al. [27] stated that they would prevent us from evaluating parameter-free data mining through our own assumptions and proposed a compression-centered data mining paradigm [27]. Alhammady [2] proposed a new approach to analyze patterns (EPs) that appeared in the data flow. EPs have experimentally proved that there are clusters of items in one class whose frequencies are significantly higher than those in other classes, and that they have an excellent effect on the classification process of data streams [2].

Venkatadri and Reddy [65] compared data mining trends from past to present in terms of techniques or algorithms used, data formats, computing resources and application areas [65].

1.2 Industry 4.0, Big Data and Service System

Today, Industry 4.0, Society 5.0 and similar concepts are known as the last stage of economic production and service models. They are poised to ensure digital transformation of businesses in terms of intelligent production and service economy of the future.

Technological elements, including Industry 4.0, are known to encompass items such as Big Data and Data Mining, Internet of Things, Cloud, 3D Printing, Simulation, Smart Robots, Augmented Reality Glasses, Cyber Security, Horizontal and Vertical Integration. According to the 2016 Global Industry Survey, with Industry 4.0, it is predicted that there will be revolutionary changes in industrial service and production. And thanks to the increase in productivity, trillions of dollars of economic contribution will be made to the world economy in the future. The key to realizing this potential is seen as digital modeling of the entire value chain and running scenarios that will produce the most ideal business results. After the digital supply chain is modeled and the applications that can be done are determined, it is also important to take note of how the process will be followed in the enterprise [44]. The fields related to Industry 4.0 are given below in Fig. 2.

Taking advantage of big data can be an important point when researching the impact of consumers' products on firms in production and reproduction processes. Service systems should be developed to determine which pricing mechanisms are preferred [67] by considering full or partial reproduction strategies under different conditions, determining the remaining life of the product utilizing fuzzy methods

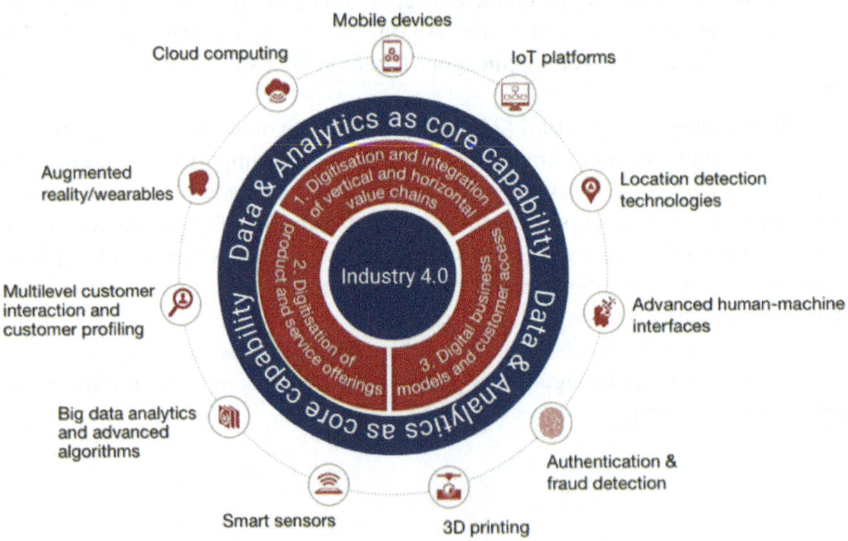

Fig. 2 The fields related to Industry 4.0. [19]

[31]. In order to create value in the company by managing customer processes with CPM models [34], big data analysis will enable more accurate decision making.

In addition, it makes use of big data analysis in survey studies [71] about the transformation of service systems in smart factories that have come up with the development of technology [59]. Examples of various service systems with big data are given below.

1.3 Data Mining and Service Systems

With the development of information and communication technologies, information, efficiency, customer and value oriented approaches gain importance in the service sector. Data mining is in businesses; it further provides information discovery in the service sector, such as finance, retail, communications, e-commerce, analysis for customer-oriented businesses, changing customers' purchasing habits, marketing or creating public relations campaigns, etc.

Data mining can be defined as a process of analyzing data stacks from different and diverse sources and converting this valuable information into business intelligence. With data mining applications; the emergence of meaningful relationships between meaningless data, to predict the impact of the business's sales, profit and customer satisfaction, and detailed and knowledge-based information such as customer profile analysis are produced. Data Mining services can help organizations interpret data accurately, which in turn help the organizations make informed business decisions, improve processes, improve quality standards, and increase customer satisfaction.

Firm using data mining can also increasing customer needs and expectations ensure accurate analysis of the products covered by the service systems within the scope of warranty, determine product quality and reliability, obtain information about maintenance and faults and assist design changes [60]. Data mining techniques bring a huge difference to the service systems of businesses and solve their problems. Some of the remarkable researches in this area include:

- ***Customer value analysis***: In the automotive maintenance industry, customer value analysis was performed using K-Means and SOM networks [33]. In addition, subjective perceptions of users, in order to analyze the relationship between service characteristics, emotional orientations, and intended use, decision trees were used to contribute to the development of product delivery services and cancer engineering studies [69]. In reproduction processes, customer demands were estimated using decision tree regression models (CART, M5, RF) [64].
- ***Design***: Ruvald et al. [52] showed that it can help in the development of a new machine with human–robot interaction with data mining and augmented reality methods used in product prototype design stage in the construction sector [52].
- ***Determination of product quality and reliability***: In the drop tests, which ensure the reliability of portable electronic products, intelligent classification methods

have been used to reduce the design cycles hence accelerating the introduction of the product [72].

- ***Maintenance and troubleshooting***: Preventive maintenance activities in factories; equipment maintenance was estimated by using decision trees, clustering, regression and artificial neural network techniques [5], customer service records were investigated by data mining methods, and intelligent machine faults diagnostics [23], data mining in naval avionics maintenance of navy aircraft. The working principle of the system and the ways of interaction with the other systems were determined by using the methods. Thus, the operational availability of a single fleet was increased by 22% [45].

1.4 Electronics Service System and Data Mining Modeling

Over the years, intense competition and short-lived competitive advantage have forced several businesses and industries to pay close attention to every stage in the lifecycle of a product. On any given day, manufacturing firms produce big amounts of data that ought to be in some way extracted, cleaned and processed in order to make value-adding decisions. Big data is analyzed using the various concepts on data analytics and data-based scientific management. One of the widely used methods is data mining. Many researchers and data experts strongly believe that data mining can transform manufacturing economies across the world.

Big data and the need for data mining are predicted to increase enormously in both private and public institutions, as well as in other peripheral organizations worldwide. One such industry that is poised to get immense benefit from data mining is the electronic industry. In 2006, according to Harding et al., from the design and manufacturing phase to the repair and service phases, electronic producers have come to rely heavily on data mining and system development for important decisions. A comprehensive research study on the use of data mining in electronic repair and service systems will contribute to the solution of many problems. As a result, better decisions will ultimately provide better customer services when it comes to repairing the products under warranty [21]. In 2014, Bastos et al. stated that with data mining, firms can convert their voluminous data into important information that will feed the decision-making process. The writers opined that data mining is a very important tool when it comes to maintenance as well as predicting machine breakdowns in the future. By doing so, the firms will be able to have a robust maintenance team capable of carrying out maintenance in an effective and efficient manner [6]. The manufacturer or a professional repair specialist is required to carry out repair services. However, non-professional repair applications are also available. In non-professional repair applications, it is important to identify the most common faults faced by consumers and to identify any difficulties that may arise in repair applications. For this purpose, regression analysis by taking information from consumers through web sites, researching consumer's purchasing behavior and redesigning repair manuals [49], operational problems of service providers of civil air engines, service risk, and service problems were identified [46]. In addition, it has been determined that the loss of value caused by consumers' unwillingness to repair their mobile phones may reach the level of 331 million dollars [54]. In 2013, according to the studies of Cox et al., the electronic repair industry faces a huge challenge due to the high cost of labor and spare parts. These factors are in turn fueled by the very short lifespan of the electronic products and over-consumption from the aspect of electronic consumers. Studies show that the financial cost element associated with electronic repairs forces the consumers to opt for disposal instead of repairs [14, 42, 49]. In 2018, Lv et al. called for a comprehensive review for more comprehensive research works in data mining and big data management, especially in the electronics industry. In 2018, [39] opined that our knowledge of data mining will be advanced, if researchers begin to pay very close attention to data mining application techniques across an electronic product's lifecycle [38]. Sabbaghi et al. [55] explored factors that impeded or dissuaded electronic consumers from repairing their broken electronics. After exploring 2170 repair technicians that belonged to third-party repair service providers, the authors concluded that there are three main barriers to electronic repair services, which are economic, technical and operational barriers. The operational barriers cover the firm's lack of proper repair and service systems. The technical barriers include the absence of repair manuals, tools, and spare parts [55].

Similarly, King et al. [28] stated that consumer's behavior (social factors) and manufacturers' lack of responsibility stifle repair and service systems. In Boon-Long and Wongsurawat [8], low electronic patronage has been attributed to repairer's inability to deal properly with the processes involved in the repairs and the outcome of those repairs. The data mining approach in the repair and service system is estimated to focus on scientific analysis of a firm's process that adds value to consumers of electronic products. The system should be capable of generating and analyzing data from thevarious phases in the product lifecycle. The electronic repair services in Turkey's industry consists of consumer electronics including radio, TV and VCR repairs, computer and office machine repairs, communications equipment repairs. Companies in this industry are involved in repairing and maintaining one or more of the following: consumer electronic equipment, computers, office machines, communication equipment and other electronic and precision equipment.

Today, it is critical for companies to use big data and analytical approaches, to make strategic and right decisions and to plan. These are statistics, data mining, simulation, supply chain planning, software technologies, cost approaches in production and service systems and many other analytical approaches are used. These techniques are scheduled for reporting, data mining, forecasting, model visualization, and graphical trend analysis.

Business Intelligence Architecture is given below in Fig. 3.

Data mining has become a key science that exposes meaningful information from big data piles in all business processes and human resources management of companies growing and globalizing. It provides the opportunity to see the whole process from product lifecycle software to product-service design and production to after-sales services. In addition, customer relationship management software allows

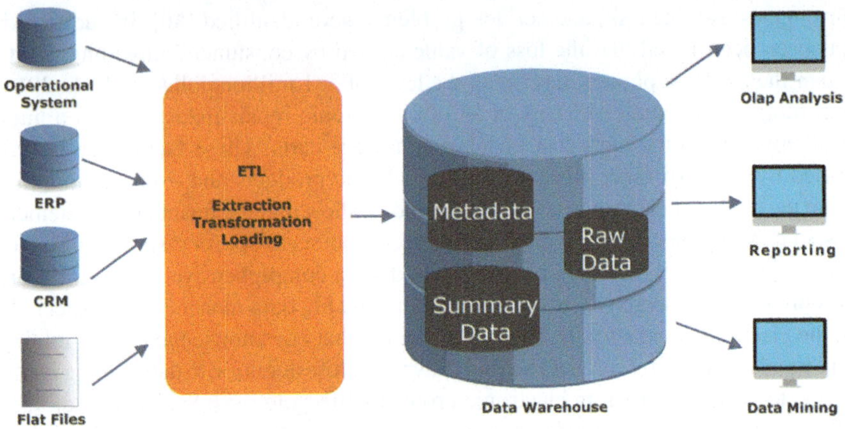

Fig. 3 Business intelligence architecture [62]

customers to analyze the demographic characteristics, purchasing habits, marketing effectiveness, customer loyalty and many more current and new performance criteria in detail. With this information, strategic roadmaps can be drawn to decision-makers and managers, future forecasts can be made, and more accurate decisions can be taken.

1.5 *Inventory Management for Servicing Electronic Products Under Warranty*

Many electronic products manufacturing firms continue to invest huge amounts of resources to help them create or integrate systems such as enterprise resource planning, customer relationship databases, supply chain and distribution systems, and other knowledge transfers or databases discoveries. At the heart [49] of all these systems is a robust and fluid information technology system—the internet of things or big data management systems. The goal is to create an effective and efficient feedback loop that moves along the design, production, distribution, repair, and delivery services of the products. The system makes sure that every phase is properly optimized. This enables the firm to save big on products that would otherwise be returned, which could cause a sense of dissatisfaction among the customers.

The electronic sector is a dynamic sector that is developing rapidly and can provide more product range than the expectations of customers. Since the product structure in the electronics sector is composed of many sub-components, even if one of the sub-components of the product fails, the product becomes unusable or the quality of use decreases.

Therefore, in order to reuse the product, to reduce production costs and to ensure customer satisfaction, companies provide repair services within their bodies. By meeting this need in the electronic repair service sector, the industrial income can be increased by 0.1% [25]. For example, it is not possible to use a phone that has a charging port input fault after the battery is discharged. However, changing the charging port input will increase the usable life of the product and prevent the product from becoming idle by providing customer satisfaction. When these issues are taken into consideration, it is important to follow up and analyze the processes of service where repairing activities are performed. Indecision support processes, analysis can be utilized in many aspects such as demonstrating customer value, determining product quality and reliability, obtaining information on maintenance and faults and assisting with design changes.

The after-sales/ spare parts industry has to become a significant one in the current business world. Many companies are facing worldwide competition; customer satisfaction has become crucial recently. A significant means to keep customers satisfied is the quick repair of a product or system that failed. After-sales services, such as the maintenance of warranty products, are all situations which contribute to the increase of returns within an organization. Many companies have focused on improving their product manufacturing and product delivery processes for improving after-sales service and customer support. To provide the necessary after-sales service, enough parts, often referred to as service parts, have to be available at appropriate points within the supply chain to guarantee the desired service level [10].

Customers expect fast and high quality after-sales services. For this reason, companies, after-sale services without losing the trust of customers by planning the good parts of the repair is to find and supply quickly [10, 47]. Technically, in selecting suppliers of advanced systems, the importance of high quality after-sales services, system availability and increasing the value of inventory investment of service parts has increased the importance of after-sale service parts management in recent years [47]. Therefore, good inventory planning of service parts is important both in terms of cost and customer satisfaction.

Repair service companies are looking for technological and logistical systems that can reduce both cost and time loss at the same time. One of the most important aspects of customer service is satisfying the repair requests for a product already used by a customer. The repair under warranty of products are concerns which increasingly affect organizations, it to improve customer service. There are few tools to standardize the spare parts in the inventory according to its price, needs and importance. These tools help to classify replacement parts to use the repair inventory system more effectively.

In 2000, Jack and Schouten defined [26] warranty as a contractual agreement between a manufacturer (seller) and a consumer (buyer) which requires the manufacturer to either rectify or compensate all faults occurring within the warranty period. In addition, the scope of warranty is defined as an assurance policy for repair and replacement of all customers to whom the goods or services are represented. This assurance applies to a product sold for a specified period and performs all technical, administrative and administrative functions during the warranty period. A warranty is a contract between the consumer and the manufacturer. Warranty management is important that

Fig. 4 Service management [15]

many of today's companies must confront. Warranty management deals with decision making at strategical and operational levels. This requires taking into account the interactions between engineering, marketing and many of the manufacturing firms. Service management is given in Fig. 4.

Product warranty is defined as the contracts entered by customers and companies during the purchase of electronic devices. Return of a product under warranty is considered defective and the customer can then request a new product again.

General warranty management process model is given in Fig. 5.

Fig. 5 General warranty management process model [15]

It is expected that the spare parts stocks of the product/s will be in a supply chain in order to ensure the follow-up of the after-sales services and to ensure the achievement of the predetermined service-service objectives [10]. To reduce the cost of product service under warranty, businesses keep the system as a spare part for use in product returns and follow up.

Service parts; in order to ensure the continuity of production, can be classified as repairable and consumables parts, as well as those that are required to repair the returned product from the end user in technical service. Consumables, in other words, disposable parts/products are spare parts that become idle by replacing them with a new part in case of faults. Repairable products are technically feasible and economically profitable [10, 47]. Returnable repairable products are sent to the maintenance and repair point (O&R) and, after repair, the product is kept in the inventory ready for delivery until requested. For repairable products, inventory policies are determined by factors such as product repair time, quantity of product to be repaired, number of stations to be repaired [40].

Therefore, this research focused on a niche area of the product lifecycle of electronic products—repair and service systems. The repair and service phase in the lifecycle hardly gets enough coverage in data mining literature. The majority of the data mining research works concern themselves primarily with the manufacturing phase. Furthermore, as the electronic products industry gets more and more competitive day by day, it will be very crucial that the firms pay close attention to how they manage their repair and service systems [48].

In the literature, it has been seen that the majority of the studies in the electronics industry are related to the production phase and those related to the reproduction and repair processes address issues such as the preferences and effects of the customers to use the service, customer purchasing behaviors, the main obstacles in the electronic repair service, and the technologies used for fault diagnosis. In this study, after all the services that can be done for an electronic product are determined, it is aimed to optimize the number of repair benches in order to contribute to the decision support processes by clustering according to the types of breakdowns and repairs. Considering these aspects, the study is the only study in the literature. However, factors such as repair time and costs were not taken into consideration in the study. Future researchers may be advised to conduct a similar investigation, taking into account supply time, repair time or costs. In this way, the impact of part supply in the sector and at the customer return rate can be determined. In addition, the work was done through a single technical service and a single product. The study can also have effective results in the construction of all products in a service or working on a single product in multiple services. This work aims to increase customer satisfaction and facilitate their specialization by matching fault types to the staff. However, the employees were separated as experts and beginners in the study and were not evaluated according to their performance. It is also possible to do the study taking into account the working performance. In addition, it is expected to improve the quality of service with feedback to the production and sales units of the research.

2 Material and Methods

This study was conducted in one of the technical services of an electronics firm operating in a province of Turkey in the international arena. The information about 852 televisions belonging to nine different models under the warranty, which came to the technical service between 01.06.2015 and 01.06.2016, was collected and analyzed with the data mining modeling approach. K-Means clustering algorithm, which is one of the clustering methods, was used in the modeling of the data. K-Means clustering method is an advantageous method because it provides faster results in the solution of hierarchical algorithms and makes the characteristics of clusters different [66] (see Table 3). Data were analyzed by IBM SPSS Modeler (Clementine 18.1). In this research, it is aimed to make operational and tactical improvement decisions which can help decision support processes in the repair service system. Analysis by K-Means clustering method,

- Identifying the models with the most repair operations,
- Determining which electronic faults occurred in which month and in which models, giving feedback for production,
- Clustering of repair and repair services are aimed at creating meaningful relationships in the electronic repair service system and identifying the repair processes of the employees and providing support to service, production, and management processes.

Data mining is a business process that explores big amounts of data to discover meaningful patterns and rules [37]. In summary, data mining can be defined as a way to transform data into qualified information. Data mining is associated with big data. When data mining is performed, it can generate new and future related information by using the big recorded data. Today, data mining is also known as Business Intelligence and Business Analytics. Other definitions, Knowledge mining in databases, Knowledge extraction, Data and pattern analysis, and Data archeology [17].

In data mining, the data is passed through several stages before modeling, with an examination aimed at achieving specific results from big and meaningless data stacks. After introducing the problem and the system, the first step is to clear the data before modeling. After the detection of outliers and extreme values, clean and high-quality data is obtained, and the data are combined to enable the same language to be spoken. Here, the selection of relevant and important variables and dimension reduction are carried out. As a result of the analysis, converting the data into a format suitable for re-use, evaluating the importance of data and relations and presenting the results to the decision-makers are the processes that complete the data mining [17].

In the data mining knowledge process, many methods that use statistical techniques and algorithms are used. Data mining models can be divided into three groups as prediction, association rules, and clustering models. In predictive models, classification is based on a dependent variable [17]. The most commonly used of these methods are, Artificial Neural Networks, Decision trees (C5.0, CART, Chaid, Quest, etc.), linear and logistic regression. In the association rules model, a certain result is tried to be associated with a set of rules. Commonly used association rules algorithms; Apriori, Gri, and Carma. In clustering models, it is aimed to find similar and different groups in multidimensional data. Common clustering methods; Cohonen Networks are K-Means and Two-Step [24].

Clustering analysis is the division of each data into groups based on similarity or difference/distance criteria. It is desirable that each sample in the cluster be close to each other and that the distance between the clusters is far from each other. TwoStep and K-Means methods are widely used in clustering analysis [66].

After the cluster centers are determined in the K-Means model, each data is assigned to the cluster that minimizes the distance to the nearest cluster center. The number of clusters in the model is determined by the analyst [37].

3 Results

The dynamic structure of the electronic service sector makes it important to make fast and accurate decisions in this sector. Therefore, data mining techniques are widely used in the service sector. In this study, related data of 852 televisions to 96 different models were collected by using K-Means clustering method in a repair service system. All analysis processes with the Data Mining approach are given in the following sub-headings.

3.1 Data Understanding

At this stage, the system problems and the characteristics of the system are determined. The current structure of the electronic technical service discussed in the study is given below:

- The service is carried out both under the warranty and for electronic products that do not have any guarantee.
- In the electronic service, service activities are carried out for ten different products under the guarantee. These products; built-in products, dishwasher, air conditioning, sound system, camera, broom, printer, refrigerator, television, washing machine.
- The products coming to the technical service are delivered to the customer after the necessary service activities or canceled due to the customer request, the decision of the company center, the cost not desired to be covered by the customer / company center, the fault not found, and the same form entered into the system more than once.

When the general structure of the technical service is taken into consideration, it is desirable to analyze which type to fault and repair televisions belong to similar groups and how many clusters are formed in terms of determined characteristics. In this way, it is aimed to determine the product characteristics and which models are in similar fault groups.

3.2 Data Preparation

At the stage of model preparation; data collection, data integration, clearing outlier and extreme values of data, converting data into data mining language and reducing the data by removing unnecessary variables entered the model [17]. Limitations such as analysis objectives, data quality, and volume limitations, or data types are also made at this stage [43]. The stages and characteristics of the data collected from the technical service are determined as follows.

3.2.1 Data Collection and Integration

Collecting the data accurately and determining the volume of the data is very important for the validity of the analysis. The data used in the analysis should reduce the complexity of the data and be able to represent the population.

In the electronic service, repair service activities are carried out for 10 different electronic products under warranty. These products; built-in products, dishwasher, air conditioning, sound system, camera, broom, printer, refrigerator, television, washing machine. After the data collected on different databases were combined for the study, the most accurate product that could be used for analysis was searched. Figure 6 shows the distribution chart of the products coming to the technical service during the one year of the study.

When the distribution of the products coming to the technical service given in Fig. 6 is examined, it is seen that the service activities mostly related to television (55.30%–903 units) are realized in the technical service. For this reason, the research was conducted between 01.06.2015 and 01.06.2016 with the television data covered by the guarantee.

3.2.2 Data Cleaning and Spotting Outliers

To improve data quality; and identifying, cleaning or identifying missing, noisy and inconsistent data in the data. To determine whether there is a deviation or abnormal value in the study data, the data was examined. After cleaning the data that was not covered by the warranty first, it was checked for outliers and missing data with extreme values and frequencies. As a result of the examination of the data used in the study, it was found that there were no noisy data with errors and outliers, but

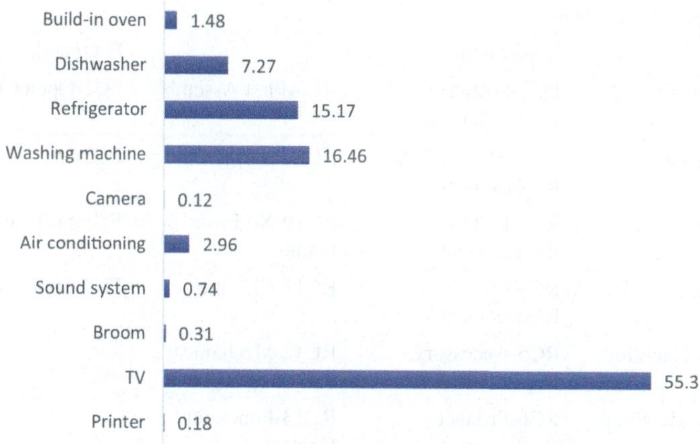

Fig. 6 Distribution chart of products coming to service (%) (2015–2016)

missing data. Due to the lack of many variables belonging to the same customer in the lost data, these data were not identified and cleaned. Since only the products covered by the warranty were considered and the lost data were removed, 852 of the 903 televisions were analyzed.

3.2.3 Data Transformation and Reduction

To prepare the data, it is necessary to remove (reduce) the variables that will not be used in the analysis. In Fig. 7, the variables used and reduced from the analysis are given.

To make an accurate and meaningful analysis, variables such as fault, repair time, repair completion time and dates showing the time and date for the operation of the service activity have been omitted. Repair code, fault and operation status information were collected. Table 1 gives the variables used for analysis and their explanations.

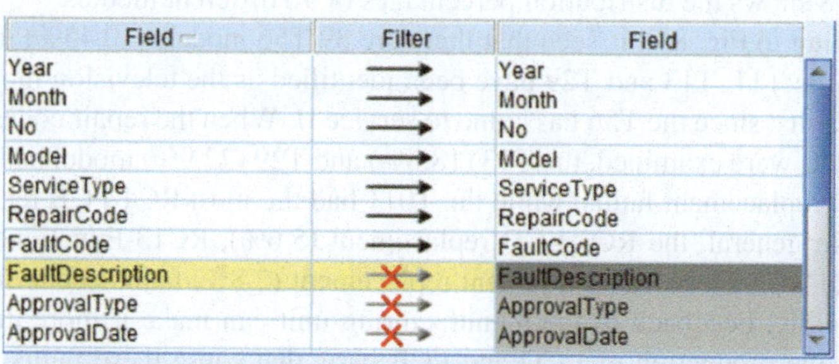

Fig. 7 Data reduction

Table 1 Data characteristics

Service type	Repair code		Fault code
IH—In Home	RC1-Software Reinstallation	RC8-First Assembly	FC1-Open Circuit
II-Initial Installation	RC2-Electrical Parts Replacement	RC9-Product Change	FC2-Dirty
CI-Carry In	RC3-LCD Replacement	RC10-No Error Found	FC3-Software Error
RS-Inventory Device Repair	RC4-PCB Replacement	RC11-Cleaning	FC4-Short Circuit
RH-Return Handling	RC5-Accessory Replacement	RC12-Mechanical Parts Replacement	–
SH-Service Handling	RC6-Product Replacement as Requires DNA Part	RC13-Functional Control	–
–	RC7-Software Upgrade	RC14-Refund	–

The repair codes made as a result of the fault occurring in the product and the class codes are indicated by the fault codes. The televisions coming to the technical service according to the service type; product repair (IH), installation (II), transport (CI), repair of stock devices (RS), product return / exchange (RH) or not (SH) can be classified into six groups.

3.3 Data Analysis and Data Modelling

Before the data model is established, it is necessary to determine the characteristics of the data used in the analysis that means to perform descriptive statistical analysis. The variables used, the number of repairs and the percentage of repairs are given in Table 2.

Figure 8 shows the distribution percentages of 96 different models.

According to Fig. 8, it is seen that there are 89 T56 models (10.45%) in service. However, the T11, T13 and T29 have been identified as the television models with the most faults, since the T56 has come to service II. When the repair codes of these three models were examined, the T13 (18.8%) and T29 (22.9%) model had the most RC3-LCD replacement faults, while the T011 had the most RC4-PCB replacement (10.4%). In general, the RC3-LCD replacement (5.6%), RC13-Functional control (4.6%) and RC2-Electrical component replacement (2.8%) faults occur after RC8-First assembly. Feedback to the manufacturing unit can make it more durable by reviewing the production of LCD and PCB parts that cause these faults. Figure 9 shows the distribution percentages of the televisions coming to the service according to months.

Table 2 Descriptive statistics

Variables	Sub-variables	Count (n)	Percentage (%)
Service Type	IH—In Home	102	12.00
	II—Initial Installation	701	82.30
	CI—Carry In	7	0.80
	RS—Stock Device Repair	13	1.50
	RH—Return Handling	8	0.90
	SH—Service Handling	21	2.50
	Total	**852**	**100.00**
Repair Code	RC1-Software Reinstallation	3	0.40
	RC2-Electrical Parts Replacement	24	0.10
	RC3-LCD Replacement	48	0.10
	RC4-RC4- PCB Replacement	14	0.40
	RC5-Accessory Replacement	5	4.60
	RC6-Product Replacement as Requires DNA Part	4	0.20
	RC7-Software Upgrade	3	2.80
	RC8-First Assembly	701	5.60
	RC9-Product Change	4	1.60
	RC10-No Error Found	1	0.60
	RC11-Cleaning	1	0.50
	RC12-Mechanical Parts Replacement	3	0.40
	RC13-Functional Control	39	82.30
	RC14-Refund	2	0.50
	Total	**852**	**100.00**
Fault Code	FC1-Open Circuit	803	94.20
	FC2-Dirty	42	4.90
	FC3-Software Error	6	7.00
	FC4-Short Circuit	1	0.10
	Total	**852**	**100.00**
Month	January	64	7.50
	February	55	6.50
	March	73	8.60
	April	44	5.20
	May	48	5.60
	June	108	12.70
	July	79	9.30
	August	75	8.80
	September	67	7.90

(continued)

Table 2 (continued)

Variables	Sub-variables	Count (n)	Percentage (%)
	October	96	11.30
	November	79	9.30
	December	64	7.50
	Total	**852**	**100.00**

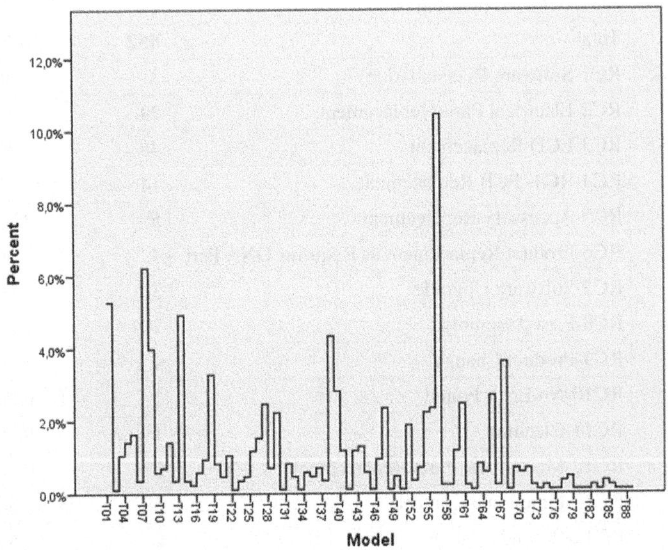

Fig. 8 Distribution of models coming to repair service

According to Fig. 9, it is observed that the product the most comes to technical service in June (12.88%) and the lest in April (5.16%). Considering that the initial installation of service activities (II) includes, it can be said that there was a decrease in television sales in April and May. For this reason, it may be suggested to put a promotion in April and May with the feedback given to the sales units.

Finding the most suitable model for the defined problem is possible by installing and testing as many models as possible. For this reason, data preparation and model building is a recurring process until the model is considered to be the best [17].

Clustering analysis is a technique that groups data of similar characteristics and is used successfully in decision processes [29]. Two-step and K-Means methods are widely used in clustering analysis. The K-Means clustering method is advantageous because hierarchical algorithms yield faster results in the solution and make the characteristics of the clusters different [66]. Furthermore, it was observed that the success of the Two-step model was 60% in data with the same variable and the same set number as a result of the analysis experiments, while the K-Means model success was 90%.

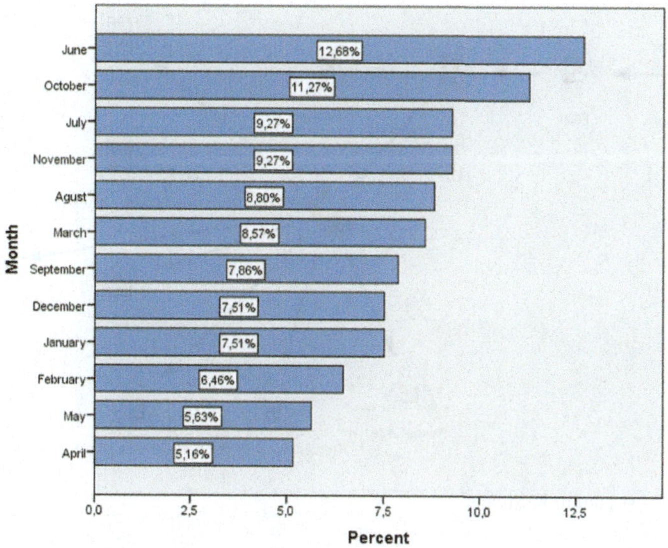

Fig. 9 Distribution of repaired products by months

Table 3 Model success rate of clustering methods (cluster quality)

Number of clusters	K-Means (%)	Two step (%)
2	0.90	0.20
3	0.90	0.60
4	0.90	0.30
5	0.90	0.30

Model success according to different cluster numbers and algorithms is given in Table 3.

When Table 3 is examined, it is seen that the model success of the K-Means method is higher. Therefore, K-Means clustering method was used in the analysis.

After the cluster centers are determined in the K-Means model, each data is assigned to the cluster that minimizes the distance to the nearest cluster center. The number of clusters in the model is determined by the analyst [37]. In the analysis, the formula ($\sqrt{\frac{numberofproject}{2}}$) proposed by Mardia et al. [39] was used to determine the number of clusters [39, 66].

The number of clusters that can be used in the analysis was found to be 21. However, when the model success rate is given in Table 3 and the percentages of the clusters are taken into consideration, it is concluded that analysis with 3 clusters is appropriate.

Figure 10 shows the three-clusters for K-Means model made using fault code, repair code, and service type variables.

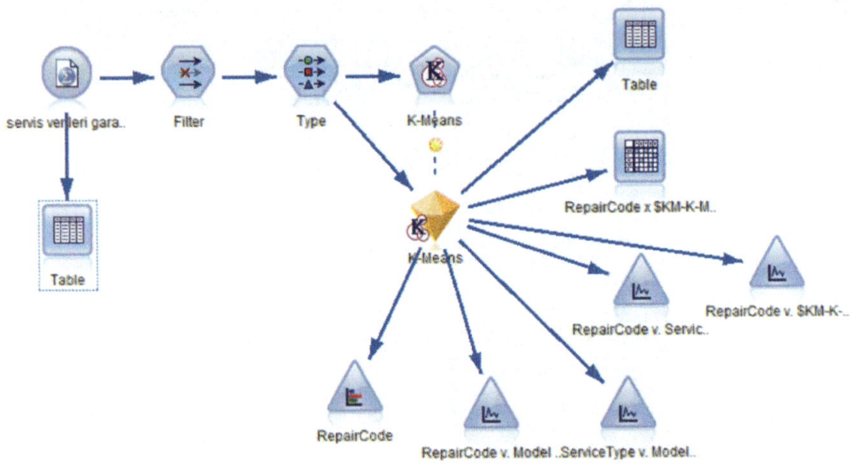

Fig. 10 Final stream to K-Means

Figure 10 shows the modules used during the data preparation process, the modules used for modeling and the modules used for analysis when modeling with Clementine 18.1.

In Fig. 11, using the K-Means method, a clustering model with three inputs and three clusters is given.

Cluster quality of the K-Means clustering model is 90%. When the cluster characteristics are examined, the televisions coming to the service;

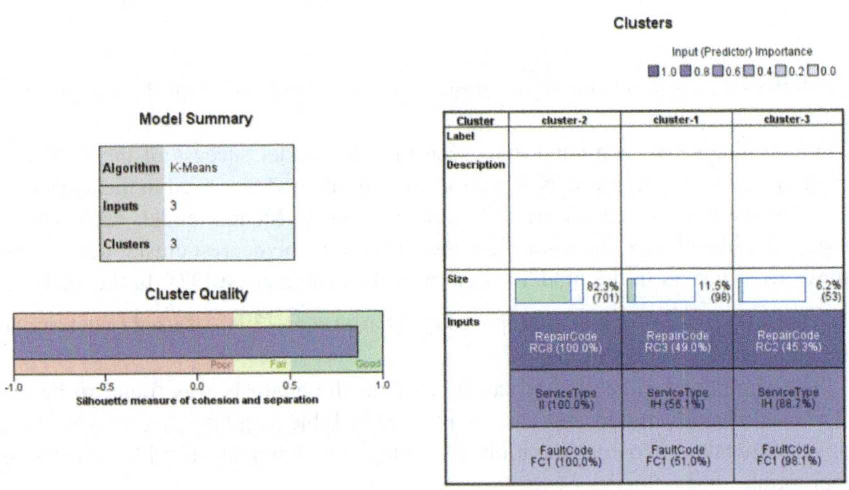

Fig. 11 Cluster model success rate and characteristics of clusters

- 11.5% are in the first cluster. In the first cluster, the repair code was found to be 49% RC3-LCD replacement, 56.1% service type IH-In home and 51% FC1-Open circuit faults.
- 82.3% of the data is in the second cluster. In the second set, the repair code is 100%; RC-Initial installation, service type 100%: II-Initial installation and fault 100%: FC1-Open circuit detected.
- 6.2% of the data is in the third cluster. The repair code of the data in the third cluster was determined as 45.3% RC2-Electrical parts replacement, 88.7% IH-In home and Fault code 98.1% FC1-Open circuit fault.

When the analysis results were examined, IH-In home service was handled in two separate clusters. This result allows two experts to accelerate the repair activities at the same time. By looking at similar data character-directions, determining which repair activities take place in which cluster, the repair activities expected from these experts can be determined. Figure 12 shows the repair codes showing the operation performed because of fault in each cluster.

According to Fig. 12; The first cluster includes the repair codes RC1, RC3, RC6, RC7, RC9, RC11, RC13, while the second cluster contains only the RC8 and the third cluster includes the repair codes RC2, RC4, RC5, RC6, RC9, RC10, RC12, RC14. When taking tactical and operational decisions, the operations performed because of fault in the first and third clusters can be evaluated. As a result of these evaluations, it is possible to employ two experts at the same time or to reorganize the fault codes by arranging the faults according to the results of the analysis, to increase customer satisfaction and provide fast return.

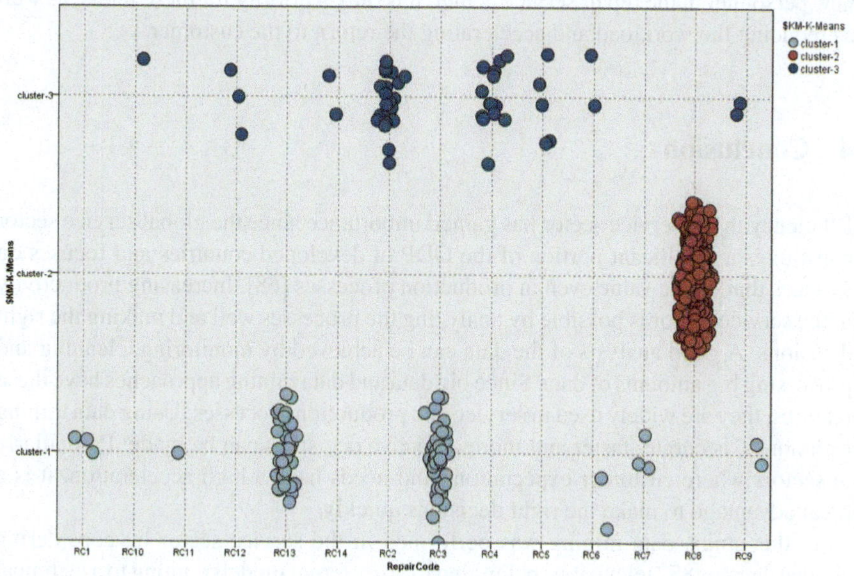

Fig. 12 Repair codes in each cluster

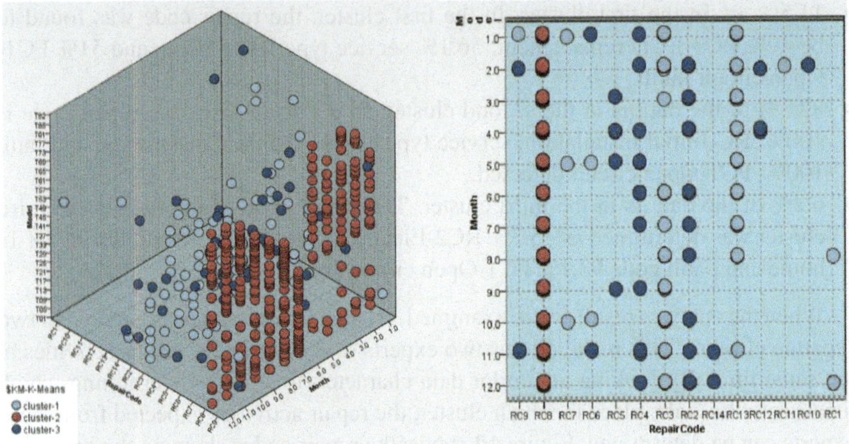

Fig. 13 Distribution of electronic repair models by months

Besides, it will assist in decision-making processes by determining the months in which the faults occurring based on model mostly occur and guide both production feedback and sales.

Figure 13 shows the types of repairs by model monthly.

According to Fig. 13, the RC8 model is first assembled every month and RC2, RC3, RC4, and RC13 faults are observed for at least 9 months. Since the faults related to these models occur frequently, it is important to ensure that during the training of new personnel in the repair service center, it is only a priority for these faults, as well as reducing the workload and accelerating the return to the customer.

4 Conclusion

Efficiency in the service sector has gained importance since the global service sector constitutes a significant portion of the GDP of developed countries and focuses on services that create value even in production processes [68]. Increasing productivity in the service sector is possible by analyzing the processes well and making the right decisions. A good analysis of the data can be achieved by monitoring, cleaning and processing big amounts of data. Since big data and data mining approaches have these features, they are widely used in service and production processes. Using data mining techniques, accurate, faster and more objective decisions can be made. Particularly in sectors where customer expectations and needs have a high acceleration, it is a great advantage to make the right decisions quickly.

In this study, data mining was performed in the service sector by considering service data for 852 televisions belonging to 96 different models coming to a technical service under warranty. After the pre-processing of the data, the K-Means clustering method was used. As a result of the analysis, it was determined that LCD and PCB replacement was made mostly in the televisions that were repaired and the production processes of these parts could be reviewed, and the possibility of fault could be reduced. However, in the study of [51], it is necessary to investigate the consumer behavior as well as reviewing the production, as some parts such as the screen are shown as the parts that are more defective due to the misuse of the consumers. Besides, due to the decrease in television sales in April and May, it was seen that sales volume could be increased by promotion or similar decisions to be made in the said months. Also, if it is desired to increase the number of experts to increase the customer return rates, it is determined that two experts are enough and which repair processes they can look for.

The results of the research are expected to contribute to the evaluation of the processes of firms operating in a similar field. Besides, this study is the only study that evaluates the operational and tactical processes in the repair service system by using data mining methods according to the types of repairs. Instead of expanding the study on a single product, it may be suggested to make changes on all products and to see the changes according to product variety and to reorganize the fault classes. In this study, elements such as supply time and repair completion time were not taken into account. The time spent in the system for the repaired products can be optimized with a system design made using the mentioned periods. In addition, the reasons and effects of the cancellation of the canceled products are analyzed according to the characteristics of the customers (age, gender, place of residence, etc.). In addition, an improvement can be made in the system to prevent time loss due to repetitive records. Similar to the survey study of [56], the effect of repair costs in the electronics industry on the attitude of customers can be investigated. In the research conducted, it was determined in which months the sales decreased, but the reasons for this were not investigated. Researching this decline can also play a major role in determining the volume of sales and related market share. Again, during the current research, the employees are in only two categories, The employees who are newcomers and the evaluated as experts and the performances of the same category are considered equal. A similar research to be conducted by evaluating employee performance may also be part of future studies. In addition, given that processes in the manufacturing and service sector are starting to be digitalized, the applicability of technologies such as wearable technologies, virtual reality, augmented reality can be investigated in electronic repair processes.

References

1. Agrawal, R., Imieliński, T., & Swami, A. (1993). Mining association rules between sets of items in large databases. In: *Proceedings of the 1993 ACM SIGMOD International Conference on Management of Data—SIGMOD '93* (pp. 207–216). New York, USA: Association for Computing Machinery (ACM). https://doi.org/10.1145/170035.170072.
2. Alhammady, H. (2007). A novel approach for mining emerging patterns in data streams. In: *2007 9th International Symposium on Signal Processing and its Applications*, ISSPA 2007,

Proceedings.
3. Alt, F. L. (1972). Archaelogy of computers: Reminiscences, 1945–1947. *Communications of the ACM, 15*, 693–694. https://doi.org/10.1145/361454.361528.
4. Ashayeri, J., Heuts, R., Jansen, A., & B. . (2006). Inventory management of repairable service parts for personal computers. *Int J Oper Prod Manag, 16*, 74–97. https://doi.org/10.1108/01443579610151760.
5. Bastos, P., Lopes, R., Pires, L., & Pedrosa, T. (2009). Maintenance behaviour-based prediction system using data mining. In *IEEM 2009—IEEE international conference on industrial engineering management* (pp. 2487–2491). https://doi.org/10.1109/IEEM.2009.5373535.
6. Bastos, P., Lopes, I., & Pires, L. (2014). Application of data mining in a maintenance system for **failure** prediction. In *Safety, Reliability and Risk Analysis. Beyond the Horizon—Proceedings of the European Safety Reliability Conference ESREL 2013* (pp. 933–940). ISBN: 978–1–138–00123–7.
7. Bayes, T. P. (1763). An essay towards solving a problem in the doctrine of chances. by the late rev. Mr. Bayes, F. R. S. communicated by Mr. Price, in a letter to John Canton, A. M. F. R. S. Philos Trans R Soc. London *53*, 370–418. https://doi.org/10.1080/14786445908642737.
8. Boon-Long, S., & Wongsurawat, W. (2015). Social media marketing evaluation using social network comments as an indicator for identifying consumer purchasing decision effectiveness. *Journal of Direct, Data and Digital Marketing Practice, 17*, 130–149. https://doi.org/10.1057/dddmp.2015.51.
9 Boser, B. E., Guyon, I. M., & Vapnik, V.A. (1992) A training algorithm for optimal margin classifiers. In: *Fifth Annual Workshop on Computational Learning Theory.* Pittsburgh: ACM.
10. Botter, R., & Fortuin, L. (2000). Stocking strategy for service parts—A case study. *International Journal of Operations and Production Management, 20*, 656–674. https://doi.org/10.1108/01443570010321612.
11. Bryson, E., & Jr. Yu-Chi, H. (1975). Applied optimal control: Optimization, estimation and control—Taylor & Francis, Blaisdell, Waltham, Massachusetts. ISBN 0–89116–228–3.
12. Chon, K. W., & Kim, M. S. (2018). BIGMiner: A fast and scalable distributed frequent pattern miner for big data. *Cluster Computing, 21*, 1507–1520. https://doi.org/10.1007/s10586-018-1812-0.
13. Codd, E. F. (1970). Information retrieval a relational model of data for large shared data banks. *Communications of the ACM, 13*, 377–387.
14. Cox, J., Griffith, S., Giorgi, S., & King, G. (2013). Consumer understanding of product lifetimes. *Resources, Conservation and Recycling, 79*, 21–29. https://doi.org/10.1016/j.resconrec.2013.05.003.
15. Díaz, V. G., Campos, M. L., Fernández, J. F. G., & Márquez, A.C. (2010). Warranty cost models state-of-art: A practical review to the framework of warranty cost management. Reliab Risk Saf Theory Appl—Briš, Guedes Soares Martorell, Taylor & Francis Group, London. 2051–2059 ISBN 978–0–415–55509–8.
16. Domingos, P., & Hulten, G. (2000). Mining high-speed data streams. *Proceeding of the Sixth ACM SIGKDD International Conference on Knowledge Discovery and Data Mining* (pp. 71–80). New York, USA: ACM Press.
17. Ersöz, F. (2019). Veri madenciliği teknikleri ve uygulamaları: Kavram-teori-modeller-yöntem. Seçkin Yayıncılık.
18. Fayyad, U., & Stolorz, P. (1997). Data mining and KDD: Promise and challenges. *Future Generation Computer Systems, 13*, 99–115. https://doi.org/10.1016/s0167-739x(97)00015-0.
19. Geissbauer, R., Vedso, J., & Schrauf, S. (2016) Industry 4.0: Building the digital enterprise. London: PWC.
20 Goebel, M., & Gruenwald, L. (1999). A survey of data mining and knowledge discovery software tools. *ACM SIGKDD Explorations Newsletter, 1*, 20–33. https://doi.org/10.1145/846170.846172.
21. Harding, J. A., Shahbaz, M., & Srinivas, K. A. (2006). Data mining in manufacturing: A review. *Journal of Manufacturing Science and Engineering Transactions ASME, 128*, 969–976. https://doi.org/10.1115/1.2194554.

22. Holland, J. (1992). *Adaptation in natural and artificial system* (p. 232). Massachusetts: The MIT Press. ISBN: 978–0262581110.
23. Hui, S. C., & Jha, G. (2000). Data mining for customer service support. *Information and Management, 38*, 1–13. https://doi.org/10.1016/S0378-7206(00)00051-3.
24. Hurwitz, J., Nu-gent, A., Halper, F., & Kaufman, M. (2013). *Big data for dummies*. New Jersey: John Wiley & Sons. Inc. ISBN: 978-1-118-64417-1.
25. IBISWorld. (2019). Electronic & computer repair services industry in the US - market research report: How is the industry trending in the US?
26. Jack, N., & Schouten, F. (2000). Optimal repair-replace strategies for a warranted product. *International Journal of Production Economics, 67*, 95–100. https://doi.org/10.1016/S0925-5273(00)00012-8.
27. Keogh, E., Lonardi, S., & Ratanamahatana, C.A. (2004) Towards parameter-free data mining. In: *KDD-2004—Proceedings of the Tenth ACM SIGKDD International Conference on Knowledge Discovery and Data Mining* (pp 206–215). New York, USA: ACM Press. https://doi.org/10.1145/1014052.1014077.
28. King, A. M., Burgess, S. C., Ijomah, W., & McMahon, C. A. (2006). Reducing waste: Repair, recondition, remanufacture or recycle? *Sustainable Development, 14*, 257–267. https://doi.org/10.1002/sd.271.
29. King, R. S. (2014). *Cluster analysis and data mining: An introduction.* Mercury Learning and Information, Dulles, Virginia. ISBN: 978-1-938549-38-0.
30. Koza, J. R. (1992). *Genetic programming: On the programming of computers by means of natural.* Cambridge: MIT Press. ISBN: 0-262-11170-5.
31. Kumar, A., Shankar, R., & Thakur, L. S. (2018). A big data driven sustainable manufacturing framework for condition-based maintenance prediction. *Journal of Computational Science, 27*, 428–439. https://doi.org/10.1016/j.jocs.2017.06.006.
32. Legendre, A. M. (1805) Nouvelles méthodes pour la détermination des orbites des comètes. Chez Firmin Didot.
33. Liang, Y. H. (2010). Integration of data mining technologies to analyze customer value for the automotive maintenance industry. *Expert Systems with Applications, 37*, 7489–7496. https://doi.org/10.1016/j.eswa.2010.04.097.
34. Lim, C., Kim, M. J., Kim, K. H., et al. (2019). Customer process management: A framework for using customer-related data to create customer value. *Journal of Service Management, 30*, 105–131. https://doi.org/10.1108/JOSM-02-2017-0031.
35. Lin, M. Y., Lee, P. Y., & Hsueh, S. C. (2012). Apriori-based frequent itemset mining algorithms on MapReduce. In: *Proceedings of the 6th International Conference on Ubiquitous Information Management and Communication, ICUIMC'12* (p. 1). New York, USA: ACM Press. https://doi.org/10.1145/2184751.2184842.
36. Ling, C. X., & Li, C. (1998) Data Mining for direct marketing: Problems and solutions. KDD-98. 1:7.
37. Linoff, G., Berry, M. J. A. (2011). *Data mining techniques: for marketing, sales, and customer relationship management*. Kanada: Wiley. ISBN: 978-0-470-65093-6.
38. Lv, S., Kim, H., Zheng, B., & Jin, H. (2018). A review of data mining with big data towards its applications in the electronics industry. *Applied Sciences, 8*, 1–34. https://doi.org/10.3390/app8040582.
39. Mardia, K. V., Kent, J. T., Bibby, J. M. (1979). *Multivariate Analysis*. London: Academic Press.
40. Matta, K. F. (1985). A simulation model for repairable items / spare parts inventory systems. *Computers & Operations Research, 12*, 395–409. https://doi.org/10.1016/0305-0548(85)90037-1.
41. McCulloch, W. S., & Pitts, W. (1943). A logical calculus of the ideas immanent in nervous activity. *Bulletin of Mathematical Biophysics, 5*, 115–133. https://doi.org/10.1007/BF02478259.
42. McCollough, J. (2009). Factors impacting the demand for repair services of household products: The disappearing repair trades and the throwaway society. *International Journal of Consumer Studies, 33*, 619–626. https://doi.org/10.1111/j.1470-6431.2009.00793.x.

43. McCormick, K., Abbott, D., & Brown, M. S., et al. (2013). *IBM SPSS modeler cookbook*. Packt Publishing. ISBN: 978–1–84968–546–7.
44. Merdin, D., & Ersöz, F. (2019). Evaluation of the applicability of industry 4.0 processes in businesses and supply chain applications. In: *3rd International Symposium on Multidisciplinary Studies and Innovative Technologies, ISMSIT 2019—Proceedings* (pp. 1–10). Institute of Electrical and Electronics Engineers Inc.
45. Meseroll, R. J., Kirkos, C. J., & Shannon, R. A. (2007). Data mining navy flight and maintenance data to affect repair. In *AUTOTESTCON (Proceedings)* (pp. 476–481).https://doi.org/10.1109/AUTEST.2007.4374256.
46. Morar, N., Roy, R., Mehnen, J., et al. (2013). Data mining and knowledge reuse for the initial systems design and manufacturing: Aero-engine service risk drivers. *Procedia CIRP, 11*, 130–134. https://doi.org/10.1016/j.procir.2013.08.002.
47. Mostafa, N. A., Ibrahim, S. A., Elayat, H. A., & Khater, M. M. (2011). Data analysis for inventory management in a multi-echelon supply chain. *International Journal of Economics and Research, 2*, 138–150. ISSN: 2229–6158.
48. O'Donovan, P., Leahy, K., Bruton, K., & O'Sullivan, D. T. J. (2015). Big data in manufacturing: A systematic mapping study. *Journal of Big Data 2*. https://doi.org/10.1186/s40537-015-0028-x.
49. Raihanian Mashhadi, A., Esmaeilian, B., Cade, W., et al. (2016). Mining consumer experiences of repairing electronics: Product design insights and business lessons learned. *Journal of Cleaner Production, 137*, 716–727. https://doi.org/10.1016/j.jclepro.2016.07.144.
50. Reinsel, D., Gantz, J., & Rydning, J. (2017). Data age 2025: The evolution of data to life-critical don't focus on big data. IDC Analyze the Future.
51. Rosner, D. K., & Ames, M. (2014). Designing for repair?: Infrastructures and materialities of breakdown. In: *Proceedings of the 17th ACM Conference on Computer Supported Cooperative Work & Social Computing* (pp. 319–331). https://doi.org/10.1145/2531602.2531692.
52. Ruvald, R., Frank, M., Johansson, C., & Larsson, T. (2018). Data mining through early experience prototyping—A step towards data driven product service system design. *IFAC-PapersOnLine, 51*, 1095–1100. https://doi.org/10.1016/j.ifacol.2018.08.458.
53. Rygielski, C., Wang, J. C., & Yen, D. C. (2002). Data mining techniques for customer relationship management. *Technology in Society, 24*, 483–502. https://doi.org/10.1016/S0160-791X(02)00038-6.
54. Sabbaghi, M., & Behdad, S. (2018). Consumer decisions to repair mobile phones and manufacturer pricing policies: The concept of value leakage. *Resources, Conservation and Recycling, 133*, 101–111. https://doi.org/10.1016/j.resconrec.2018.01.015.
55. Sabbaghi, M., Cade, W., Behdad, S., & Bisantz, A. M. (2017). The current status of the consumer electronics repair industry in the U.S.: A survey-based study. *Resources, Conservation and Recycling, 116*, 137–151. https://doi.org/10.1016/j.resconrec.2016.09.013.
56. Sabbaghi, M., Esmaeilian, B., Cade, W., et al. (2016). Business outcomes of product repairability: A survey-based study of consumer repair experiences. *Resources, Conservation and Recycling, 109*, 114–122. https://doi.org/10.1016/j.resconrec.2016.02.014.
57. Savasere, A., Omiecinski, E. R., & Navathe, S. B. (1995). An efficient algorithm for mining association rules in large databases. In *21st International Conference Very Large Data Bases*, Swizerland.
58. Shapiro, G. P. (2000). Knowledge discovery in databases: 10 years after. *ACM SIGKDD Explorations Newsletter, 1*, 59–61. https://doi.org/10.1145/846183.846197.
59. Shukla, N., Tiwari, M. K., & Beydoun, G. (2019). Next generation smart manufacturing and service systems using big data analytics. *Computers & Industrial Engineering, 128*, 905–910. https://doi.org/10.1016/j.cie.2018.12.026.
60. Srinivasan, R., Manivannan, S., Ethiraj, N., et al. (2016). Modelling an optimized warranty analysis methodology for fleet industry using data mining clustering methodologies. *Procedia Computer Science, 87*, 240–245. https://doi.org/10.1016/j.procs.2016.05.155.
61. Stigler, S. M. (1986). *The history of statistics: The measurement of uncertainty before 1900*. The Belknap Press of Harvard Universty Press.

62. Tanphet, S., & Wanchai, P. (2018). Applying business intelligence technology for equipment maintenance and repair plan of telecommunications services provider. In *International Conference on Advanced Communication Technology ICACT 2018* (pp. 448–453). https://doi.org/10.23919/ICACT.2018.8323790.
63. Turing, A. (1936). On computable numbers, with an application to the entscheidungsproblem. *Proceedings of the London Mathematical Society, 42*, 230–265.
64. Van Nguyen, T., Zhou, L., Chong, A. Y. L., et al. (2019). Predicting customer demand for remanufactured products: A data-mining approach. *European Journal of Operational Research*. https://doi.org/10.1016/j.ejor.2019.08.015.
65. Venkatadri, M. (2011). A review on data mining from past to the future. *International Journal of Computers and Applications, 15*, 975–8887.
66. Wendler, T., & Gröttrup, S. (2016). Data mining with SPSS Modeler. Springer International Publishing AG Switzerland. ISBN: https://doi.org/10.1007/978-3-319-28709-6.
67. Xu, F., Li, Y., & Feng, L. (2019). The influence of big data system for used product management on manufacturing–remanufacturing operations. *Journal of Cleaner Production, 209*, 782–794. https://doi.org/10.1016/j.jclepro.2018.10.240.
68. Yada, K. (2014). *Data mining for service. In the series of "Studies in Big Data"*. Berlin, Heidelberg: Springer. ISBN: 978–3–642–45251–2.
69. Yeh, C. T., & Chen, M. C. (2018). Applying kansei engineering and data mining to design door-to-door delivery service. *Computers & Industrial Engineering, 120*, 401–417. https://doi.org/10.1016/j.cie.2018.05.011.
70. Zadeh, L. A. (1965). Fuzzy sets. *Information and Control, 8*, 338–353. https://doi.org/10.1016/S0019-9958(65)90241-x.
71. Zhong, R. Y., Newman, S. T., Huang, G. Q., & Lan, S. (2016). Big Data for supply chain management in the service and manufacturing sectors: Challenges, opportunities, and future perspectives. *Computers & Industrial Engineering, 101*, 572–591. https://doi.org/10.1016/j.cie.2016.07.013.
72. Zhou, C., Nelson, P. C., Xiao, W., et al. (2001). An intelligent data mining system for drop test analysis of electronic products. *IEEE Transactions on Electronics Packaging Manufacturing, 24*, 222–231. https://doi.org/10.1109/6104.956808.

Filiz Ersoz is a Professor at Karabük University, Department of Industrial Engineering, Turkey. Filiz Ersoz holds B.Sc. degree (1989) in Statistics from Anadolu University, M.Sc. (1992) and Ph.D. (1998) degrees in Biostatistics from Ankara University. She also received the rank of Associate Professor in Quantitative Decision-Making Methods from the Turkish Inter-University Council (UAK) in October 2011 and the rank of Professor in Industrial Engineering from Karabük University (2017). Her current research deals with decision making, statistics, data mining, simulation and modelling, statistical quality control, multi-criteria decision-making techniques. Dr. Ersoz has published widely in the field, with more than 100 publications and 5 books. She acted as a Project Manager and Advisor in more than 15 projects. Furthermore, she regularly works as reviewer for about 10 different scientific journals.

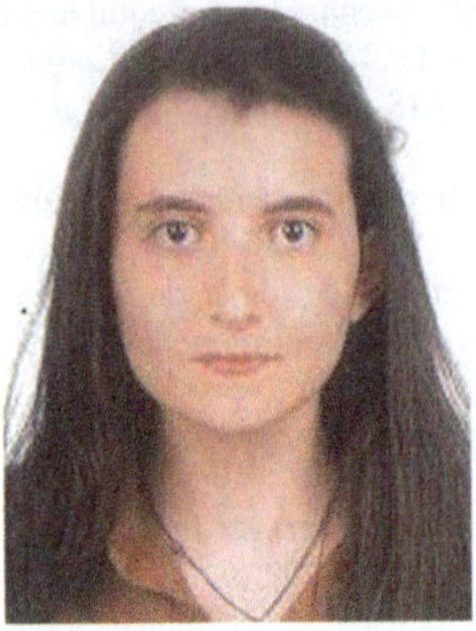

Deniz Merdin is Ph.D. student at the Department of Industrial Engineering, Karabük University, Turkey. In 2009, she began her tertiary education in the Department of Industrial Engineering in Karabük University, and in 2014 she graduated as the top student of the Department and the Faculty. In addition, the graduation thesis (An Evaluation of ISO 9001:2008 and ISO 9001:2015 Standard Changes in Quality Management System) received support within the scope of TÜBİTAK Industry Focused Graduation Thesis Support Program. Deniz holds MSc. Degree (2017) in Industrial Engineering from Karabük University. She has articles and papers in national and international journals. She continues her studies on simulation modelling, data mining and digital transformation.

Integrative Applications of Blockchain and Contemporary Technologies from a Big Data Perspective

Amit Karamchandani, Samir K. Srivastava, and Abha

Abstract Blockchain is not a standalone technology. It is a ledger that ensures the trustworthiness and traceability of data stored on it. Given the Industry 4.0 environment, it will work in integration with state-of-the-art technologies such as the internet of things (IoT), 3D printing (3DP), fog computing, geotagging, artificial intelligence (AI), and swarm robotic systems (SRS). These technologies' combined function would create large amounts of data on blockchain, popularly known as big data. This chapter contributes by highlighting the role of contemporary Industry 4.0 technologies in creating big data. The chapter features the managerial and policy implications of big data from service operations management perspective.

1 Industry 4.0 and Big Data

The digitalization of the supply chain is a field with promising opportunities, which led to the rise of Industry 4.0. Initially introduced at the Hannover fair in 2011, Industrie 4.0 was officially introduced as a German high-tech strategy initiative in 2013 aimed at leading the manufacturing industry. Industrie 4.0, also known as Industry 4.0, represents the commencement of the Fourth Industrial Revolution [1, 20, 26]. Industry 4.0 is characterized by its ability to ensure flexible products and services shared on blockchain by promoting resilience, security, autonomy, timestamped data, and scalability [34]. In Industry 4.0, it is critical to model business processes in synchronization with the digital technologies adopted in the organization. Industry

A. Karamchandani (✉)
Great Lakes Institute of Management, Gurgaon, India
e-mail: fpm17013@iiml.ac.in; amit.k@greatlakes.edu.in

S. K. Srivastava · Abha
Indian Institute of Management, Lucknow, India
e-mail: samir@iiml.ac.in

Abha
e-mail: efpm08013@iiml.ac.in

A. Emrouznejad and V. Charles (eds.), *Big Data and Blockchain for Service Operations Management*, Studies in Big Data 98,
https://doi.org/10.1007/978-3-030-87304-2_9

4.0 can be adopted by implementing various technologies such as artificial intelligence, blockchain, cloud computing, cyber-physical systems (CPS), internet of things (IoT), machine learning, 3D printing, cloud computing, geotagging, artificial intelligence, and robotic process automation [41]. Industry 4.0 promises responsiveness and customization by ensuring flexible response to supply chain disruptions allowing product customization, ensuring visibility and transparency through the integration of blockchain and IoT (BIoT), and meeting customer requirements. World Economic Forum predicts that 10% of world GDP will be put on blockchain by 2027 [37]. Industry 4.0 is anticipated to attain its decentralization characteristic through blockchain, which is one of its core technologies [35]. The role of blockchain has been increasing in recent years by its increasing applications in the manufacturing and service industry worldwide [41]. Given the Industry 4.0 environment, blockchain can integrate with other Industry 4.0 technologies, which would create large amounts of data, popularly known as big data. Big data refers to large datasets that require storage larger than a traditional Structured Query Language (SQL) database, preferably a NoSQL database that may handle large amounts of data. The big data is represented by three V's, namely Volume, Velocity, and Variety. The adoption of Industry 4.0 technologies results in an overwhelming increase in the velocity, volume, and variety of data creation [10]. Several open-source tools are currently available for big data, such as Bigtable, Hadoop, and MapReduce. These tools efficiently store and query large amounts of data [9]. Big data can be generated from multiple sources, including the following [6]:

- Enterprise systems such as supplier relationship management (SRM), enterprise resource planning (ERP) and customer relationship management (CRM).
- Online social networks such as Twitter, Instagram and Facebook.
- Industrial data collected by sensor systems such as Internet of Things (IoT) and Radio Frequency Identification (RFID).
- Open access data such as financial data, geographical data, housing data, traffic data, weather data, etc.
- Mobile data generated from mobile devices, mobile network and internet connection.

Today, big data is in great demand in most industry sectors. Organizations that can utilize data to answer nagging questions about their operations can gain from big data. Experts claim that most of the operational decisions would be based on insights derived from big data in the coming future. Nowadays, it is very common to find a department of data science experts in most organizations including startups [7]. According to Mckinsey's research, there is an excessive shortage of talent capable of deriving insights from big data. Organizations intend to use past customer data to offer better-customized services to existing customers. The same data can also be used to attract new similar customers. Customers engage with companies through various online channels such as Twitter, Facebook, Instagram, Whatsapp, Telegram, and company website. It is essential to aggregate the data from multiple sources to analyze consumer preferences. The next task is to derive decision-making insights from the data to help the organization achieve a competitive advantage. The above exercise is meaningful only if the data collected is trustworthy and secure. Centralized databases have not been efficient in fulfilling these two objectives. The alternative to centralized databases is decentralized databases, popularly known as Blockchain, which will be discussed in the next section.

The chapter is organized in the following manner: Sect. 2 introduces the concept of blockchain. Section 2.1 highlights the integrated applications of blockchain, IoT, fog computing, and AI. Next, Sect. 2.2 focuses on 3D Printing based on blockchain. Section 2.3 introduces the integrative applications of blockchain and swarm robotic systems. Further, Sect. 2.4 features the composite application geotagging and blockchain. Section 3 highlights the managerial and policy implications of managing big data from a service operations perspective. Finally, Sect. 4 concludes the chapter by appending the contributions, limitations, and future research directions.

2 Blockchain

The concept of bitcoin was introduced to the world in 2008 by Satoshi Nakamoto [24]. The blockchain is a decentralized, encrypted electronic distributed ledger or database that can maintain the records immutably without involving a trustworthy third party [40]. Each data record in a blockchain is timestamped, identity stamped, and has a hash code associated with it. The hash code of each data block is generated by cryptographic integrated transformation of multiple data such as the hash code of the previous block, current block number, data, and nonce. A nonce is an arbitrary number mused just once for cryptographic transformation. Economic experts and researchers started to find blockchain interesting after a few years of bitcoin introduction. Blockchain further evolved from bitcoin technology to believe that "what internet did for information, blockchain may do for transactions" [4].

The blockchain is of three types: Unpermissioned public, permissioned public, and permissioned private [5]. Reputed organizations like Microsoft use the term 'Enterprise Blockchain' for blockchain applications beyond cryptocurrency applications [16]. The permissioned public and permissioned private blockchain are popularly known as Enterprise Blockchain (EBC) [18]. EBC involves the use of multiple algorithms to establish consensus. Identities of all participants are known on EBC. Read and write permissions are role-based in an EBC. An example of permissioned public EBC is a government blockchain with limited writers (government officers) and unlimited readers (country population) who may access EBC by logging in with a national identification number. Examples of permissioned private EBC can be a private EBC within a single organization or a consortium EBC formed by the collaboration of multiple organizations in a supply chain. The current blockchain market is getting a tremendous amount of investments, and it is expected to grow from $3.0 billion in 2020 to $39.7 billion by 2025. EBC is expected to be an integral part of many businesses in the coming years and is proving to be the most significant technological breakthrough since the evolution of the internet. EBC is found to have transformative potential in the manufacturing and service industries [17].

2.1 Blockchain, IoT, Fog Computing, and AI

AI is observing progressive growth for knowledge discovery by employing intelligent algorithms to make data-driven decisions. AI applications are visible across sectors such as financial services, defense, healthcare, and transportation. It can learn from existing data and improve its learning from new data [13]. The immutable data on blockchain makes it suitable to feed in AI systems and record the AI output. Microsoft, IBM, and Intel are a few organizations working on integrating AI applications with blockchain. IBM is piloting with the idea of blending its blockchain (based on hyperledger fabric) service with its Watson AI platform. Everledger is using the IBM blockchain platform to track the traceability of the diamond supply chain. Similarly, IBM Watson AI applies knowledge of thousands of regulations and ensures that diamonds meet UN mandates preventing the sale of undesirable minerals [27]. Using AI systems helps to (1) process data into useful information for decision-makers, (2) learn from historical information and new information, (3) identify interesting trends that aid in making decisions, and (4) generate automated recommendations for decision-makers [28].

Most IoT devices today are connected via highly vulnerable public networks prone to hacking. BIoT helps address the data trust issues with IoT sensors and hence is likely to drive its adoption. The current technological paradigm involves data creation through machine-to-machine communication by integrating IoT sensors with blockchain (BIoT). The use of BIoT will give rise to new kinds of IoT applications. For example, insurance companies could use smart contracts to pay a certain amount based on triggers captured through IoT sensors on blockchain. If a car driving at speed below 60 km/h faces an accident, the insurance company could automatically pay 25% of the insurance claim to the policyholder [19, 36]. Alternatively, automakers may utilize BIoT for authentication of the interactions between connected vehicles and road infrastructure. Figure 1 indicates the amount of data created and

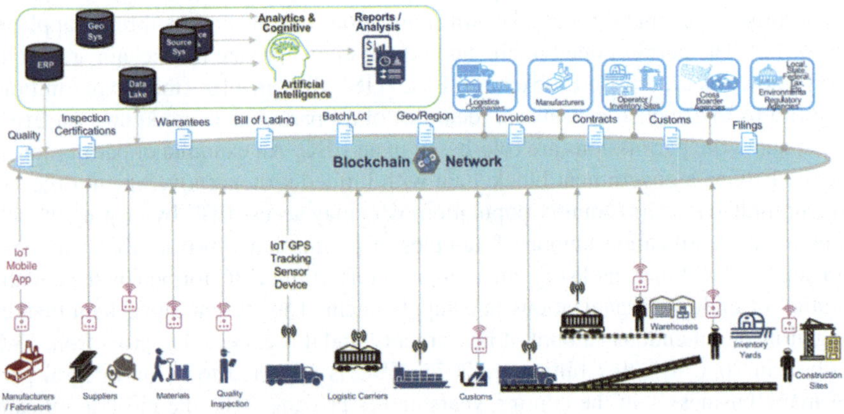

Fig. 1 Application of Blockchain, IoT, and AI in a supply chain [8]

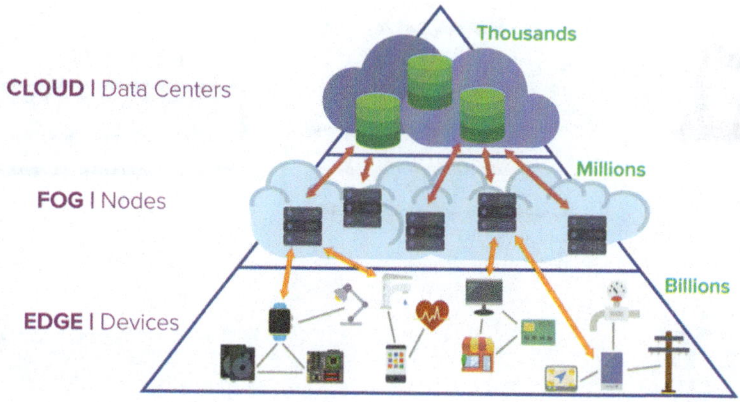

Fig. 2 Fog computing for IoT devices [25]

stored on the blockchain in a supply chain. Hence, we can see that the integration of IoT, AI, and blockchain would create big data.

Fog computing refers to another layer of network fabric for IoT applications that are bandwidth-intensive and require real-time data processing locally. It extends the cloud computing capacity to the edge of the enterprise's network. An example of a fog computing application is an offshore oil rig that generates data in terabytes daily. Leveraging the integration of blockchain, IoT, AI, and fog computing can bring about a breakthrough in technology. For example, traditional cloud computing systems are not intelligent or efficient enough to drive an autonomous vehicle. However, the AI engines can make decisions based on the immutable IoT data stored on blockchain and processed by fog nodes. Hence, the vehicle would require little to no human intervention to navigate the roads. A self-driving truck that delivers goods to distribution centers using AI is another such application. It can use the smart contract framework on blockchain to bill and charge automatically at a distribution center. The retailer can then transport the goods delivered to the customers through drones [27, 30]. Figure 2 provides a comparative overview of the number of IoT devices, fog nodes, and cloud data centers.

2.2 *Blockchain and 3D Printing (3DP)*

3DP refers to additive manufacturing processes that (1) build products layer by layer, and (2) build products based on digital models [3, 15]. 3DP facilitates supply chain agility and flexibility in product customization. The digital product can be sent to the customer, who can print it at the office, home, or via an online 3DP service provider and get it delivered at home. 3DP is likely to disrupt the traditional manufacturing industry by encouraging make-to-order production. The 3DP production initiates through a 3D CAD model divided by software into layers, which helps the printer

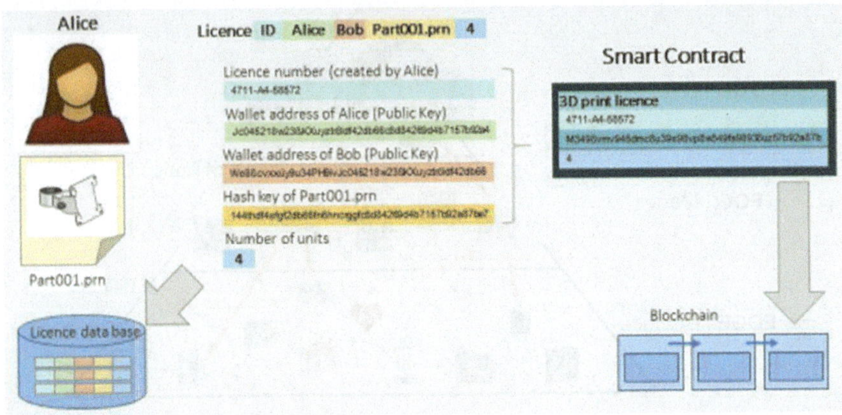

Fig. 3 Distributing license of 3DP through blockchain [32]

deposit or sinter the material accordingly [30]. The expiry of printer patents has lowered the costs, and hence 3DP has observed exponential growth in the industry. The 3D desktop printer costs have gone down from 100,000 dollars to less than 1000 dollars. The 3DP market is expected to increase from $8.35 billion in 2019 to $23.75 billion in 2027 [29]. The 3DP designs when stored on blockchain ensure their secure transfer to a verified 3D printer, which can save production, inventory and logistics costs. Storing them on blockchain also ensure a traceable history of document revisions [23]. If 3DP obsoletes the traditional manufacturing industry, we can imagine the amount of data that would be created by storing 3DP designs of various kinds of products on blockchain. Figure 3 shows the application of smart contract on blockchain to distribute 3DP license.

2.3 Blockchain and Swarm Robotic Systems

Swarm robotic systems have become popular in the last few years and are expected to disrupt the market in the near future [21]. The SRS applications include precision farming, material transportation, unmanned aerial vehicles (UAVs), nanorobotics and entertainment systems (forming interactive displays). The applications are increasing with a decrease in the costs of robotic platforms. One of the main challenges faced by SRS is security and trust. It is found that the inclusion of faulty robots (with malicious intent) can be a great risk for the swarm's goals. The security in the SRS context consists of data integrity, data confidentiality, traceability, and identity authentication. Security for SRS has been challenging to achieve due to several factors such as decentralized control, collective swarm behavior, and a huge number of robots. Integrating SRS with blockchain can make the robotic operations flexible, autonomous, and secure through reliable peer-to-peer communication. As shown in

Fig. 4, a robot looking to send messages to other robots can use their private keys to encrypt the message, share their public keys with the intended robot, and send it to the robot's address. The robot possessing the matching private key would only access the message [11]. The potential SRS applications that can benefit from its integration

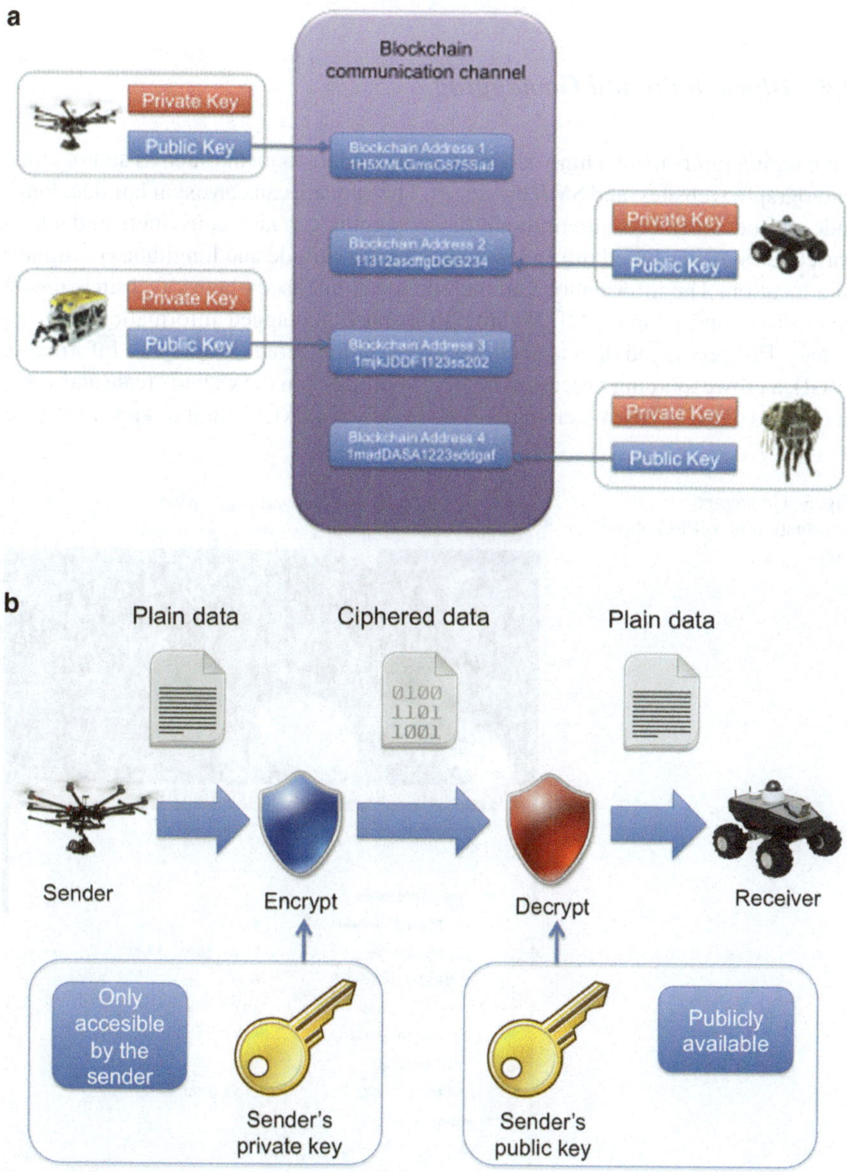

Fig. 4 **a** SRS Communication on blockchain [11]. **b** SRS Communication on blockchain [11]

with blockchain are military (require trustworthy and secure systems), disaster relief (ensures that the intended recipient gets the aid), and health care (SRS aid healthcare workers in the logistics of medicines). Given the breadth of SRS applications, the data created by SRS on blockchain will be huge and heterogeneous in the context of volume, velocity and variety.

2.4 *Blockchain and Geotagging*

Geotagging refers to attaching geographical metadata to media such as audio, video, photograph, websites, and SMS messages. The geodata can consist of latitude, longitude, altitude, distance, and timestamp. Geotagging can help consumers find a large variety of location-based information by entering latitude and longitude coordinates of a location. The information can include location-based photos, websites, media, enterprises, and people [22]. Figure 5 illustrates geotagged information in a .jpg image. The geotagged data is also created by Volunteered Geographic Information (VGI), a crowdsourcing process to involve volunteers on the web to create and verify geotagged data [14]. Few examples of VGI are Geo-Wiki, Columbus app, and flicker

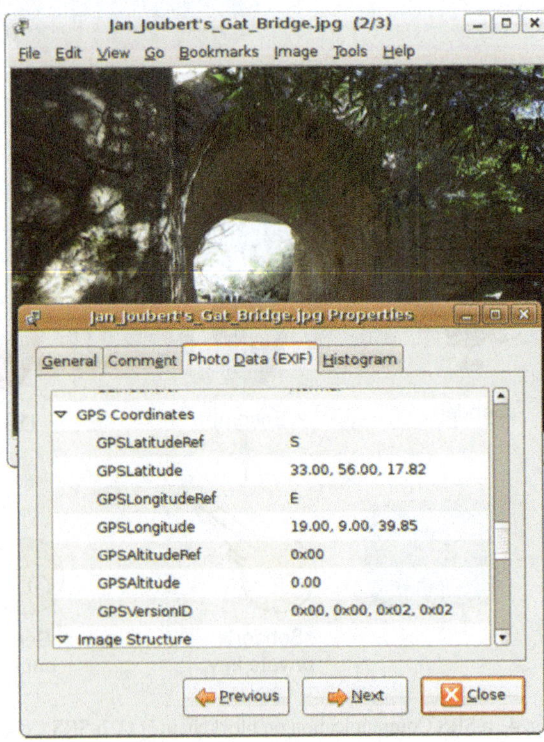

Fig. 5 Geotagged information in a JPEG photo [39]

app [31]. Adding geotag to blockchain data can provide a traceable history of events in a temporal sequence. One of the applications of geotagged data on blockchain is the inhibition of criminal activity. The criminal can be tracked easily and quickly if the crime committed is recorded on a geotagged blockchain. The crime investigation agencies can flag the blockchain hash code associated with the crime and assign it a 'hot' block. A 'hot' block means an alert would be broadcasted across the network and would be actively monitored by other blockchain nodes. The IoT devices, security cameras, and other assets would converge, keeping a close eye on the target, accelerating the process of apprehending the criminal. The number of crime events registered globally can help anticipate the amount of data that would create on a geotagged blockchain, resulting in the creation of big data [12].

3 Managerial Implications and Policy Implications

Today, firms are collecting a tremendous amount of data in the service sector. The sector includes transportation, hospitality, e-commerce, and online transaction platforms (such as Amazon, Airbnb, Uber, and Ola) [7]. Based on historical data, managers can learn about the customer's buying behavior and preferences to provide highly customized services, thereby improving service quality, profitability, and customer satisfaction. The big data collected by firms generally include sensitive personal information of the customers such as age, location, education, income, social networks, past searches, past transactions, credit card details, and purchase preferences. Such vital data on centralized databases has raised cybersecurity concerns through increased fraud, identity theft, and information leakage by hackers. The latest example is Air India, whose servers were hacked, leading to leakage of critical data of 45 lac customers globally [33]. The leaked data included the credit card and passport details of the customers. Hence, storing such data on blockchain can significantly enhance data security. Moreover, as blockchain is the chain of data connected through hash codes leading to immutability, it ensures the availability of trustable data. Integrating big data with fog computing can provide real-time personalization and customization on a large scale. Firms can use the derived insights against the customers' interest by charging higher prices, which can raise legal issues [7]. Policymakers can address such issues by implementing business-friendly policies, which can cater to all the stakeholders. The policymakers need to ensure that the privacy and confidentiality of individuals are guaranteed while reaping the benefits from shared information.

The integration of technologies discussed in this chapter can lead to improved lean and agile service operations. Big data stored on blockchain would have high predictive power, which would be helpful in crucial predictions such as crime prediction, traffic prediction, health predictions, etc. [6]. Blockchain would act as an underpinning technology, enabling better contract management, superior visibility, effective quality control, authenticated IoT data, improved accountability, and compliance in service SCM. 3DP can stimulate agile onsite production. E.g., Onsite healthcare equipment manufacturing can help healthcare facilities to be self-dependent, especially during a pandemic when the availability of healthcare equipment is scarce globally. AI can strengthen and develop the decision-making capabilities of managers, while blockchain can ensure trust in AI-derived decisions based on the trustworthiness of data. Geotagging on blockchain can augment supply chain resilience by providing alternatives to existing supply chain stakeholders such as suppliers, manufacturers, distributors, and customers.

Blockchain-based swarm robotic systems can help managers realize increased efficiency in task allocation, trusted information support, effective consensus, and distributed decision-making among robots [27]. Firms can realize such benefits, especially in healthcare and logistics sectors that involve significant movement of tangible products, which enable them to provide their respective services. For example, healthcare facilities involve substantial movement of health care equipment, beds, patients, medicines, and healthcare staff. The collective result of generating and using big data in service supply chains based on industry 4.0 technologies would be reduced congestion, resilience, and cost-efficiency with improved collaboration and trust among stakeholders. Industry 4.0 technologies share a lot in common, and hence they would depend on each other for leveraging big data. The crucial aspect that matters the most is their transformative power in improving service operations, such as customer service, sales, marketing, and human resources. The insights derived from big data can help address crucial issues in service operations such as conflicting information among stakeholders, redundant information, lack of traceability of resources, inadequate transparency, high customer acquisition cost (CAC), low customer lifetime value (CLTV), increased customer churn, etc.

4 Conclusion

This chapter contributes by highlighting the role of contemporary Industry 4.0 technologies in creating big data. First, the role of blockchain was elaborated, which is poised to play the role of foundation technology to store big data produced by the interactions of other Industry 4.0 technologies. Second, we examined the role of integrative applications of blockchain, IoT, Fog computing and AI in creating big data. The decisions derived from a dataset are as reliable as the data itself. The data stored on blockchain ensures that the data stored on it remains trustworthy, immutable, and traceable. The sensors such as IoTs collect the data from the various stages of the supply chain, stored on blockchain. Fog computing takes care of the immediate data processing requirements at the ground level. The AI systems propose decisions based on the existing data and keep learning and improving the proposed decisions with new data. The AI output is also stored on blockchain, which forms the part of the current data.

Third, the chapter highlights the role of integrated applications of blockchain, SRS, 3DP, and geotagged data in creating big data. The AI decisions can guide swarm robotic systems, and blockchain can help them credibly communicate with each other.

The AI systems can also potentially help improve the designs used for 3DP. Storing 3DP designs on blockchain, apart from saving time and reducing costs, would ensure that their original creator is incentivized, eliminating the non-value-adding intermediaries. The geotagged data stored on blockchain can help law enforcement agencies and the general population at large. In this manner, the interactions of industry 4.0 technologies would create structured and unstructured data in large volumes with high velocity and huge variety. The chapter finally highlights the managerial and policy implications of big data from service operations management perspective.

There can be several areas for future research on this subject. New optimization models can be developed to explore the potential of big data in service supply chains. Additionally, more research is required to examine the suitability of 3DP for the manufacturing of non-perishable equipment [30]. Many firms have made considerable investments in collecting and storing data, and it is found that less than 0.5% of the data has ever been analyzed. Hence, context-specific expert systems and decision support systems are required to generate useful insights from big data [2]. As the industry 4.0 technologies work in tandem with each other, it will be helpful to explore the kind of blockchain (public or private) that should be used and the type of access required by various stakeholders. Adopting industry 4.0 technologies would involve societal and legal aspects to be considered by adopting organizations, which is not yet studied in the extant literature. Though blockchain research has attracted much attention, we find very few studies of the post-adoption effects of industry 4.0 technologies on service operations management [38]. The expanse of the types of the service industry is itself very huge, and hence we can imagine the number of studies that can be carried out in this context. The variation in geography (such as India, US, Europe, etc.) and research methods (such as case studies, cross-sectional studies, longitudinal studies, modeling-based studies, etc.) would further enhance the variety of studies that can be conducted.

References

1. Alexopoulos, K., Makris, S., Xanthakis, V., Sipsas, K., & Chryssolouris, G. (2016). A concept for context-aware computing in manufacturing: The white goods case. *International Journal of Computer Integrated Manufacturing, 29*(8), 839–849.
2. Baryannis, G., Validi, S., Dani, S., & Antoniou, G. (2019). Supply chain risk management and artificial intelligence: State of the art and future research directions. *International Journal of Production Research (Taylor & Francis), 57*(7), 2179–2202.
3. Beltagui, A., & Candi, M. (2018). Revisiting service quality through the lens of experience-centric services. *International Journal of Operations & Production Management*, Emerald Publishing Limited.
4. Bloomberg. (2017). *Why IBM Is Investing in Blockchain Technology?* Retrieved February 22, 2018, from https://www.youtube.com/watch?v=J5CD29LaW7c.
5. Brennan, C., & Lunn, W. (2016). *Blockchain: The trust disrupter*. Credit Suisse Securities (Europe) Ltd.
6. Charles, V., & Emrouznejad, A. (2019). Big data for the greater good: An introduction. In A. Emrouznejad & V. Charles (Eds.), *Big data for the greater good* (pp. 1–18). Springer International Publishing.

7. Cohen, M. C. (2018). Big data and service operations. *Production and Operations Management, 27*(9), 1709–1723.
8. Daniels, J., Sargolzaei, S., Sargolzaei, A., Ahram, T., Laplante, P. A., Amaba, B., Laplante, P. A., et al. (2018). The internet of things, artificial intelligence, blockchain, and professionalism. *IT Professional, 20*(6), 15–19.
9. Emrouznejad, A., & Marra, M. (2016). Big data: who, what and where? Social, cognitive and journals map of big data publications with focus on optimization. In A. Emrouznejad (Ed.), *Big data optimization: Recent developments and challenges* (pp. 1–16). Springer International Publishing.
10. Feng, Q., & Shanthikumar, J. G. (2018). How research in production and operations management may evolve in the era of big data. *Production and Operations Management, 27*(9), 1670–1684.
11. Ferrer, E. C. (2018). The blockchain: A new framework for robotic swarm systems. In *Proceedings of the Future Technologies Conference* (pp. 1037–1058). Springer.
12. GSS. (2021). Bitcoin, blockchain and GIS could change the world – Geospatial Solutions: Geospatial Solutions. Retrieved April 19, 2021, from http://geospatial-solutions.com/bitcoin-blockchain-and-gis-could-change-the-world/.
13. Haenlein, M., & Kaplan, A. (2019). A brief history of artificial intelligence: On the past, present, and future of artificial intelligence. *California Management Review (SAGE Publications Sage CA: Los Angeles, CA), 61*(4), 5–14.
14. Halemah, N. A., & Qayem, A. (2021). *Fields of geo-data and blockchain.*
15. Holmström, J., & Partanen, J. (2014). *Digital manufacturing-driven transformations of service supply chains for complex products.* An International Journal (Emerald Group Publishing Limited).
16. Junco, P. (2017, November). Accelerating the adoption of enterprise blockchain | Blog | Microsoft Azure. Retrieved February 22, 2018, from https://azure.microsoft.com/en-in/blog/accelerating-the-adoption-of-enterprise-blockchain/.
17. Karamchandani, A., Srivastava, S. K., Kumar, S., & Srivastava, A. (2021). Analysing perceived role of blockchain technology in SCM context for the manufacturing industry. *International Journal of Production Research* (Taylor & Francis). https://doi.org/10.1080/00207543.2021.1883761.
18. Karamchandani, A., Srivastava, S. K., & Srivastava, R. K. (2019). Perception-based model for analyzing the impact of enterprise blockchain adoption on SCM in the Indian service industry. *International Journal of Information Management*. https://doi.org/10.1016/j.ijinfomgt.2019.10.004.
19. Kshetri, N. (2017). Will blockchain emerge as a tool to break the poverty chain in the Global South? *Third World Quarterly, 38*(8), 1710–1732.
20. Li, L. (2018). China's manufacturing locus in 2025: With a comparison of 'Made-in-China 2025' and 'Industry 4.0.' *Technological Forecasting and Social Change, 135*, 66–74.
21. Lopes, V., & Alexandre, L. A. (2018). An overview of blockchain integration with robotics and artificial intelligence. ArXiv Preprint arxiv:1810.00329.
22. Lunden, I. (2012). 'Find Friends Nearby': Facebook's new mobile feature for finding people around you [updated]. *TechCrunch*. Retrieved April 19, 2021, from https://social.techcrunch.com/2012/06/24/friendshake-facebooks-new-mobile-feature-for-finding-people-nearby-and-a-highlight-killer/.
23. Mandolla, C., Petruzzelli, A. M., Percoco, G., & Urbinati, A. (2019). Building a digital twin for additive manufacturing through the exploitation of blockchain: A case analysis of the aircraft industry. *Computers in Industry, 109*, 134–152.
24. Nakamoto, S. (2008). *Bitcoin: A peer-to-peer electronic cash system.*
25. OmniSci. (2021). *What is fog computing? Definition and FAQs | OmniSci.* Retrieved April 20, 2021, from https://www.omnisci.com/technical-glossary/fog-computing.
26. Qin, J., Liu, Y., & Grosvenor, R. (2016). A categorical framework of manufacturing for industry 4.0 and beyond. *Procedia Cirp, 52*, 173–178.

27. Rabah, K. (2018). Convergence of AI, IoT, big data and blockchain: A review. *The Lake Institute Journal, 1*(1), 1–18.
28. Radianti, J., Dokas, I., Boersma, K., Noori, N. S., Belbachir, N., & Stieglitz, S. (2019). Enhancing disaster response for hazardous materials using emerging technologies: The role of AI and a research agenda. In J. Macintyre, L. Iliadis, I. Maglogiannis, & C. Jayne (Eds.), *Engineering applications of neural networks* (pp. 368–376). Springer International Publishing.
29. RND. (2020). *Additive manufacturing market analysis*. Retrieved April 15, 2021, from https://www.reportsanddata.com/report-detail/additive-manufacturing-market#:~:text=The%20global%20Additive%20Manufacturing%20Market,growth%20opportunities%20for%20the%20industry.
30. Rodríguez-Espíndola, O., Chowdhury, S., Beltagui, A., & Albores, P. (2020). The potential of emergent disruptive technologies for humanitarian supply chains: The integration of blockchain, Artificial Intelligence and 3D printing. *International Journal of Production Research (Taylor & Francis), 58*(15), 4610–4630.
31. Rost, M., Cramer, H., & Holmquist, L. E. (2012). Mobile exploration of geotagged photographs. *Personal and Ubiquitous Computing (Springer), 16*(6), 665–676.
32. Scott, C. (2018). The value of blockchain for securing 3D printed intellectual property. *3DPrint.Com | The Voice of 3D Printing / Additive Manufacturing*, 5 September. Retrieved April 20, 2021, from https://3dprint.com/224165/blockchain-3d-printed-ip/.
33. Sen, M. (2021). Air India servers hacked; passport, credit card details of 45L customers leaked. *Mint*, 21 May. Retrieved May 21, 2021, from https://www.livemint.com/companies/news/air-india-servers-hacked-passport-credit-cards-details-leaked-airline-says-11621613402288.html.
34. Tama, B. A., Kweka, B. J., Park, Y., & Rhee, K.-H. (2017). A critical review of blockchain and its current applications. In *2017 International Conference on Electrical Engineering and Computer Science (ICECOS)* (pp. 109–113). IEEE.
35. Viriyasitavat, W., Da Xu, L., Bi, Z., & Sapsomboon, A. (2018). Blockchain-based business process management (BPM) framework for service composition in industry 4.0. *Journal of Intelligent Manufacturing*. https://doi.org/10.1007/s10845-018-1422-y.
36. Wang, K., & Safavi, A. (2016). Blockchain is empowering the future of insurance. *TechCrunch*. Retrieved February 20, 2018, from http://social.techcrunch.com/2016/10/29/blockchain-is-empowering-the-future-of-insurance/.
37. WEF. (2015). Deep shift: Technology tipping points and societal impact. *World Economic Forum*. Retrieved April 13, 2019, from https://www.weforum.org/reports/deep-shift-technology-tipping-points-and-societal-impact/.
38. Wen, Z., Liao, H., & Emrouznejad, A. (2021). Information representation of blockchain technology: Risk evaluation of investment by personalized quantifier with cubic spline interpolation. *Information Processing & Management, 58*(4), 102571.
39. Wiki. (2021). Geotagging. *Wikipedia*, 27 February. Retrieved 19 https://en.wikipedia.org/w/index.php?title=Geotagging&oldid=1009256465.
40. Wright, A., & De Filippi, P. (2015). *Decentralized blockchain technology and the rise of lex cryptographia*.
41. Xu, L. D., Xu, E. L., & Li, L. (2018). Industry 4.0: State of the art and future trends. *International Journal of Production Research, 56*(8), 2941–2962.

Amit Karamchandani is a faculty of Operations Management at Great Lakes Institute of Management, Gurgaon. He has a Ph.D. from Indian Institute of Management, Lucknow (India). His research interests include blockchain, sustainability, decision support systems, supply chain innovations and technology adoption. He is a graduate in Mechanical Engineering from IIT (BHU). He has seven years of industrial experience with Bharat Heavy Electricals Limited (BHEL) in roles such as Quality Assurance Engineer and Senior Design Engineer. He worked in several projects at BHEL aimed at business process digitization. He has published in peer-reviewed journals such as International Journal of Information Management (IJIM), International Journal of Production Research (IJPR) and International Journal of Technology, Intelligence and Planning (IJTIP). He has presented his research in refereed international conferences such as POMS, ISDSI and SOM.

Samir K Srivastava is a Professor and currently Dean (Research) at the Indian Institute of Management Lucknow (India). He is a prolific researcher. His work has about eleven thousand citations in Google Scholar with a g-Index of 104, more than three thousand citations in Scopus with an H-Index of 12 and more than two thousand citations in Publons. He serves on the Editorial Board of several International Journals and is an active referee for many journals of repute. His name figures among the top 2% scientists in the world based on career-long impact in the list created by Stanford University and his biographic profile appears in Marquis Who's Who. Prof Samir has been bestowed with many recognitions and awards. These include four Best Paper awards, Emerald/IAM Indian Management Research Fund Award 2015 and Prof. Manubhai M. Shah Memorial Award for Excellence in Commerce and Business Management (MMSMA) 2015.

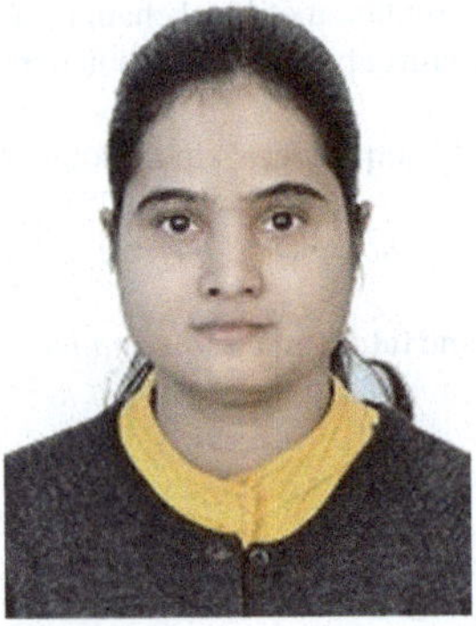

Abha is a doctoral student in the area of Operations management at Indian Institute of Management, Lucknow (India). Her research interests include sustainability, public procurement and application of emergent technology areas in energy sector. She is a postgraduate in management from S.P. Jain Institute of Management and Research, Mumbai. She has five years of experience in public procurement and six years of industrial experience with Bharat Heavy Electricals Limited (BHEL). She has presented her work in refereed international conferences such as POMS and SOM.

Blockchain for Disaster Management

Kyle Hunt and Jun Zhuang

Abstract Natural disasters, acts of terrorism, and man-made disasters have affected our planet in an unpredictable way. These events prove to cost nations around the world billions of dollars annually, and the destruction that they bring is merely impossible to forecast. Disaster victims may not receive timely help due to poor coordination among responding agencies, late disaster response, and inadequate distribution of resources such as food, water, medical supplies, and transportation. This chapter explores the application of blockchain in disaster management to address these issues, and in turn, reduce the loss of time, money, and human life. Disaster management is a unique form of operations management in which emergency service providers and humanitarian agencies provide resources (e.g., medical services, logistical services, food, shelter, etc.) to support the public before, during, and after disasters. As disaster management operations unfold, massive amounts of data are produced as a result of the service transactions and aid efforts that take place to help support and restore society. The adoption of a blockchain approach in disaster management can reduce corruption, facilitate and accelerate the formation of partnerships between disaster relief agencies, deliver verified and timely disaster communications, improve the allocation of vital resources, and enable secure access to the valuable data that is produced during response and recovery operations. We offer a comprehensive blockchain framework for disaster management which includes the following stakeholders in its design: governments, residents, telecommunication providers, shelter providers, food service providers, medical service providers and suppliers, transportation providers, and non-governmental relief organizations.

Keywords Blockchain technology · Big data · Disaster response · Disaster recovery · Operations management

K. Hunt (✉) · J. Zhuang
Department of Industrial and Systems Engineering, University at Buffalo, New York, USA
e-mail: kylehunt@buffalo.edu

A. Emrouznejad and V. Charles (eds.), *Big Data and Blockchain for Service Operations Management*, Studies in Big Data 98,
https://doi.org/10.1007/978-3-030-87304-2_10

1 Introduction

Many people around the world are affected by disasters every year; whether it be a natural disaster in the form of a hurricane, an act of terrorism, or a man-made disaster such as a nuclear leak or an oil spill. An immediate effect of disasters in the modern era is the displacement of huge populations. According to the World Health Organization, close to 160,000,000 people around the world are affected by natural disasters every year (including earthquakes, tsunamis, volcanic eruptions, landslides, hurricanes, floods, wildfires, heat waves, and droughts) (World Health Organization, 2019). In the wake of disasters, many people are forced to leave their homes and communities in order to seek safety and vital resources, which can greatly disrupt their accessibility to healthcare, education, and income.

Without proper disaster management efforts, death tolls may rise even after the immediate threat has passed, as people are without access to medical supplies, nutritional necessities, and a livable residence. According to the International Federation of Red Cross and Red Crescent Societies (IFRC), disaster management is defined as "the organization and management of resources and responsibilities for dealing with all humanitarian aspects of emergencies, in particular, preparedness, response and recovery in order to lessen the impact of disasters" (International Federation of Red Cross and Red Crescent Societies. (2019)). In other words, disaster management is a unique form of operations management in which emergency service providers and humanitarian agencies provide resources (e.g., medical services, logistical services, food, shelter) to support the public before, during, and after disasters. Disaster management includes the development of disaster recovery plans and the implementation of such plans across all emergency events. Due to the extreme costs associated with disasters, such as the loss of life, destruction of land and properties, and disruptions to business and society, it is of extreme importance to continuously improve and adapt disaster management strategies to meet the demands of growing populations and improving technologies. Many agencies view and facilitate disaster management in four stages; namely, mitigation, preparedness, response, and recovery (Coppola, 2006). Improving functions within any of these phases can save billions of dollars, while also creating a safer pre- and post-disaster environment. Although the current disaster management approaches around the world are significant in reducing the effects from disasters, many problems still arise when disasters strike.

In recent years, a plethora of research has studied the spread of misinformation during disasters, and the lack of timely, moderated, and verified emergency communications (Hunt et al., 2020a; b; c; Wang and Zhuang, 2018). These faulty communications can directly change the public's opinions and beliefs in some of the most vulnerable of times, and the research indicates a need for more efficient and reliable emergency communications. Additionally, the allocation of food, water, medical services, and other vital resources following disasters are issues that have been researched widely. The works of (Sabbaghtorkan et al., 2020; Galindo and Batta, 2013; Altay and Green, 2006; Hoyos et al., 2015) have focused on issues in disaster management from an operations research and management science perspective. These studies address problems in resource allocation, emergency logistics, and facility location, along with developments in the broad fields of disaster mitigation, preparedness, and response. In Sobel and Leeson, (2006), the authors studied the United States Federal Emergency Management Agency's (FEMA's) role following Hurricane Katrina, and found that their efforts were inefficient. FEMA took charge of the distribution of relief supplies, and to be effective, FEMA had to know what was needed (and in what amounts), who needed it, and when and where it was needed. Along with the demand side, FEMA also had to know the supply side, such as who had the resources and where they were located. These broad responsibilities that FEMA undertook led to an ineffective disaster response (Sobel and Leeson, 2006). Further research has studied corruption following natural disasters, especially in the public sector. Yamamura, (2014) found that corruption is increased with more considerable and frequent disasters. In Sobel and Leeson, (2006), the authors provide that state and local officials, like other public agents, have "an incentive to request a larger than needed amount of resources" when they are not the ones assuming the cost. Like other government agencies, this is generally done by exaggerating the state of the problem at hand (Sobel and Leeson, 2006).

As studied in Akter and Wamba, (2019), some of the major issues that society faces during disasters are caused by the massive amounts of data that exist during response and recovery operations (e.g., aid distribution, emergency logistics). These data may reflect demands for services and commodities, supply chain transactions, emergency communications, service transactions, requests for rescue, among others. In order to have an effective disaster response and recovery, these data must be processed and analyzed efficiently to ensure that resources and services are being allocated properly. The big data Ishikiriyama and Gomes, 2019 that is generated can hinder the performance of traditional data storage and processing systems, while limiting the potential for swift collaboration among responding agencies (Akter and Wamba, 2019). Data access and data security are critical considerations when addressing the problems that the big data era has introduced within disaster management.

Limited research has studied the higher level need for a better system to efficiently and securely facilitate the transfer of funds, goods, services, and emergency communications during disasters. On the same note, blockchain technology has recently received an increased attention by researchers and practitioners regarding its potential to benefit disaster management (Hunt et al., 2021). Blockchain has the potential to be more effective than the traditional systems that are deployed in disaster management operations, especially in storing big data, fostering trust and collaborations among stakeholders, and improving logistical processes that enable effective relief operations. This chapter focuses on blockchain's potential to notably improve disaster management through a multifaceted approach. By proposing a blockchain framework consisting of governmental agencies, private agencies, non-governmental organizations, and residents, this chapter offers a novel and detailed framework that can be applied throughout the world to improve and aid in disaster management.

In Sect. 2, we introduce existing frameworks for blockchain in humanitarian aid and disaster management. We include the field of humanitarian aid given that humanitarian operations and disaster operations are very similar in that they bring together multiple stakeholders to provide support and aid for a vulnerable community. In Sect. 3, we present our approach for improved disaster management via a robust blockchain framework. In Sect. 4, we present the conclusions and limitations of this work, along with the promising future research directions.

2 Existing Blockchain Frameworks

2.1 Current Frameworks for Blockchain in Humanitarian Aid and Disaster Management

Here we present state-of-the-art frameworks for blockchain in humanitarian aid and disaster management. Most notably, we (i) cover a deployed blockchain solution in the humanitarian sector that has received significant global attention, (ii) discuss how the United States government has recently considered adopting blockchain in their disaster management approach, and (iii) discuss a novel blockchain framework for disaster management which brings together multiple key stakeholders. While the literature we cover in this section provides a strong background for the application of blockchain technology in humanitarian aid and disaster management, we note that there exists a greater body of literature in this domain. For a more detailed and complete overview of blockchain technology in humanitarian operations management, we direct our readers to the literature review offered in (Hunt et al., 2021).

Over the last few years, the world has witnessed progress in the research and application of blockchain technology in humanitarian aid projects, and many of these works offer insights that can be translated directly to disaster management. The World Food Programme (WFP) is a global humanitarian organization who delivers food assistance to refugees and those affected by disaster situations. Starting in 2017, the WFP launched "Building Blocks," which is a blockchain enabled project that facilitates cash transfers to beneficiaries while protecting their personal data (World Food Programme, 2019). In 2018, Building Blocks was used to transfer the equivalent of 1.6 billion USD, while eliminating 98% of third-party transaction fees (World Food Programme, 2019). Money is transferred directly from the WFP to participants, and the project runs on a private, permissioned blockchain. Due to this initiative, over 100,000 people in Jordan refugee camps can purchase groceries by simply scanning their iris (World Food Programme, 2019). In Ko and Verity, (2016), the authors identify different ways in which blockchain can improve humanitarian operations, including protected data sharing, improved logistics, and identity verification. The authors also mention the use of blockchain for distributing funding, but the applied work of World Food Programme, (2019) is much more comprehensive. Ko and Verity, (2016) described the needs for future research into the applications and use cases of blockchain in global humanitarian operations. A more recent work from

Zwitter and Boisse-Despiaux, (2018) expands on blockchain's potential in humanitarian action and development. The authors propose the use of smart contracts to facilitate automated funding, and they claim that utilizing blockchain technology and smart contracts will result in more efficient and less politically-based funding. Smart contracts can also be used to improve supply chains in development and humanitarian projects. The authors provide that donations can be digitally marked for certain services or to reach certain communities, mitigating the potential of corruption and fraud Zwitter and Boisse-Despiaux, (2018). In McIsaac et al., (2019), the authors explore the issues of identity management and payment systems in disaster and refugee relief operations. Their research offers insights into the use of blockchain to prevent fraud, offer cryptocurrency to aid recipients, and provide a single system to validate the identification of refugees and disaster victims in order to allow them access to the relief supply chain.

Few researchers and practitioners have developed high level frameworks for integrating blockchain in disaster management. In 2018, the United States Defense Logistics Agency (DLA) hosted a meeting in Philadelphia regarding the use of blockchain to improve disaster management. The meeting was orchestrated in direct response to the Hurricane Maria recovery operations (Dwyer, 2018). As a result of the 2017 hurricane season, the DLA provided over 50,000 power poles, 88,000,000 meals, and over 1,000 generators in partnership with FEMA. All of the associated logistical and service processes were tracked through systems and databases managed by different agencies, and as a result, these agencies had to synchronize their data with each other to ensure that their information was up to date (Dwyer, 2018). The meeting resulted in plans for the DLA and other agencies to conduct research and increase knowledge in the possible applications of blockchain to assist in the information sharing and supply chain efforts following disaster events. In 2019, potential applications for blockchain in disaster management were released by the DLA in a technical report (Keenaghan et al., 2019). Their work implies that blockchain may enhance communications, improve logistics planning, ensure compliance with regulatory applications, support the monitoring of third party deliveries, and reduce the duplication of relief efforts. Their work also highlights that each of these improvements can provide significant savings in time and dollars. After Hurricane Maria alone, there were over 36,000,000 power grid items, 20,000 handheld radios, 344,000 water jugs, 12,000 cots, and 14,000 sandbags that were distributed to the citizens and regions in need. The inclusion of a blockchain network to facilitate these transactions and logistics would improve relief efforts, and the United States government is continuing to explore blockchain for potential adoption by federal agencies and their partners (Keenaghan et al., 2019).

The most comprehensive study regarding the applications of blockchain in disaster management is that of (Mohan, 2018). In this work, the author proposes a blockchain network coupled with a mobile application to connect citizens, service providers, and government agencies during crisis situations. Although residents and shelters are considered as stakeholders in disaster management, they are not included in the final proposed blockchain network design (Mohan, 2018). In this proposed framework, the government would be responsible for leading the initiative to bring all key players into the network, and also for overseeing all transactions within the network. The framework requires the use of cryptocurrencies and mobile wallets as opposed to traditional currency such as the USD. The author provides that the blockchain network and mobile application would cost an estimated 100,000 USD per month during disaster periods, and 40,000 USD per month during non-disaster periods.

2.2 *Gaps in Current Frameworks*

In the work of McIsaac et al., (2019), the authors propose a worldwide blockchain relief framework, which is infeasible to implement in the current age, as there is no central organization to facilitate the adoption and implementation of the proposed network around the world. Although the WFP is proving to extend their work within different countries, their system is still localized and focused on specific cohorts of affected residents World Food Programme, (2019). Neither McIsaac et al., (2019) nor World Food Programme, (2019) offer guidelines or ideas on how to train residents and agencies on the use of blockchain. Additionally, to the best of our knowledge, no research has taken general disaster communications into account, such as emergency declarations that are released to affected residents, and the use of social media and other online platforms to disseminate credible disaster-related updates and news. Likewise, no prior work has integrated non-governmental relief organizations (NGOs), such as the American Red Cross, into the blockchain networks proposed. Previous research from Mondal et al., (2015) has provided the significant roles that NGOs play in disaster management and relief, and these agencies should be considered as key players in the blockchain network. NGOs can offer their services throughout the network, and also identify where help is needed in order to allocate their services and resources accordingly.

Requiring the government to be responsible for the complete oversight of the blockchain network, as proposed in Mohan, (2018), is not an efficient approach based on previous problems the government has faced in holding too much responsibility post-disaster Sobel and Leeson, (2006). The government should have power in disseminating and approving monetary transactions, as well as adding new members to the network, but other service and resource related transactions can be approved through delegated agencies and organizations. Lastly, the process for residents and agencies to register in the network is not addressed in previous research to ensure the security of the nodes in the blockchain network. In Table 10.1, we provide details covered in previous research, along with the comprehensive solution that our framework offers. We contribute to the existing literature in the following ways: (i) we include residents and NGOs in the blockchain framework, (ii) we discuss and address the issue of secure registration in the blockchain network, (iii) we provide general strategies to support the adoption of a blockchain solution within disaster management, while exposing the realistic need for training and marketing, and (iv) we relax the need for the government to have total oversight of the disaster management processes by enabling them to delegate responsibilities to other nodes in the network.

Table 10.1 A comparison between previous research and this work regarding the details covered for the application of blockchain in disaster management. The works from World Food Programme (2019); Ko and Verity (2016); Zwitter and Boisse-Despiaux (2018); McIsaac et al. (2019) are concentrated on the more general field of humanitarian aid, but we employ their contributions exclusively for disaster management

Frameworks for Blockchain in Disaster Management

Details covered	World Food Programme 2019)	Ko and Verity 2016)	Zwitter and Boisse-Despiaux 2018)	McIsaac et al. 2019)	Keenaghan et al. 2019)	Mohan 2018)	This work
Supply Chain/Resource Logistics		x	x	x	x	x	x
Financial Transactions	x	x	x	x	x	x	x
Secure Registration of Nodes							x
Identity Verification	x	x	x	x		x	x
Strategy for Adoption							x
Smart Contracts			x		x	x	x

3 A Novel Blockchain Framework for Disaster Management

3.1 Framework Overview

We propose a consortium, permissioned blockchain solution (Zheng et al., 2018) which will empower government agencies, NGOs, private organizations, and residents to transact and communicate efficiently in the disaster management process. Since all transactions are approved within the network, and visible by the entire network, information cannot be falsified. This allows blockchain to provide a secure environment, which leads to trust and supports governance and accountability (Casey and Vigna, 2018). Blockchain's shared distributed ledger ensures that transactions and data reach all parties as soon as they are created, leading to the instantaneous transmission of vital information in the context of disaster management. The main stakeholders in the proposed blockchain network for disaster management are the government, telecommunication providers, shelter providers, food service providers, medical service providers and suppliers, transportation providers, non-governmental relief organizations, and residents, as provided is Fig. 10.1. These stakeholders are selected based on their central roles in disaster management operations, as studied in previous research (Coppola, 2006; Mohan, 2018; Mondal et al., 2015; Panesir, 2018).

Within the consortium blockchain system, the government will be in charge of creating the set of nodes who will be responsible for validating specific transactions (e.g., shelter providers validate a new resident joining their shelter, NGOs validate the receipt of resources from food and medical service providers). Smart

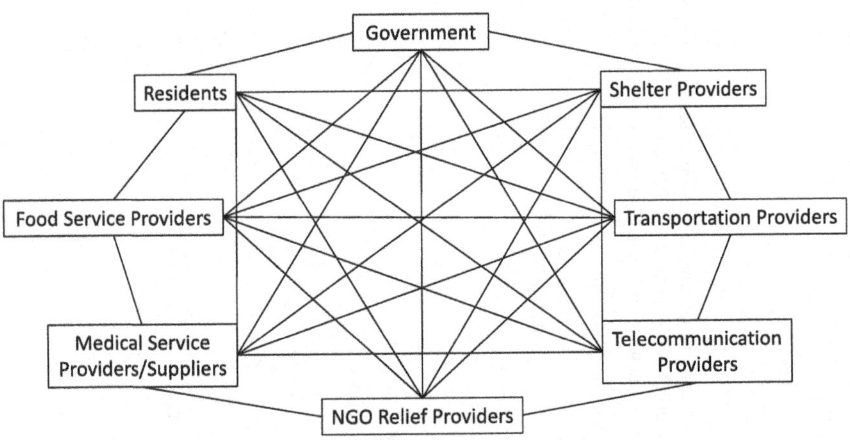

Fig. 10.1 SEQ Figure * ARABIC 1 The key players of the blockchain network. The information in the blockchain is distributed to all nodes, and the government grants permission for nodes to enter the consortium

contracts are used for the automatic disbursement and payment for goods and services between the government and all other key players (Zheng et al., 2018). For example, resources and funds can be automatically distributed to agencies (e.g., shelters, food providers, medical servicers, etc.) based on their activity and needs represented in the ledger. Through its ability to foster trust, enable collaborations among relief agencies, provide a secure and auditable transaction ledger, and enable real-time information sharing, blockchain has the potential to significantly improve disaster operations. Our proposed framework is motivated by the vast potential for blockchain technology to benefit current disaster management approaches (Zwitter and Boisse-Despiaux, 2018; McIsaac et al., 2019; Dwyer, 2018; Hunt et al., 2021; Farooq et al., 2020), along with the perpetual need for such improvements. This framework is novel in its approach and detail throughout the various components, but we acknowledge the significant work of previous researchers in their initial approaches and fundamental frameworks (World Food Programme, 2019; Zwitter and Boisse-Despiaux, 2018; Mohan, 2018; Farooq et al., 2020; Panesir, 2018). Their work serves as an inspiration in the following solution.

3.2 *Key Players*

3.2.1 Government and Residents

In this framework, the government will be responsible for the development (or outsourcing of the development) of a web and mobile compatible application with the blockchain network functioning as the back end of the service. The government should also be responsible for facilitating smart contracts (Zheng et al., 2018) between their agencies and service providers, where the government supplies funding and/or resources based on the demands from external agencies. For example, since FEMA is currently the backbone of U.S. disaster management, most of the partnerships and smart contracts should be formed and executed by their leadership and administration. If an agency requests governmental assistance without a smart contract in place, the funding and/or resources can still be approved and recorded on the ledger. On the mobile application, residents will apply for registration by providing personal information such as their name, age, birth date, and national identification number (such as a Social Security number in the U.S.). This information will be approved and validated by governmental agencies, such as the U.S. Social Security Administration, to create a secure blockchain network. Following approval, a user will be able to access the mobile application and blockchain network. Parents or guardians should be responsible for registering their children, and when children reach a certain age, such as 18 or 21 in the U.S, they should then be responsible for their own registration. Additional user information, such a medical records (Azaria et al., 2016), can be added at the users' discretion after they have been added to the network. This will allow medical service providers and suppliers to identify the medical needs of affected citizens. Although registration should be completed pre-disaster, residents must also be able to apply during and after disasters. All residents will only need to be granted permission to the blockchain network once, and this will not need to be renewed. Therefore, if their personal information is ever attempted to be validated again after they have already been granted permission to the network, it will flag the system and require investigation, similar to the credit card fraud process. All external agencies will be added to the blockchain network by the government via a request made from the agency to be added, or the government taking the initiative to partner with the agency. The government should then delegate validation rights to certain users within agencies so that non-financial transactions can be facilitated securely and efficiently. With the help of this application, the government can analyze the conditions and needs of the disaster affected citizens and organizations and provide timely communications, funding, and resources. Following a disaster, all of the information and transactions stored in the blockchain can be analyzed by the government using big data approaches (Charles and Emrouznejad, 2019), and the insights gathered can support the improvement of future disaster management operations.

3.2.2 Telecommunication Providers

The telecommunication providers are one of the most important players in advanced disaster management, as they offer the communication networks and infrastructure to make the blockchain service functional before, during, and after disasters (Patricelli et al., 2009). Using the blockchain-enabled application, residents or agencies can request to locate a missing person. If a missing person request is made, the telecommunication providers can identify the GPS location or IP address of the individual based on the coordinates of the device that is hosting their blockchain network connection. If the missing person is registered in a shelter, telecommunication providers can resolve the missing person request, signifying that the person is safe. If the GPS location shows that the individual is in a high risk area, agencies can request for immediate rescue efforts, and the responding rescue agency or person can update the blockchain to notify the network that they are pursuing the missing person(s). This information will be recorded in real time on the blockchain, and all nodes within the network will have direct updates and access to the information. Additionally, the telecommunication companies can partner with trusted weather and news outlets in order to get the most timely emergency communications and updates. The updates and emergency communications from weather and news outlets can be immediately broadcasted to the network, so that all key stakeholders have the information they need to make informed decisions.

Given the significant use of social media during disasters and its vast potential to benefit emergency management efforts (Houston et al., 2015), emergency information from social media platforms can be validated and added to the network. For example, if the telecommunication providers and/or their partner agencies identify important emergency communications on social media platforms, this information can be validated and added to the network. This will significantly lower the chances of misinformation being spread to the network. Moreover, FEMA has offered rumor control pages for many recent hurricanes, such a Hurricane Florence in 2018. These pages, along with similar efforts from agencies around the world, should also be offered on the application and network in order to offer more sources of validated emergency information. All of these measures can ensure the safety of the information and communications that residents and organizations receive online.

3.2.3 Shelter Providers

An individual, a group of individuals, or an organization that wants to provide their services by offering shelter to affected people can apply to be registered on the application before, during, or after disasters. Upon being granted permission to join the blockchain network, they can give information about the type of services they are offering. Examples would include information such as the number of beds they have, the types and amounts of food, and the type of medical services available (if any). The application can then match the people seeking shelter to nearby shelters according to their need(s), and these actions will be reflected and updated in the blockchain. Vacancy information can be automatically updated on the network as new people check into and out of shelters. This will ensure that the affected population has knowledge about the availability of services. Shelters can directly communicate with the government regarding needed funds, or with the other service providing organizations regarding needed resources. Smart contracts should be established and utilized to facilitate such transactions. Since the communication would be direct and without involvement from a third party, needs will be fulfilled efficiently.

3.2.4 Food Service Providers

An individual or organization who is willing to provide food services to the affected community can apply to be registered on the application. Upon being added to the network, they can provide information to the network regarding where they are located and what kind of food supplies that they have in stock. The application will help people and/or shelters who are seeking food supplies by connecting them directly with the providers. The blockchain network helps secure these relations by having the delivery of food published to the network, and subsequently validated to ensure the resources are being received and no fraudulent activity occurs. Likewise, when a delivery has been validated, the food stock of the service provider can be automatically updated so other nodes have accurate information regarding what is available. Any provider who has the inventory to provide the requested food can directly be in contact with the people in need. The food service providers can also request for food and ingredients from the government or other agencies in order to have a continually stocked inventory.

3.2.5 Medical Service Providers and Suppliers

As medical service providers and suppliers are added to the network, affected residents in need of medical care or medications can request services directly through the network. Medical service providers can update the information regarding medications and available personnel, and can appropriately deploy personnel to the requested area. The medical service providers can also work closely with the telecommunication providers to work on the basis of geographical priority. Since providers and suppliers are validated by the government, the medical personnel can be trusted. Likewise, the supply chain of the medications and medical supplies will be tracked on the blockchain, ensuring that no fraudulent transactions occur.

3.2.6 Transportation Providers

Transportation providers can offer information regarding their coverage and locations supported. They can also offer information regarding the type of transportation they can provide (and capacity) and publish such data on the application, along with the people they have moved from the disaster site to a shelter. The shelters and residents can contact these organizations directly through the application and request for the necessary transportation services. If residents are being transported to shelters, the shelters (or residents themselves) can validate the transportation block to signify that they have safely reached the destination. Likewise, if medical providers are being transported to a residence or business structure, the medical personnel can validate that they have reached their destination.

3.2.7 Non-Governmental Relief Organizations

The role of NGOs during a disaster is to support quick response and to help as many affected citizens as possible with their resources. NGOs are involved in the rescue and relief operations by arranging shelters and providing medical, food, transportation, and communication resources (Mondal et al., 2015). Due to the lack of transparency and efficient communication during a disaster, it is difficult for NGOs, as well as governmental organizations, to find the right service providers in times of need. If the NGOs decide to be a part of the blockchain network, they can respond to the victims' requests and provide direct information regarding the services and help they can provide. The proposed blockchain network also allows the other service providers to communicate and partner directly with the NGOs and work towards the common goal of efficient and effective disaster response.

3.3 Reaching Citizens and Providers

One of the largest unaddressed aspects of adopting a blockchain approach in disaster management is the training and information that would need to be disseminated to residents and agencies in order to ensure that all stakeholders understand how the application works, and how it aids their safety and efforts when disasters strike. Upon the development of the application and blockchain network, and after formal approval of the system, the government should begin training agencies and residents on the use of the proposed solution. These trainings should consist of how to register on the network, how to use the application's user interface, how the underlying framework works, and how this solution benefits all stakeholders. Trainings should be organized to take place in disaster prone areas at accessible locations, such as malls, gymnasiums, and other community gathering sites. Trainings should also be offered online to ensure that all residents have access to the information and resources that are needed for successful implementation. Simulations should be created showing the use of the application and blockchain network in a disaster situation in order to inform all potential users how the proposed framework will benefit them and support a safer and more efficient disaster management process. Trainings, simulations, video support, and documentation should be easily accessible for residents of different race and genders, and well as residents with disabilities, impairments, or illnesses. All materials should be offered in multiple languages to ensure accessibility among all residents. The application and framework should be broadly marketed in order to ensure its adoption around the nation. Via press conferences and news programs, many residents would be able to get the information needed in order to follow up with the respective trainings and registration.

4 Conclusions and Future Research Directions

4.1 Concluding Discussion

Blockchain allows organizations to perform services and publish records within the network. Upon validation, the records are reflected on the network as a block which cannot be modified. Since blockchain technology functions on a distributed ledger system, it ensures that the data and transactions reach all nodes as soon as they are created. This study proposed how blockchain technology, coupled with a mobile application, can be employed to provide an enhanced disaster management solution. While the application of blockchain technology has been studied in the operations management literature (Babich and Hilary, 2020), there is a limited literature focusing on how blockchain can be employed in disaster management, where affected residents are in need of emergency services and resources. The key players in our consortium, permissioned blockchain network are the government, telecommunication providers, shelter providers, food service providers, medical service providers and suppliers, transportation providers, non-governmental relief organizations, and residents. The gains in efficiency, transparency, and security obtained by using this blockchain framework have the potential to save lives and decrease financial losses and corruption. Instead of having a centralized disaster management approach, the proposed solution can decentralize the disaster response and recovery operations, while maintaining a secure and trustworthy system that is approved of and monitored by the government. The big data that is generated during disaster management operations can be securely stored and disseminated to stakeholders to enable the timely sharing of crisis information. All stakeholders can come together and communicate with each other, while recording all transactions and related data within the distributed blockchain network. Given this, all stakeholders have the most accurate and up to date information needed to make informed and safe decisions. The framework will deliver real time emergency communications and weather updates to users based on their locations, and also aid in the resolution of missing person requests, the effective allocation of resources, and the effective distribution of vital services such as rescue operations and medical expertise.

4.2 Limitations and Future Research Directions

The cost of developing and integrating the blockchain network and mobile application have not been studied in this work. In the related literature, there is no insight into how much money needs to be invested to hire the personnel to create and maintain the framework. Similarly, this study does not contain a cost analysis on the organization and implementation of trainings. Besides the financial investment, many employees and/or volunteers would be needed to train the public on the use of the blockchain system. In future research, these shortcomings should be addressed, and the framework should be tested in a simulated disaster scenario before being deployed in a real disaster scenario. Such simulations will provide an opportunity for researchers and disaster management stakeholders to analyze the performance and functionality of the framework. Furthermore, testing the framework in future research will allow for the iterative updating of the solution based on recognized shortcomings or opportunities for improvement. As discussed in Hunt et al., (2021), there is currently a lack of empirical evidence for the use of blockchain in disaster management. While few frameworks have been developed, such as the one we present here, there is a dearth of field deployments showing the utility that blockchain can offer stakeholders in the midst of a disaster event.

Future research should also focus on the integration of optimization models within the framework that we developed in this work to offer dynamic recommendations for the allocation of resources and services. As mentioned throughout this chapter, the blockchain solution would host a great deal of data which could be effectively used to drive decision support models. Such models would create an even more efficient framework, and assist providers in their decisions regarding response and resource disbursement.

References

Akter, S., & Wamba, S. F. (2019). Big data and disaster management: A systematic review and agenda for future research. *Annals of Operations Research, 283*(1), 939–959.

Altay, N., & Green, W. G., III. (2006). OR/MS research in disaster operations management. *European Journal of Operational Research, 175*(1), 475–493.

Azaria, A., Ekblaw, A., Vieira, T., & Lippman, A. (2016). Medrec: Using blockchain for medical data access and permission management. In *Proceedings of 2016 2nd International Conference on Open and Big Data* (pp. 25–30).

Babich, V., & Hilary, G. (2020). OM Forum—Distributed ledgers and operations: What operations management researchers should know about blockchain technology. *Manufacturing & Service Operations Management, 22*(2), 223–240.

Casey, M. J., & Vigna, P. (2018). In blockchain we trust. *MIT Technology Review, 121*(3), 10–16.

Charles V., Emrouznejad A. (2019). Big data for the greater good: An introduction. In A. Emrouznejad & V. Charles (Eds.), *Big Data for the Greater Good. Studies in Big Data* (Vol. 42). Springer.

Coppola, D. P. (2006). *Introduction to international disaster management*. Elsevier.

Dwyer III, J. (2018). Troop support event poses question: How and where can blockchain help? https://www.dla.mil/AboutDLA/News/NewsArticleView/Article/1720207/troop-support-event-poses-question-how-and-where-can-blockchain-help/

Farooq, M. S., Khan, M., & Abid, A. (2020). A framework to make charity collection transparent and auditable using blockchain technology. *Computers & Electrical Engineering, 83*, 106588.

Galindo, G., & Batta, R. (2013). Review of recent developments in OR/MS research in disaster operations management. *European Journal of Operational Research, 230*(2), 201–211.

Houston, J. B., Hawthorne, J., Perreault, M. F., Park, E. H., Hode, M., Halliwell, M. R., McGowen, S. E., Davis, R., Vaid, S., McElderry, J. A., & Griffith, S. A. (2015). Social media and disasters: A functional framework for social media use in disaster planning, response, and research. *Disasters, 39*(1), 1–22.

Hoyos, M. C., Morales, R. S., & Akhavan-Tabatabaei, R. (2015). OR models with stochastic components in disaster operations management: A literature survey. *Computers & Industrial Engineering, 82*, 183–197.

Hunt, K., Agarwal, P., & Zhuang, J. (2020.a). Monitoring misinformation on twitter during crisis events: A machine learning approach. *Risk Analysis*. https://doi.org/10.1111/risa.13634

Hunt, K., Agarwal, P., Al Aziz, R., & Zhuang, J. (2020.b). Fighting fake news during disasters. *OR/MS Today, 47*(1), 34–39.

Hunt, K., Wang, B., & Zhuang, J. (2020.c). Misinformation debunking and cross-platform information sharing through twitter during Hurricanes Harvey and Irma: A case study on shelters and ID checks. *Natural Hazards, 103*, 861–883.

Hunt, K., Narayanan, A., & Zhuang, J. (2021). *Blockchain in humanitarian operations management: A review of research and practice*. Working paper.

International Federation of Red Cross and Red Crescent Societies. (2019). About disaster management. https://www.ifrc.org/en/what-we-do/disaster-management/about-disaster-management/

Ishikiriyama C. S., & Gomes C. F. S. (2019). Big data: A global overview. In A. Emrouznejad & V. Charles (Eds.), *Big Data for the Greater Good. Studies in Big Data* (Vol. 42). Springer.

Keenaghan, D. J., Londo, E., King, R. M., Herzer, J. W., Ayala Ortiz, M., & Simerly, M. T. (2019). *Use case for blockchain technology: Supply chain response to humanitarian assistance/disaster relief*. Defense Logistics Agency.

Ko, V., & Verity, A. (2016). Blockchain for the humanitarian sector: Future opportunities. *Digital Humanitarian Network*, 1–19.

McIsaac, J., Brulle, J., Burg, J., Tarnacki, G., Sullivan, C., & Wassel, R. (2019). Blockchain technology for disaster and refugee relief operations. *Prehospital and Disaster Medicine, 34*(s1), s106–s106.

Mohan, P. (2018). Disaster management solution, part 3: Reference architecture. https://developer.ibm.com/articles/disaster-management-using-blockchain-iot-part-3-reference-architecture/
Mohan, P. (2018). Blockchain disaster management solution. https://developer.ibm.com/series/blockchain-disaster-management-solution/
Mondal, D., Chowdhury, S., & Basu, D. (2015). Role of non governmental organization in disaster management. *Research Journal of Agricultural Sciences, 6*, 1485–1489.
Panesir, M. S. (2018). Blockchain applications for disaster management and national security (Master thesis, University at Buffalo, State University of New York)
Patricelli, F., Beakley, J. E., Carnevale, A., Tarabochia, M., & Von Lubitz, D. K. (2009). Disaster management and mitigation: The telecommunications infrastructure. *Disasters, 33*(1), 23–37.
Sabbaghtorkan, M., Batta, R., & He, Q. (2020). Prepositioning of assets and supplies in disaster operations management: Review and research gap identification. *European Journal of Operational Research, 284*(1), 1–19.
Sobel, R. S., & Leeson, P. T. (2006). Government's response to Hurricane Katrina: A public choice analysis. *Public Choice, 127*(1–2), 55–73.
Wang, B., & Zhuang, J. (2018). Rumor response, debunking response, and decision makings of misinformed twitter users during disasters. *Natural Hazards, 93*(3), 1145–1162.
World Food Programme. (2019). Blockchain for zero hunger. https://innovation.wfp.org/project/building-blocks
World Health Organization. (2019). Environmental health emergencies: Natural events. https://www.who.int/environmental_health_emergencies/natural_events/en/
Yamamura, E. (2014). Impact of natural disaster on public sector corruption. *Public Choice, 161*(3–4), 385–405.
Zheng, Z., Xie, S., Dai, H. N., Chen, X., & Wang, H. (2018). Blockchain challenges and opportunities: A survey. *International Journal of Web and Grid Services, 14*(4), 352–375.
Zwitter, A., & Boisse-Despiaux, M. (2018). Blockchain for humanitarian action and development aid. *Journal of International Humanitarian Action, 3*(1), 1–7.

Kyle Hunt is a Ph.D. student in the Department of Industrial and Systems Engineering at the University at Buffalo (UB), focusing his studies in operations research. At UB, he conducts his dissertation work in the Decision, Risk and Data Laboratory, where he applies game theory, decision analysis, operations research, and data science techniques across a wide range of problem settings. His research focuses primarily in national security (with applications in aviation security, border security, and illicit networks), emergency management, and humanitarian operations. He is a U.S. National Science Foundation Graduate Research Fellow (NSF GRF), and has received many awards for his research. Mr. Hunt has published his research in peer-reviewed journals such as Reliability Engineering and System Safety, Risk Analysis, Natural Hazards, and Decision Analysis, among others.

Jun Zhuang is a Professor, Director of Graduate Studies, and Director of the Decision, Risk and Data Laboratory, Department of Industrial and Systems Engineering, School of Engineering and Applied Sciences (SEAS) at the University at Buffalo (UB). Dr. Zhuang has a Ph.D. in Industrial Engineering in 2008 from the University of Wisconsin-Madison. Dr. Zhuang's long-term research goal is to integrate operations research, big data analytics, game theory, and decision analysis to improve mitigation, preparedness, response, and recovery for natural and man-made disasters. Dr. Zhuang has been a principal investigator of over 30 research grants funded by the U.S. National Science Foundation (NSF), by the U.S. Department of Homeland Security (DHS), by the U.S. Department of Energy, by the U.S. Air Force Office of Scientific Research (AFOSR), and by the National Fire Protection Association (NFPA). Dr. Zhuang has published 100+ peer-reviewed journal articles in Operations Research, IISE Transactions, Risk Analysis, Decision Analysis, and European Journal of Operational Research, among others.

Jun Zhuang is a Professor, Director of Graduate Studies, and [illegible]

Blockchain Production Planning in Mass Personalized Environments

Fernando Tohmé, Daniel Alejandro Rossit, Mariano Frutos, Óscar Vásquez, and Andrea Teresa Espinoza Pérez

Abstract Industry 4.0 has substantially increased the degree of coordination and autonomy of production systems, lending manufacturing systems the ability to respond flexibly to the changing conditions of the market. Additive technologies enhance still further the flexibility of production processes. This increasing flexibility will allow the implementation of business strategies oriented towards satisfying "long tail" demands (mass customization and mass personalization). This poses the question of how to manage efficiently the production of goods that incorporate intangibles, like the preferences of individual customers. This is a novel problem in the field of Service Operations Management (SOM), namely the design of a management system able to elaborate in the aforementioned autonomous and decentralized way the production plans in mass customized/personalized environments in which the goods are individual requests incorporated in physical objects. This system, which draws on the capacities of cyber-physical systems (CPS), will be able to generate individual prototypes based on the specification of customers and, if it responds to their demands, plan the operations for its fabrication. The result will be a data structure autonomously elaborated by the network of CPS. This database must be freely available to all the CPS involved in the production process. This large data structure can be configured as a blockchain. That is, the intervening CPS will sequentially generate the blocks codifying the information needed to elaborate a production

F. Tohmé · D. A. Rossit
Departamento de Matemática, INMABB UNS CONICET, Av. Alem 1253, 8000 Bahía Blanca, Buenos Aires CP, Argentina

F. Tohmé
Departamento de Economía, Universidad Nacional del Sur, Bahía Blanca, Argentina

D. A. Rossit (✉) · M. Frutos
Departamento de Ingeniería, Universidad Nacional del Sur, Bahía Blanca, Argentina

M. Frutos
Departamento de Economía, IIESS UNS CONICET, Av. Alem 1253, 8000 Bahía Blanca, Buenos Aires CP, Argentina

Ó. Vásquez · A. T. E. Pérez
Industrial Engineering Department, Universidad de Santiago de Chile, Santiago, Chile

A. Emrouznejad and V. Charles (eds.), *Big Data and Blockchain for Service Operations Management*, Studies in Big Data 98,
https://doi.org/10.1007/978-3-030-87304-2_11

plan for a customized/personalized good. This structure will allow the intervention of "expert CPSs" at each step, signing their contributions as well as the hashing of the previous blocks. The use of Big Data methods will provide the grounds for applying Proof-of-Stake tests that will disallow revisiting the previous stages and will induce a chain of blocks capturing, in the end, the entire plan. The blueprint for this architecture is our original contribution to the SOM literature.

Keywords Blockchain · Cyber-physical systems · Mass personalization

1 Introduction

Recent advances in Smart Manufacturing and Industry 4.0 have substantially increased the degree of coordination and autonomy of smart production systems connected through the Internet of Things (IoT) and linked to decision-making processes [22, 41]. These advances lend manufacturing systems the ability to respond flexibly to the changing conditions of the market [17]. Technologies such as Cyber-Physical Systems (CPS) and Additive technologies, for instance 3D printing, enhance still further the flexibility of production processes, inducing a potential new industrial revolution. CPS are systems consisting of physical components with computational functionalities, for instance mechanisms that are controlled or monitored by algorithms running on computers [26]. Advanced CPS architectures link CPS to platforms in the cloud [22]. These systems constitute the adequate tools for Additive Manufacturing (AM, a process in which layers of material are joined as to generate products following the patterns found in databases of 3D models [32, 36]. In comparison to traditional fabrication processes, AM exhibits significant advantages, for instance the reduction of delays between designing, producing in a just-in-time way and delivering products to consumers. AM can produce functional pieces with complex internal structures and optimized physical properties, with a lesser amount of waste than subtractive production [52].

These technologies contribute to improve the flexibility of the production process, leading to a larger variety of products available to customers. In this sense, [54] claim that the synergy between Industry 4.0 environments and those technologies will intensify business strategies of mass customization (customers will face a large variety of options) and mass personalization (each customer defines her own product, [24]. The wide possibilities of increasing the variety of products and their characteristics pose technological challenges which require an increasing complexity of the information systems associated to the production processes. Managing the production of personalized or customized goods will be quite demanding, given their different requirements [58].

Since a substantial component of customized goods will involve the idiosyncratic preferences of customers, which are by nature intangible and thus make the final product a physically embodied service-like good. This issue of managing the distributed production of customize goods poses a new problem for Service Operations Management (SOM). The goal is the dual of that of reducing the uncertainty of customers [14] by seeking to address the requirements of individual customers. Since the production process is distributed, it requires a network approach similar to that of Service Supply Networks [46]. This makes the problem close to one the main problems in SOM enumerated by Fields et al. [12]. The analysis of the problem requires assessing the alternative costs, the different production stages, the dispatch of finished products, as well as the materials involved and the due dates of the processes. CPS are fundamental pieces in this structure, since they have access to all this information and can play some roles proper of DSS, involving many business functions related to production [43]. Therefore, it is natural to think that the entire flow of tasks in these mass customized/personalized production processes, involving the planning of operations and their implementation, can be handled in a decentralized and autonomous way by the CPS themselves. This means that CPS could collect information and specifications from the clients and, on the basis of records of processes and operations carried out in the past and with the aid of 3D design tools, develop a production plan for every specific product. Of course, the efficiency of the plan requires the interoperability among the different business processes involved [11, 62].

In order to ensure that production planning gets carried out autonomously the structures and systems able to do this must be developed. This chapter is intended as a contribution towards this goal. We first start with a question not yet treated in the literature, namely that of identifying the problems that arise in production planning of mass customized/personalized activities in Industry 4.0. The different participants in the production process of the service-like manufactures generate large reams of data. To coordinate them we propose to solve this Big Data issue by means of the blockchain technology [23]. Basically, blockchain is a distributed database (a *ledger*) in which the records correspond to all the transactions or digital events that have been executed and shared among the participant entities. Each transaction in the ledger is approved consensually by a majority of the participants of the system [7]. The use of this technology ensures the transparency and reliability of transactions [3, 4].

The rest of this chapter is organized as follows. In Sect. 2 we introduce basic characterizations of the technologies that participate in mass customization /personalization production processes. Section 3 presents such a scenario and the business strategy that supports it. In Sect. 4 we present and formalize decision-making processes in mass customization/personalization environments. In Sect. 5 we present an efficient solution to the problems of Sect. 4 based on the use of blockchain. Finally, Sect. 6 concludes the chapter.

2 Advanced Technologies: Concepts

In this section we present the advances in Industry 4.0 that amount to a paradigm change in the way in which production is carried out, allowing the extend the traditional business model to mass customization/personalization settings. This requires reviewing some of the key technologies of Industry 4.0 and AM. Next, we will describe the information technologies based on blockchain.

2.1 The Industry 4.0 Environment

The novelty of Industry 4.0 environments is mostly due to the increasing penetration of internet connectivity in production systems, fundamentally through the Internet of Things (IoT). The data collected by production machines fed into physical control units through a SCADA (Supervisory Control and Data Acquisition) or similar system. With the enhanced connectivity of production systems this information can be transmitted to processing centers with higher capacity, availing Decision Support Systems (DSS) with data traditionally restricted to control levels and thus becoming useful for CPS [41, 50]. Data collected through a Wireless Sensor Network (WSN) gets compiled by IoT, yielding a better picture of the state of the system. This assessment gets even more precise if a connection with Cloud Computing [34] is established, since then data processing capacities increase considerably [49]. Notice that manufacturing generates large volumes of data, which require, in turn, of large data processing [29, 60].

An interesting aspect of the penetration of internet into Industry 4.0 systems, is that it goes beyond it use for the production flow process. The Research and Innovation funding program for 2007–2013 (FP7) of the EU established that a future networked society had to be grounded on four feet: Internet by and for People (IoP), Internet of Contents and Knowledge (IoCK), Internet of Things (IoT), and Internet of Services (IoS) [55].

IoT links cyber and physical systems making fabrication processes intelligent. IoP connects all the participants, eliminating the barriers between producers and consumers, creating online communities for the design, creation and sale of products. IoS uses the internet as a medium for the exchange of services by applying technologies like Service Oriented Architectures (SOA) o Cloud Computing. Finally, IoCK transforms data (generated, for instance, by intelligent objects connected through the IoT) into information/knowledge that can be used in manufacturing systems [54]. While all these uses of internet are currently in use, they will become even faster and cheaper facilitating closer interactions between customers and production units, connected in platforms [39].

2.2 Additive Manufacturing

AM is a group of additive fabrication technologies, in which a 3D object is created by superimposing layers of material [52]. AM allows developers the capacity of printing and assembling parts fabricated with different materials and having different physical and mechanical properties. Advanced AM technologies can even help to create prototypes, improving the process of designing products [32].

There exist different AM processes like Stereolithography (also known as optical fabrication, photo-solidification or resin printing) which is the oldest AM technique, using a laser to create polymers in a liquid resin and generating the final product [20, 25, 28]. Another AM technology is Fused Deposition modeling, which uses a continuous filament of a thermoplastic material which a printing head melts and deposits on a piece of work under the control of a computer, defining the printed shape (Zein et al. 2001). In the last decade these techniques became combined in new and improved 3D printers. These, embedded in CPS systems seem poised to become a crucial component of fabrication processes.

2.3 Blockchain Technology

Blockchain can be seen as a data structure which the information is grouped in blocks, which are chained and secured through cryptographic method. Each block contains a cryptographic hash of the previous block in the chain, as well as a time stamp and the data on some transactions [47]. Blockchain is "an open, distributed ledger that can record transactions between two parties efficiently and in a verifiable and permanent way" [19]. Being a distributed ledger, blockchain is handled in peer-to-peer[1] networks, adhering to a collective protocol for the communication among nodes and validation of new blocks. Once recorded, the data in a block cannot be modified without altering the next blocks in the chain, which would require the consensus of most of the users of the network. The most popular application of blockchain is Bitcoin [33], a peer-to-peer version of electronic money, facilitating online transactions without requiring the intervention of financial institutions. Bitcoin highlights one of the main features of blockchain, namely the veracity of transactions, based on the consensus of the majority without requiring the intervention of third party acting as an authority [15, 56].

In this work we intend to use the decentralized nature of blockchain in order to handle the demands to a mass customization/personalization fabrication system in an Industry 4.0 environment as to generate valid production plans in an autonomous and decentralized way.

[1] A peer-to-peer network is an architecture for distributed applications that divides tasks or workloads among participants. The latter are all considered to be on a par and act both as servers and clients.

There are five main features that [19] identify in this technology. The first one is that it is a distributed database in which all the users have access in the same footing. Each user can verify the transactions of any other without going through intermediaries. The second feature is the peer-to-peer linkage between users, again with regulators or intermediaries. The third feature is the transparency of the system since, as said, each transaction and its associated value is visible for anyone with access to the system. Each user has an identification address. Each of them can decide to remain anonymous or reveal her identity, but in transactions only matter the addresses. The fourth feature is the irreversibility of records. Once a transaction is recorded and all the hash values are updated, the record cannot be modified because this would affect all the other blocks in the chain. The fifth feature is the digital nature of the ledger, which allows the use of algorithms and computer procedures to carry out automatically any transaction.

We can illustrate how blockchain works with an example drawn from [7]. Suppose that an user A wants to make a payment to B. This constitutes a transaction from A to B, described online by a block which is sent to every user in the network. They have to verify whether this block corresponds to a valid operation. If there is a consensus on that, the block is incorporated in the chain. Once this happens, A's money is sent to B. Notice that then, nobody can claim that A did not pay to B or that B did not receive the full amount sent by A.

This example shows that there is a stage of validation in order to accept a block. There are different possible procedures of validation. Bitcoin, for instance, uses proof-of-work (PoW), in which each user can only validate a block by solving complicated puzzles, requiring a heavy computation load and thus very demanding in electrical power [47, 61]. This protects the system of cyberattacks, which the openness and deregulation of the system seem in principle prone to incentivize. Another validation procedure is by proof-of-stake (PoS), which unlike PoW, which bases the validation on the capacity or willingness to do heavy work, facilitates the validation of blocks by users that have more valid transactions, being the effort to be carried out inversely proportional to the latter. The assumption behind PoS is that those who have more at stake (having, say, more transactions registered) are particularly interested in the survival of the system and thus are more willing to shield it from attacks. PoS requires much less computation (and electrical power), optimizing the use of resources. In our proposal we will use this type of validation procedure.

3 Mass Customization/personalization Production Processes

Mass customization and personalization have become characteristic features of the fourth industrial revolution. These strategies intend to bring the products closer to the requirements and preferences of customers [8, 24, 38]. The traditional mass production processes limit the variety of products from which customers can choose.

Fig. 1 Long tail market

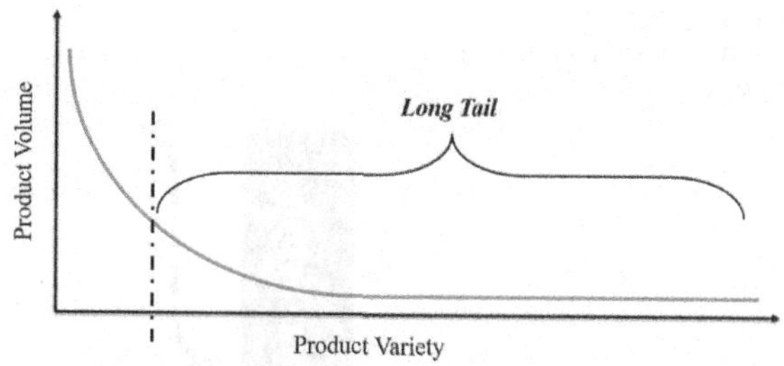

Industry 4.0 is able to cover the demands of traditional markets [17], as well as to face new "long tail" markets [1, 2]. According to [2], "long tail" markets are those in which the goal is to sell smaller volumes but of a larger number of products, unlike the traditional objective of selling large amounts of a few products. The expression "long tail" is used to represent probability distributions in which observations in the tails have still a fair amount of weight. Anderson uses this term to distinguish the traditional market strategies oriented to generate a very concentrated production, from the new approaches in which a demand for variety can be satisfied, as shown in Fig. 1 [1]. Historically, industrial activities focused on market sectors in which mass production generated higher yields and returns to investments [5]. In the last two decades the widespread use of the internet (IoP, IoT, IoS, IoCK) and the digitalization of retail markets have laid the grounds for long tail productions systems [2].

Industry 4.0 or Smart Manufacturing environments address these business strategies in an efficient way, based on the use of CPS and AM [35, 37]. They collect updated information about physical objects and processes through IoT/CPS, improving the productivity and flexibility of already existing mass production processes. This empowers the customers to choose or contribute to the attributes of the product or service in which they are mostly interested [48, 54]. These closer interactions with customers can be handled thanks to the larger flexibility of Industry 4.0 environments, endowed with intelligent production systems that can be autonomously configured to optimize production, yielding a higher level of service and larger returns. Besides, AM systems are perfectly able to make single-unit products o industrial productions with very low volumes [32]. Therefore, mass customization and mass personalization can be achieved, and in fact, it is already in use in areas like dental implants, healthcare products, sport equipment, hearing aids, etc. [13, 18].

One of the main impacts of implementing these technologies is the reduction of costs in non-massive productions. [54] illustrate this with the example of the AM production of a plastic gear, which has a cost of 55$ per unit. The traditional injection production of these gears has a cost of only 5$ per unit, but requires a matrix that costs 10,000$. These authors calculate the equilibrium point and indicate that if less than 200 gears are needed, AM can provide them at a lower cost. This indicates that the new technologies allow producing low volumes or even single units without increasing the costs over mass production. Long tail markets can thus become commonplace.

Despite this encouraging realization we have to keep in mind that the advantages of the new production processes can be achieved only if the system is autonomous

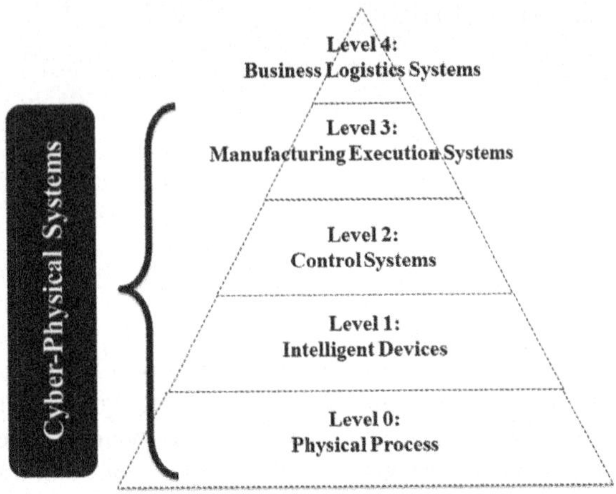

Fig. 2 Levels of ANSI/ISA95 integrated into CPS

and able to keep the fabrication process running. While, as far as we know, there have not been many contributions analyzing the production engineering basis for these business strategies, several authors have already pointed out the possibility of linking the production and decision-making functions in such a way that they could be handled autonomously in Industry 4.0 environments [30, 31, 41, 43].

3.1 Production Planning and Control: ANSI/ISA 95

To understand why Industry 4.0 environments are able to run production processes in an autonomous way, let us consider the control standard ANSI/ISA 95 or ISA 95. This standard provides a framework for an automated interface between production facilities and control systems. Officially is defined as[2]: " ISA-95 is the international standard for the integration of enterprise and control system. ISA-95 consists of models and terminology that can be used to determine which information has to be exchanged between systems for sales, finance and logistics and systems for production, maintenance and quality". It can yield a common ground for the communication of all the participants in a production process and gives a representation of how information can be modelled and used. It organizes the different levels of decision-making hierarchically. It is based on the "Purdue Enterprise Reference Architecture" (PERA) which distinguishes five levels, as shown in Fig. 2. Level 0 is associated to the physical process of manufacturing. Level 1 involves the intelligent devices that measure and manipulate the physical process are located. Typical instruments at this level are sensors, analyzers, effectors and related instruments. Level 2 represents the control and supervision of the underlying activities. Systems acting on ISA-95 Level 2 are SCADA and Programmable Logic Controllers (PLC), for instance. Level 3 involves

[2] http://www.isa-95.com/.

the management of the operations and the production work flow in the production of the desired products. Some of the systems comprised at this level are Batch Management, manufacturing execution/operations management systems (MES/MOMS), the laboratory, maintenance and plant performance management systems, data historians and related middleware. This level has special importance for our work, since it is here where the scheduling process takes place. Finally, level 4 is associated to the business activities of the entire firm. This architecture represents, in a synthetic way, the different activities and functions of a production system. Besides, it establishes the way in which the different levels communicate; in traditional productions settings in particular, each level interacts only with its adjacent levels [41].

Decision making in CPPS

As discussed in [41], CPS will have an impact on decision-making activities in the area of industrial planning and control. This will be due to the ability of CPS to carry out a wide spectrum of activities, ranging from the physical operations of production (level 0) to planning, evaluating and managing the entire production process (level 3), by controlling the actions and systems on levels 1 and 2 (i.e. the measurement and sensing instruments as well as the control systems). This approach is illustrated in Fig. 2, in which the levels of ISA 95 that should be incorporated to CPPS are highlighted.

Some of the direct benefits of this integration of functionalities are, for instance, the increased flexibility to respond to unexpected events, or the faster transmission of information through the entire system. These advantages are due to the fact that CPPS can translate the data obtained at level 1 to the higher order language used at level 3, eluding the adjacency limitations inherent in PERA, generating answers faster to unforeseen events.

This, in turn, will affect directly the way in which decisions are made in production planning, which in terms of PERA will be managed by both ERP (Enterprise Resource Planning) and the CPPS. Figure 3 shows this:

Figure 3 indicates that only the decisions at the aggregate level (for instance the goals of the company) will be handled by ERP systems, already adapted to smart manufacturing environments. All other decisions will automatically be made and executed by CPPS. In this way, current Manufacturing Execution Systems (MES) will be absorbed by CPPS, which will also take care of integrating the dispatch of work orders and their schedule in the shop floor. This will improve the quality of the information at this level, increasing the flexibility and the ability to respond to changing circumstances [41].

The ensuing structure, jointly handled by ERP and CPPS in a distributed fashion, modifies the traditional hierarchical and centralized view on production planning. Of course, this will also affect Scheduling, since decisions will now be made in a distributed and collaborative way (Wang and [51]. Despite this, the literature still treats the problem in the traditional and static way. The scheduling community will have to accept the challenges presented by the new paradigm, developing strategies and methods accordingly. The incorporation of more complex and flexible production

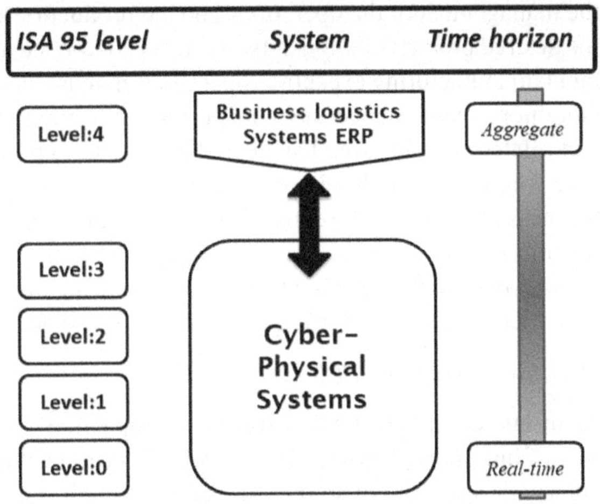

Fig. 3 Distribution of ISA 95 levels between ERP and CPPS

structures will be a substantial contribution, as in non-permutation programming of fabrication cells [40, 41].

A good deal of the decisions made by ERP systems (as inventory control, management of databases, handling information about suppliers, etc.) will be managed by CPPS. But we leave them separated as to indicate at what point the system becomes autonomous and up to which human interventions may be needed, particularly in the area of production planning. The linkage with human decision-makers will be at the aggregate or strategic level. They will define the goals and guidelines for the firm and the system. An ERP system will get them and will translate these guidelines for the rest of the system, in particular to the CPPS that handle the production system. The latter are thus not completely autonomous since they keep an open loop with the ERP system, at least on production planning [31, 44, 45].

4 Production Planning in Mass Personalized Environments

In this section, we discuss production planning in mass customized/personalized environments. We start considering the linkages between customers and manufacturing firms in which the former define the requirements and specifications. The planning process has to respect these conditions, leading to some decision-making problems that have to be solved in order to ensure a mass customization/personalization production.

4.1 The Setting

Consider a class I of customers and a firm in a special manufacturing sector. Each costumer $i \in I$ wants to acquire a new and specific product j_i, which is demanded to the firm. Let P and O be the sets of processes and operations historically used by the firm. Assume that it has a set of CPS K to handle the production of the new item. Each $k \in K$ can perform a set of operations O_k and/or a set of processes P_k. All the information associated to the processes, operations and production resources (i.e., processing times, materials, machinery and standards of quality, among others) is available from historical databases.

4.2 Decision-Making Process

The decision-making process of the firm having to plan the production in mass customized/personalized environments follows two main steps. The expected result of each step corresponds to a feasible solution to an open problem to be solved by the firm.

Open problem 1. Given a set of new products J demanded by the customers, identify a subset of processes $P_{j_i} \subseteq P$ and/or a subset of operations $O_{j_i} \subseteq O$, required to produce a prototype $t_{j_i} \subseteq T$ for each new product $j_i \in J$ according to the specifications of customer $i \in I$.

The output of this problem defines the core processes and core operations that the firm should implement to produce prototype t_{j_i}.These processes and operations are drawn from the historical records of previous production experiences of the firm. This problem can be seen as a concept test of the product demanded by the consumer. If there are not enough processes and operations in the database of past experiences to produce the prototype, the product cannot be produced (at least by autonomous processes) and this must be informed to the customer.

Open problem 2. Given a set of prototypes T and the processes and/or operations required to manufacture them, determine the class of expert cyber-physical systems (ECPS)E such that each $e \in E$ is able manufacture the elements of T. Each $e \in E$ is defined by a subset of CPS $K_e \subseteq K$. A CPS becomes an ECPS when it has an accumulated experience in running a family of processes and operations, and thus it became an "expert" in doing this. This problem amounts to associate the activities already seen as feasible solutions to the problem 1 to an CPS experienced in carrying them out.

The solution to this second problem is the class of ECPS, the network of systems that are able to manufacture the set of prototypes T according to the specifications of the class of customers I. A roadmap for the activities leading to the satisfaction of the demands of customer i must be produced. The production design so obtained is defined by the prototype t_{j_i}, the processes and operations required (P_{j_i} and O_{j_i}) and the resources able to engage in the production (E).

5 The Blockchain Solution

We have discussed the problems that have to be solved once a personalized demand for a product *j* arrives. Even if the company already produces goods similar to *j*, it has particular features that make it different, which have to be reflected in its individual roadmap. Since Industry 4.0 environments can run autonomously by using CPS [30], many of its functions can be used to generate automatically the roadmap for the production of *j*. Since this is the main goal of this chapter, let us go through the problems stated in the previous section.

Problem 1 involves the translation of the request of product *j* by customer *i* into a class of operations and processes for its production. An autonomous solution obtains thanks to the ability of CPS to go back and forth between the virtual and the physical realm [27]. A full digital model of product *j* can be obtained using the specifications provided by *i* and the 3D design tools available in the Industry 4.0 system, yielding a digital prototype t_{j_i}. This prototype can be analyzed by the production CPS using the database of the processes and operations previously executed (sets P and O) and of the resulting products. Matching the specifications of t_{j_i} to the adequate subsets $P_{j_i} \subseteq P$ and $O_{j_i} \subseteq O$, solves Problem 1. So, for instance, if the production of t_{j_i} requires hollow cylindrical shapes, the CPS in charge of lathes may identify the operations corresponding to those shapes. Even if no subtractive operations and processes can be found matching the specifications of t_{j_i}, AM technologies can be still be used, although the latter tend to be slower than the former [32], and thus the first choice should always be subtractive activities.

An important question is whether a conflict can arise among the different autonomous CPS in the selection of P_{j_i} and O_{j_i}, since they could recommend different sets, all satisfying the requirements of t_{j_i}. The proposals could be mutually exclusive making it impossible to choose in a decentralized way one of them. In our example, let us add to the recommendation of the lathe controlling CPS one of the milling cutter CPS, suggesting to drill a hole and some additional operation. While both proposals may involve making a hole, they may differ in the next operation. To let the autonomous system to negotiate which recommendation to follow can be complicated. This is why our proposal involves an information system based on blockchain, which solves in a simple way these kinds of problems. This approach will also allow solving most of Problem 2. This technology facilitates the decentralized organization and coordination of the contributions of the different CPS (the users of the blockchain network) to yield the roadmap for the production of t_{j_i}.

In order to achieve this goal we will use the Proof-of-Stake (PoS) validation method instead of that of Proof-of-Work (PoW). New blocks of processes and operations will be validated by the CPS that are better fit for the task than the others. This is less computationally demanding that the use of PoW. The aptitude of CPS will be associated to the number of times that it processed similar operations in the past. As an example consider a company specialized in the fabrication of tennis rackets receiving a request for a personalized racket. The know-how of the firm in making this sport equipment in the past can help to rank the CPS. Data analytics can indicate the CPS that have more chances to be used at each stage of the process. The blockchain protocol will give more weight, to validate each new process or operation in the ledger, to the CPS with the highest probability in the corresponding stage. This means that even if the information is distributed and all CPS can make recommendations, the only contributions that will matter are those that are deemed fitter at each stage. Interestingly, a CPS with more chances of validating the activities at stage $s + 1$ can revise those at stage *s*, already in the ledger. Changing them would be costly but this review allows it to recommend an adequate continuation to the previous stage. This, in turn, contributes to the generation of a more efficient roadmap for t_{j_i}. In the example of the lathe and milling cutter CPS, in the case that the former is the one with primacy at the first stage, it could contribute with the suggestion of the operations on external and internal turning. The milling cutter CPS, that has the stake at the second stage can, by analyzing the recommendations of the lathe CPS, select the more efficient operations to apply after those.

In this way, the CPS will contribute the processes and operations that in the end yield the selected sets P_{j_i} and O_{j_i}. Furthermore, since the blocks reflect the order in which they will be carried out, the ledger will be the roadmap for the production of t_{j_i}. Notice that the implementation of PoS validation also solves Problem 2 since it automatically identifies the CPS better fit to carry out the activities at each stage. That is, it yields the set E of ECPS.

5.1 Properties of This Blockchain Solution

Let us analyze the different aspects of the roadmap-as-blockchain as a solution to planning the production of customized/personalized goods.

Quality of the solution

While the chain of blocks generates a roadmap, nothing ensures that it will be the optimal solution that selects the best sequence of processes, setups, etc. By design, each CPS recommends some operations for the production of t_{j_i}, oblivious of the possibilities of the next CPS in the sequence. On the other hand, the assignation of the relative weights at each step for the validation by PoS responds to the historical record. We can suspect that the resulting roadmap may be at least a good solution in the light of the fabrication processes of the past, be it because it respects the idiosyncrasies of the firm or the layout of the system.

On the other hand, if we consider that many of CPS intervening in the production process may respond to different firms, each of them aiming to increase its profits, the actual outcome could be seen as a Nash equilibrium in a game-theoretic setting, if their relative weights are common knowledge. That is, if each CPS knows which one will be in charge of validating a new block at a certain stage, we must consider two cases. Either the CPS is the one that will perform the validation, in which case it will perform optimally according to the goals of the firm it represents. If the CPS, instead, will not be in charge of validating the new addition, it has no reason for not proposing to do the best for itself. Of course, the efficient (Pareto) solution may differ from the competitive (Nash) one.

Autonomy and speed of response

One of the main virtues of this proposal is that the roadmap can be generated rather easily and efficiently. Since the entries in the ledger correspond to the CPS that are already fit for each stage, the response is rapid, without depending on the other units (except for the previous one in the sequence). This contributes to providing an agile and flexible supply of customized goods.

The choice of PoS validation makes the process much faster than under PoW. Since there is no need to run complex processes, unrelated to the actual information incorporated into the ledger, the validation phase is much shorter. This, combined with the fact that information is generated autonomously by CPS that are designed to run efficiently, make the process very agile and faster than if it were managed in a centralized way.

User-friendliness

Another aspect to highlight is that the ledger, so defined, will have a format readable by human users. Once the entire blockchain is validated, the ledger can be submitted to all the workstations without the need of further processing. In order to ensure this, it is convenient to implement the system in an XML (Extended Markup Language) in such way that both the autonomous CPS and the human operatives can read it. An example of an XML that respects ISA 95 is B2MML (Business to Manufacturing Markup Language) [16]. The use of such an XML facilitates, in case of an unavoidable conflict during the generation of the ledger, the intervention of a human user that can make the decisions that lead to the generate an adequate roadmap of fabrication.

A tool for SOM

As shown in [6], a useful way of modelling supply networks is by focusing on the differentiation and the level of interrelationships between suppliers. This is exactly what our approach captures, namely a way of assigning differential weights to the different CPS participating in the production process. This allows applying a PoS validation method for the addition of new blocks to the ledger. But this strategy can be readily adapted to solve the coordination problem of service supply networks, particularly those in which "prestige" may serve as a consensual way of giving more weight to some suppliers.

Since many services are already run by the equivalents to the cyber component of CPS, the only aspect that may change in an application to SOM is the nature of the good produced. In our case the good has many service-like features, due to the fact that the final product responds to the individual desires of customers. This personalized aspect, and the decentralized production process indicate that our setting is basically a service supply network. The solution proposed is thus easy to extend to a blockchain for the supply chain of services.

A way of reducing complexity

One important issue in the literature on distributed production by means of the use of CPS is how to use the large amount of data generated by them. Traditional data managing systems seem to be poor to handle the rapidly growing databases generated by those production systems. Big Data methods are needed to extract useful information out of them summarizing this dearth of data [9]. One way, is the registration of the relevant information in the ledger. The validation by PoS makes the addition of new blocks the main component in the process of reducing the complexity of the entire system, by defining (by discriminating between what is finally included and what is left out) which information is actually meaningful.

The literature shows that information generated by the different CPS can be better processed using Machine Learning methods [10]. The best candidates can be *unsupervised learning* methods, either using *deep learning* neural networks or *K-means* clustering [21]. These methods do not assume a previous classification of data. The algorithms themselves find the pattern hidden in data, mainly the weight that should be given to each CPS at a given stage of the production process (it may vary in time). The downside of the use of these methods is the possible loss of accuracy due to the lack of centralized labelling.

6 Conclusions

In this chapter we presented the tools to generate a business strategy for mass customized/personalized production in Industry 4.0 environments. We propose an autonomous and decentralized system managed fundamentally by CPS. This allows associating the decision-making processes concerning production planning to the CPS. We formalized the problems that such a system has to address, which can be solved under different approaches. Our proposal is a blockchain-based system that profits from the autonomy and decentralization of Industry 4.0 environments. This system uses a PoS validation procedure that avoids conflicts and duplication of efforts among the CPS in the network. The ledger generated in this way provides an organized sequence of production activities for the fabrication of customized/personalized goods.

In this sense, our proposal is readily applicable to Service Operations Management, given that personalized goods embody many features shared with services. Thus, our framework describes a setting similar to a Service Supply Network. In particular, it operates on the basis of detecting among the components of the network those with more weight at a given stage of the production process to grant them the right to validate new blocks to be added to the blockchain. The detection of those salient components, CPS in our case, the distributed information must be processed with efficient and decentralized methods. The best candidates are Big Data tools of unsupervised learning.

Managers can implement this abstract model implementing it on off-the-shelf technologies.[3] In simpler contexts, a restricted version of our approach can be applied by defining in clear and precise terms the criteria according to which the weights for the exertion of the Proof of Stake validation are granted. This simplifies matters by discarding the need to implement unsupervised learning methods.

As already noted above, our approach has potential weaknesses due to the lack of objective criteria for the application of PoS validation. The decentralization of the production process also introduces many potential complications due to the physical distance among CPS and thus, even if the tasks are correctly allocated, increased costs and possible delays may arise from the need of displacing different components and the application of different processes.

Future work on this model may follow different lines. Some of them are:

- The explicit extension to SOM, in particular to the supply chain of services.
- The explicit introduction of a restriction on the physical separation of CPS, in order to minimize the costs of displacements among them.
- The incorporation of other Machine Learning methods to apply anticipatory allocation of weights, to base the PoS not only on past performance but also on future possibilities (see [59]).
- In a more speculative setting, to consider the possibility of ultimately transferring the whole control structure directly to final consumers, ensuring them access to the relevant information of a fabrication process carried out according to their specification.

References

1. Anderson, C. (2004). The long tail. Wired. http://www.wiredcom/wired/archive/1210/tailhtml
2. Anderson, C. (2008). The long tail: why the future of business is selling less of more. Hyperion Books.
3. Charles, V., & Emrouznejad, A. (2019). Big data for the greater good: An introduction. In A. Emrouznejad & V. Charles (Eds.), *Big data for the greater good*. Studies in Big Data, vol. 42. Springer.
4. Charles, V., & Gherman, T. (2019). Big data analytics and ethnography: Together for the greater good. In A. Emrouznejad & V. Charles (Eds.) *Big data for the greater good*. Studies in Big Data, vol. 42. Springer.
5. Chase, R. B., Aquilano, N. J., & Jacobs, F. R. (1998). *Production and operations management*. Irwin/McGraw-Hill.
6. Choi, T. Y., & Krause, D. R. (2006). The supply base and its complexity: Implications for transaction costs, risks, responsiveness, and innovation. *Journal of Operations Management, 24*(5), 637–652.

[3] For instance, *Statebox* or *Daml* for the blockchain and *IBM Watson Machine Learning for z/OS* for the unsupervised learning algorithm.

7. Crosby, M., Pattanayak, P., Verma, S., & Kalyanaraman, V. (2016). Blockchain technology: Beyond bitcoin. *Applied Innovation, 2*, 6–10.
8. Da Silveira, G., Borenstein, D., & Fogliatto, F. S. (2001). Mass customization: Literature review and research directions. *International Journal of Production Economics, 72*(1), 1–13.
9. Emrouznejad, A. (2016). *Big data optimization: Recent developments and challenges*. Studies in Big Data, vol. 18. Springer.
10. Emrouznejad, A., & Marra, M. (2016). Big data: who, what and where? Social, cognitive and journals map of big data publication. In A. Emrouznejad (Ed.) *Big data optimization: Recent developments and challenges*. Studies in Big Data, vol. 18, Springer.
11. Espadinha-Cruz, P., Gonçalves-Coelho, A., Mourão, A., & Grilo, A. (2015). Re-design of an interoperable buyer-seller automotive relationship aided by computer simulation. *Procedia CIRP, 34*, 98–105.
12. Field, J. M., Victorino, L., Buell, R. W., Dixon, M. J., Goldstein, S. M., Menor, L. J., Pullman, M., Roth, A.V., Secchi, E. & Zhang, J. J. (2018). Service operations: what's next? *Journal of Service Management, 29*(1).
13. Frazier, W. E. (2014). Metal additive manufacturing: A review. *Journal of Materials Engineering and Performance, 23*(6), 1917–1928.
14. Gao, F., & Su, X. (2017). Manufacturing & service operations management. *INFORMS, 19*(1), 84–98.
15. Gorkhali, A., Li, L., & Shrestha, A. (2020). Blockchain: A literature review. *Journal of Management Analytics, 7*(3), 321–343.
16. Harjunkoski, I., & Bauer, R. (2014). Sharing data for production scheduling using the ISA-95 standard. *Frontiers in Energy Research, 2*, 44.
17. Hermann, M., Pentek, T., & Otto, B. (2016, January). Design Principles for Industrie 4.0 Scenarios. In *2016 49th Hawaii International Conference on System Sciences (HICSS)* (pp. 3928–3937). IEEE.
18. Huang, S. H., Liu, P., Mokasdar, A., & Hou, L. (2013). Additive manufacturing and its societal impact: A literature review. *The International Journal of Advanced Manufacturing Technology, 67*(5–8), 1191–1203.
19. Iansiti, M., & Lakhani, K. R. (2017). The truth about blockchain. *Harvard Business Review, 95*(1), 118–127.
20. Jacobs, P. F. (1992). *Rapid prototyping & manufacturing: fundamentals of stereolithography*. Society of Manufacturing Engineers.
21. James, G., Witten, D., Hastie, T., & Tibshirani, R. (2017). *An introduction to statistical learning with applications in R*. Springer.
22. Jardim-Goncalves, R., Romero, D., & Grilo, A. (2017). Factories of the future: Challenges and leading innovations in intelligent manufacturing.
23. Karafiloski, E., & Mishev, A. (2017). Blockchain solutions for big data challenges: A literature review. In *EUROCON 2017–17th International Conference on Smart Technologies* (pp. 763–768). IEEE.
24. Kumar, A. (2007). From mass customization to mass personalization: A strategic transformation. *International Journal of Flexible Manufacturing Systems, 19*(4), 533.
25. Kucukkoc, I. (2019). MILP models to minimise makespan in additive manufacturing machine scheduling problems. *Computers & Operations Research, 105*, 58–67.
26. Lee, E. A. (2008, May). Cyber physical systems: Design challenges. In Object oriented real-time distributed computing (isorc), *2008 11th IEEE international symposium on* (pp. 363–369). IEEE.
27. Lee, J., Bagheri, B., & Kao, H. A. (2015). A cyber-physical systems architecture for industry 4.0-based manufacturing systems. *Manufacturing Letters, 3*, 18–23.
28. Melchels, F. P., Feijen, J., & Grijpma, D. W. (2010). A review on stereolithography and its applications in biomedical engineering. *Biomaterials, 31*(24), 6121–6130.
29. Moktadir, M. A., Ali, S. M., Paul, S. K., & Shukla, N. (2019). Barriers to big data analytics in manufacturing supply chains: A case study from Bangladesh. *Computers & Industrial Engineering, 128*, 1063–1075.

30. Monostori, L. (2014). Cyber-physical production systems: Roots, expectations and R&D challenges. *Procedia CIRP, 17*, 9–13.
31. Monostori, L., Kádár, B., Bauernhansl, T., Kondoh, S., Kumara, S., Reinhart, G., ... & Ueda, K. (2016). Cyber-physical systems in manufacturing. *CIRP Annals-Manufacturing Technology, 65*(2), 621–641.
32. Mueller, B. (2012). Additive manufacturing technologies–Rapid prototyping to direct digital manufacturing. *Assembly Automation, 32*(2).
33. Nakamoto, S. (2008). Bitcoin: A peer-to-peer electronic cash system. www.bitcoin.org/bitcoin.pdf
34. Nodehi, T., Jardim-Goncalves, R., Zutshi, A., & Grilo, A. (2017). ICIF: An inter-cloud interoperability framework for computing resource cloud providers in factories of the future. *International Journal of Computer Integrated Manufacturing, 30*(1), 147–157.
35. Ogunsakin, R., Marin, C. A., & Mehandjiev, N. (2021). Towards engineering manufacturing systems for mass personalisation: A stigmergic approach. *International Journal of Computer Integrated Manufacturing,* 1–29.
36. Oh, Y., Witherell, P., Lu, Y., & Sprock, T. (2020). Nesting and scheduling problems for additive manufacturing: a taxonomy and review. *Additive Manufacturing, 101492.*
37. Park, K. T., Lee, J., Kim, H. J., & Do Noh, S. (2020). Digital twin-based cyber physical production system architectural framework for personalized production. *The International Journal of Advanced Manufacturing Technology, 106*(5), 1787–1810.
38. Pine, B. J., Victor, B., & Boynton, A. C. (1993). Making mass customization work. *Harvard Business Review, 71*(5), 108–111.
39. Porter, M. E., & Heppelmann, J. E. (2015). How smart, connected products are transforming companies. *Harvard Business Review, 93*(10), 96–114.
40. Rossit, D., Tohmé, F., Frutos, M., Bard, J., & Broz, D. (2016). A non-permutation flowshop scheduling problem with lot streaming: A Mathematical model. *International Journal of Industrial Engineering Computations, 7*(3), 507–516.
41. Rossit, D., & Tohmé, F. (2018). Scheduling research contributions to Smart manufacturing. *Manufacturing Letters, 15*(B), 111–114.
42. Rossit, D. A., Tohmé, F., & Frutos, M. (2018). The non-permutation flow-shop scheduling problem: A literature review. *Omega, 77*, 143–153.
43. Rossit, D. A., Tohmé, F., & Frutos, M. (2019). Industry 4.0: Smart scheduling. *International Journal of Production Research, 57*(12), 3802–3813.
44. Rossit, D. A., Tohmé, F., & Frutos, M. (2019). A data-driven scheduling approach to smart manufacturing. *Journal of Industrial Information Integration, 15*, 69–79.
45. Rossit, D. A., Tohme, F., & Frutos, M. (2019). Production planning and scheduling in cyber-physical production systems: A review. *International Journal of Computer Integrated Manufacturing, 32*(4–5), 385–395.
46. Sampson, S. E., Schmidt, G., Gardner, J. W., & Van Orden, J. (2015). Process coordination within a health care service supply network. *Journal of Business Logistics, 36*(4), 355–373.
47. Swan, M. (2015). *Blockchain: Blueprint for a new economy.* O'Reilly Media, Inc.
48. Tan, C., Chung, H., Barton, K., Hu, S. J., & Freiheit, T. (2020). Incorporating customer personalization preferences in open product architecture design. *Journal of Manufacturing Systems, 56*, 72–83.
49. Tavana, M., Shahdi-Pashaki, S., Teymourian, E., Santos-Arteaga, F. J., & Komaki, M. (2018). A discrete cuckoo optimization algorithm for consolidation in cloud computing. *Computers & Industrial Engineering, 115*, 495–511.
50. Wang, L., Törngren, M., & Onori, M. (2015). Current status and advancement of cyber-physical systems in manufacturing. *Journal of Manufacturing Systems, 37*(Part 2), 517–527.
51. Wang, L., & Wang, X. V. (2018). Cloud-enabled distributed process planning. In *Cloud-based cyber-physical systems in manufacturing* (pp. 105–123). Springer, Cham.
52. Yampolskiy, M., King, W. E., Gatlin, J., Belikovetsky, S., Brown, A., Skjellum, A., & Elovici, Y. (2018). Security of additive manufacturing: Attack taxonomy and survey. *Additive Manufacturing.*

53. Yao, X., Jin, H., & Zhang, J. (2015). Towards a wisdom manufacturing vision. *International Journal of Computer Integrated Manufacturing, 28*(12), 1291–1312.
54. Yao, X., & Lin, Y. (2016). Emerging manufacturing paradigm shifts for the incoming industrial revolution. *The International Journal of Advanced Manufacturing Technology, 85*(5–8), 1665–1676.
55. Yao, X., Zhou, J., Lin, Y., Li, Y., Yu, H., & Liu, Y. (2017). Smart manufacturing based on cyber-physical systems and beyond. *Journal of Intelligent Manufacturing*, 1–13.
56. Yli-Huumo, J., Ko, D., Choi, S., Park, S., & Smolander, K. (2016). Where is current research on blockchain technology?—a systematic review. *PloS One, 11*(10), e0163477.
57. Zein, I., Hutmacher, D. W., Tan, K. C., & Teoh, S. H. (2002). Fused deposition modeling of novel scaffold architectures for tissue engineering applications. *Biomaterials, 23*(4), 1169–1185.
58. Zhang, L. L., Lee, C. K., & Akhtar, P. (2020). Towards customization: Evaluation of integrated sales, product, and production configuration. *International Journal of Production Economics, 229,* 107775.
59. Zhi, W., Liao, H., & Emrouznejad, A. (2021). Information representation of blockchain technology: Risk evaluation of investment by personalized quantifier with cubic spline interpolation. *Information Processing & Management, 58*(4), 102571.
60. Zhong, R. Y., Newman, S. T., Huang, G. Q., & Lan, S. (2016). Big Data for supply chain management in the service and manufacturing sectors: Challenges, opportunities, and future perspectives. *Computers & Industrial Engineering, 101,* 572–591.
61. Zhou, Q., Huang, H., Zheng, Z., & Bian, J. (2020). Solutions to scalability of blockchain: A survey. *IEEE Access, 8,* 16440–16455.
62. Zutshi, A., Grilo, A., & Jardim-Goncalves, R. (2012). The business interoperability quotient measurement model. *Computers in Industry, 63*(5), 389–404.

Fernando Tohmé is a Principal Researcher of CONICET (National Research Council of Argentina) and Full Professor at the Department of Economics of the Universidad Nacional del Sur, in Bahía Blanca, Argentina. He holds an undergraduate degree in Mathematics and a Ph.D. in Economics (under the late Rolf Mantel). A former Fulbright Scholar he held visiting positions at U.C. Berkeley, Washington University in St. Louis and Endicott College. His research has focused on decision problems, game theory and optimization in socio-economic settings. He is currently working on a category-theoretical reformulation of game theory. Tohmé has been the thesis advisor of more than fifteen doctoral students in Economics, Engineering and Mathematics. He has published research articles in peer-reviewed journals like Theory and Decision, Mathematics of Social Sciences, Journal of Applied Logic, Physica A, Artificial Intelligence, Mathematical and Computational Modeling, and Annals of Operations Research among others.

Daniel Alejandro Rossit has an Industrial Engineer degree and a Ph.D. in Engineering from Universidad Nacional del Sur. He is an Assistant Researcher at CONICET (National Research Council of Argentina) and Teaching Assistant in the Engineering Department of the Universidad Nacional del Sur, Bahía Blanca, Argentina. His research has focused on production problems, operations research and engineering systems optimization. He has published in journals such as Omega, International Journal of Production Research, European Journal of Operational, Research, The International Journal of Advanced Manufacturing, Computers and Electronics in Agriculture, International Journal of Computer Integrated Manufacturing, among others.

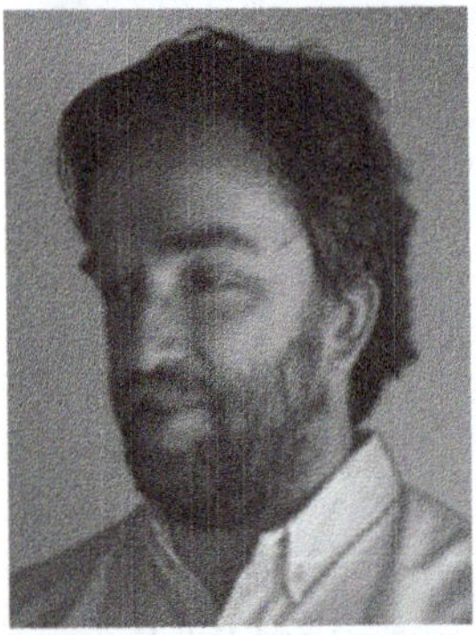

Mariano Frutos received the bachelor's degree in industrial engineering and the master's and Ph.D. degrees in engineering from his home university. He is currently an Independent Researcher of CONICET (National Research Council of Argentina) and a Teaching Assistant with the Department of Engineering, Universidad Nacional del Sur, Bahía Blanca, Argentina. His research is centered on scheduling problems in production and its treatment through metaheuristic methods. He has published in European Journal of Operational Research, Omega, Operational Research, Annals of Operations Research, IEEE access, Mathematics, Waste Management & Research, Journal of Industrial Information Integration, Computers & Operations Research and International Journal of Production Research, among others. He participates actively in Operational Research Communities in Latin America and Europe.

Óscar C. Vásquez received his MSc. Eng. mention in Industrial Engineering from the Universidad de Santiago de Chile, MSc. in Mathematics Applied from the Ecole Polytechnique, France and PhD in Computer Science from the Pierre and Marie Curie University. He is an associate professor at the Industrial Engineering Department, Universidad de Santiago de Chile. His research interests lie on decision problems from decentralized and centralized perspectives, involving combinatorial optimization, algorithmic game theory, algorithms analysis and continuous/discrete-event simulation. He has published in several international journal such as Applied Mathematical Modelling, Computers & Operations Research, Discrete Applied Mathematics, European Journal of Operational Research, Expert System with Applications, International Journal of Production Economics, Annals of Operations Research, Optimization Letters, Journal of Scheduling, Theoretical Computer Science, Waste Management, Resources, Conservation & Recycling, among others.

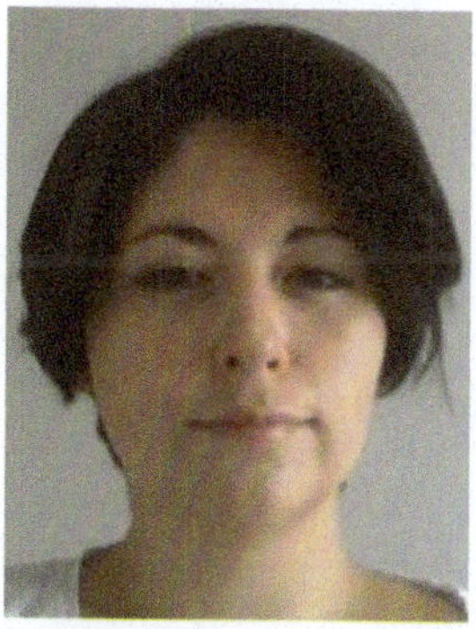

Andrea Teresa Espinoza Pérez was born in Santiago, Chile. She holds a Ph.D. in Industrial Systems Engineering and a Master degree in Innovation Management and Industrial Design from the University of Lorraine, France; a Master degree in Engineering Science, specializing in Industrial Engineering and a Diploma in Industrial Engineering from the University of Santiago, Chile. She is currently a researcher and professor at the Department of Industrial Engineering of the University of Santiago, Chile. Dr. Espinoza's research activities deal mainly with sustainable supply chain modeling and optimization. She has published her research in journals such as Cleaner Production and Renewable & Sustainable Energy Reviews. She also researches multiobjective optimization for natural resources like mining, decentralized energy and biofuels generation, and circular economy.

[illegible] Espinoza Pérez was born in Santiago, Chile. [illegible]

Frontiers of Blockchain for Railways

Mohita G. Sharma and Sachinder Mohan Sharma

Abstract The objective of this chapter is to provide the reader with the application of blockchain technology in railway system. As the focus of this book is Big Data and Service Operations Management, it is pertinent to understand Big data-blockchain relationship first before we can delve into Railway service operations and application. We begin by linking Big Data and Blockchain, followed by studying blockchain applications in various sectors, in order to provide a solid foundational understanding of the same. Thereafter, we set the context of Indian Railways, moving on to a structured understanding of the railway service and blockchain applications. followed by creating a specific provenance framework followed by conclusion, limitation. Big data signifies a novel technology paradigm for data that are generated at high velocity and high volume, and with high variety. The significance of Blockchain technology has drastically enhanced with the advent of Big Data because blockchain technology can help in recording and storing data for long duration. These technologies are envisioned as a game changer capable of revolutionizing the business drastically. Blockchain which is essentially a 'distributed ledger' has found applications in the financial and supply chain space. The value derived through this technology is in terms of transparency, veracity, track ability, traceability and auditing. The possibility of transparent information that can be traced to the origin of materials and processes and operations which are shared on blockchain ledger can provide product provenance, chain of custody and authenticity. Transport is an important service and represents the essence of service operations management. Railways is an eco-system. It provides a service that is asset intensive which has its own challenges in terms of maintenance and infrastructure. At the same time there are service concerns in terms of human safety and comfort. This provides a complex eco-system to work with. Blockchain as a digital tool intervention providing transparency, provenance and auditability can enhance the performance of the system This study attempts to create a model to understand the contribution of blockchain in the Indian Railway system.

M. G. Sharma (✉)
FORE School of Management, New Delhi, India

S. M. Sharma
GGM, DFCCIL New Delhi and Research Scholar IIM-Kozikhode, New Delhi, India

A. Emrouznejad and V. Charles (eds.), *Big Data and Blockchain for Service Operations Management*, Studies in Big Data 98,
https://doi.org/10.1007/978-3-030-87304-2_15

Keywords Blockchain · Railways · Service Operations · Provenance

1 Introduction

The Blockchain characteristics which make it a disruptive technology lies in its following characteristics. The data on Blockchain is always complete, accurate, reliable and authentic thereby providing Integrity. It is believed that the overall cost of the chain shall also decrease as many intermediaries which are into auditing, assurance and quality and thus might get supernumerary. It is Cryptographically secure which is due to the consensus mechanism. Further, Transparency can be attributed to the non- alteration feature along with the time-stamped version.

Railways is an eco-system. It provides a service that is asset intensive which has its own challenges in terms of maintenance and infrastructure. At the same time there are service concerns in terms of human safety and comfort. This provides a complex eco-system to work with. Blockchain as a digital tool intervention providing transparency, provenance and auditability and can enhance the performance of the system This study attempts to create a model to understand the contribution of blockchain in the Indian Railway system.

The chapter is organised as follows. First chapter explains the relationship between big data and blockchain. The next section looks at comprehending blockchain as a technology and its various applications in business. This is followed by the relevant background studies of blockchain to understand the spread of applications. This is followed by creating the background with the details of the context of Indian Railways. Further, mapping the provenance framework to blockchain application is accomplished followed by conclusion [15], future research and limitations.

2 Big Data and Blockchain

The era of 'Big data' has arrived. It is not limited to a mere concept, fad but has started impacting the lives and businesses significantly [15]. Big data is the compilation of huge data sets that can be analysed computationally to provide insights and patterns, trends and associations relating to behaviour and interactions [1, 2, 9, 10]. It has applications in almost all industries: retail, healthcare, financial services, government, agriculture, customer service among others.

In the initial study four specific features characterized big data [24]. These include:

- Volume which represents the large scale of big data, which requires innovative tools for their collection, storage, and analysis.
- Velocity refers to the rate at which the data are generated or updated, pointing to the real-time nature of big data.

- Variety refers to the variation in types of data. Big data can come in diverse and dissimilar forms from multiple sources, such as texts, spreadsheets, audios, videos, and sensors.
- Veracity refers to the complex structures of big data assets that make them ambiguous, imprecise and inconsistent.

The process of Big data analytics includes the process of examining large data sets to find and visualise business insights viz. hidden patterns and correlations, environmental impacts, business trends, customer choices. These insights and analytical results can help targeted approach towards marketing, better operational control and efficacy new revenue opportunities, and better customer service. The successful application of big data requires a well-researched and realistic deployment of big data analytical tools. The steps include data generation, data acquisition, data storage, advanced data analytics, and decision making for value creation. One of the ways of recording and storing data is to use cloud-based storage, but it is an expensive solution. Further, Blockchain provides a cheaper, efficient solution for storing data through its traceability and distributed ledger solution. Its ability to guarantee the accuracy and transparency of data has shifted the focus onto blockchain.

Blockchain can be an efficient mode for online data storage [11]. In addition to this, the decentralized network can grant access to many users. Various parties in a transaction store the transactional information in different ledgers. All of these parties can be granted access to one single network with the help of Blockchain. Transactions can be recorded in the network which can be further validated by all the concerned parties. Since all the information will be stored in the Blockchain it will be convenient to access these details. Because of the design of the Blockchain technology, users can view historical transactions effortlessly. Tracking the origins of a transaction becomes effortless.

3 Comprehending Blockchain Technology

Blockchain technology can be perceived as a kind of open distributed digital ledger that can record transactions between two parties efficiently and in an authentic way. Rather than relying on a third trusted party such as a bank, it uses a peer to peer approach where verification and validation are done by the nodes of the distributed network. Not only this technology eliminates the involvement of a Third Party for the purpose of authentication (because very often these third parties become the source of security breach). Using this approach, the possibility of fraudulent transaction is minimized to zero because the group of transactions called blocks is not only protected by digital signature that subsequently goes through a mathematical puzzle for time alignment, but also needs to be the longest and fastest one in the network amongst the existing good nodes to succeed. This means that all transactions are fully secured and fraudulent transactions, if any shall form a long chain of such blocks which will be transparently visible to all parties clearly identifying the perpetrators for disciplinary action and also such fraudulent transactions will be next to impossible to get overlooked. Thus this mechanism is completely secured from hacking by outsiders through spoofing, phishing, etc.

4 Recent Background Studies

Researchers have combined supply chain and block chain and analysed the potential for value creation for various stakeholders both in the upstream and downstream business ecosystems. The model builds a common IT based platform centred on distributed data storage using the node consensus algorithm to generate and update the same. The fact that the data is encrypted ensures security of transmission and restricted access. This helps in speeding up the cash flow in an enterprise and thereby improving asset creation and utilization [19]. Studies cite that Blockchain is a digitally signed account transaction with a highly trusted record [3, 4]. in some policy studies it has been shown that Transparency, traceability and auditability provided by blockchain help minimize system corruption and fraud while keeping the system under control [3, 14]. Integration of blockchain with supply chain would ensure product tracking which is a big challenge; [5, 6]. The intermediary can be eliminated and information exchange between parties is structured and controlled. It has been propounded that overall efficiency of the system is increased because of transparency provided by physical traceability in supply chains, information and financial security in management systems, leading to an increase in overall efficiency [6, 8]. Blockchain technology has the ability to address all three dimensions to address collaboration, cost optimization, and risk management. This technology allows companies to gain total access of the end-to-end supply chain data while ensuring that it can be securely stored and privately shared with authorized stakeholders [23]. There are many industry specific studies for application and challenges of block chain application like the construction industry [22]. Researchers have looked at the possibility of Distributed Ledger Technologies (DLTs) to overcome such events [16]. As has been illustrated by Muniandi blockchain can help achieving the goals of providing a platform with more transparency, fit for use by all passengers without any exclusions, prevents siloation and prevents frauds.

Researchers have looked at the potential application of block chain for digital ticketing and enhancing customer experience [20, 21]. This can be done by a mix node which may be active or passive. The passive nodes only maintain the latest data and do not perform any mining operation [18]. Since, the data is verified by simultaneously cross checking with adjoining nodes, it is fail safe, collusion resistant and resilient to external attacks. Thus this technology is useful in various facets of train operation and maintenance [20, 21]. Researchers have also found utility for block chain in ensuring safety in heavy haul operations by reducing manual decision making [12, 13].

5 Application Context of Indian Railways

Railways are the preferred choice of transportation mode in the world population due to their environment-friendly, safer, long-distance interstate, and intercity mobility options. Indian Railways (IR) is the fourth largest Railway network in the World and considered the lifeline of the country. About 22,000 trains run on the network in a day. It therefore has a large fleet of about 11,000 locomotives, 70,000 coaches and about 2.94 lakh wagons. During 2018–19, it carried 8,439 million passengers and 1221.48 million tonnes of freight traffic. The present rail share of freight is 27% and is mainly attributable to traditional bulk commodities like fertilizers, coal, iron ore and food grains. These constitute approximately ~60 % of IR's freight business. IR needs to cater to freight increase for commodities which are dominated by road movement through appropriate strategies/product offerings. Such a modal shift is possible by improving the transit through speed and predictability and by improving the capacity through better terminal and network access and improved availability of rolling stock. The service in terms of product offering and costing also has to improve to meet the customer needs in real time. Thus the objective of railway to provide a safe, secure, dependable, economical and reliable service. It represents a service which is both technology and human intensive and applies all the essential elements of service operations. This is the basis of including the chapter in this book.

The application of Blockchain in Railways can be classified into three broad categories (Fig. 1).

1. Infrastructure and Real Time Operations.
2. Equipment Maintenance Management.
3. Customer oriented Service Operations.

Fig. 1 Application of blockchain in railways

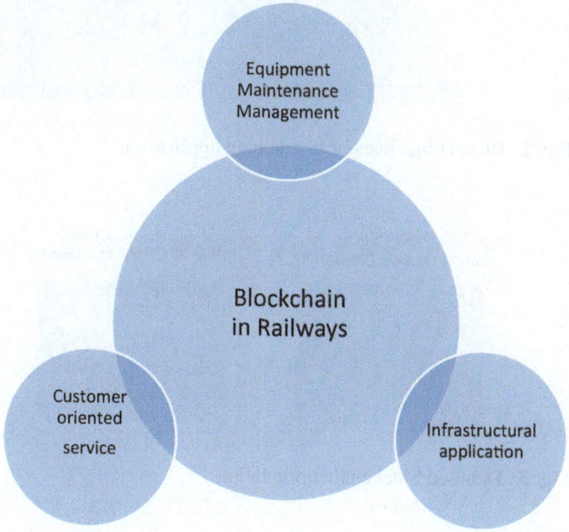

Given the objective of Indian railways to provide safe, reliable, dependable and comfortable service, the blockchain intervention shall secure both financial transactions but also train running (Fig. 2).

Railway systems are infrastructure intensive and these have to be maintained in proper fettle so that trains can ply smoothly. The fixed infrastructure consists of tracks, overhead wires for electric traction and the signaling system. The moving assets are the rolling stock consisting of the coaches, wagons and locomotives. The Fig. 3 gives a succinct details of the application which has been elaborated further.

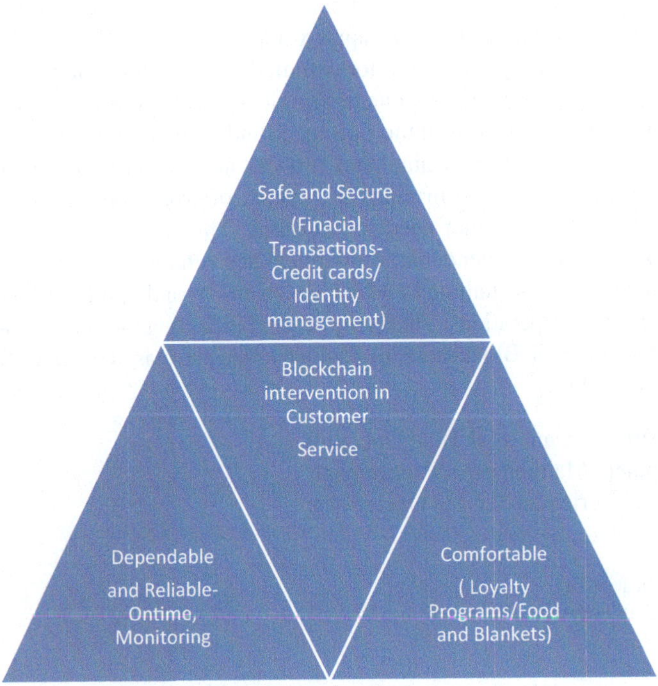

Fig. 2 Broad objectives of blockchain application

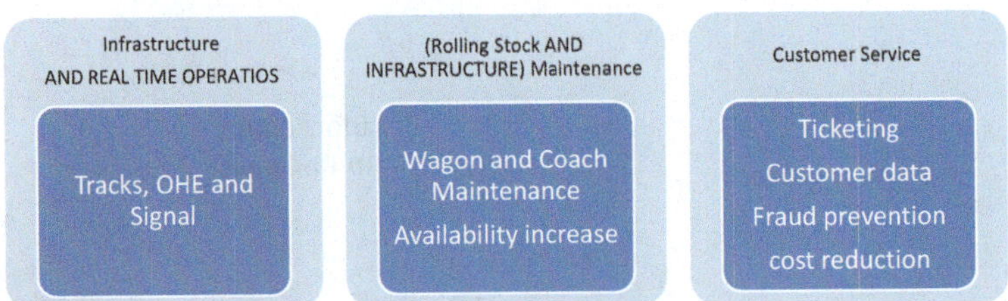

Fig. 3 Detailed blockchain application

5.1 Infrastructure

With the primary focus on safety and given the enormity, complexity, heterogeneity an interoperability issues of the present system, blockchain can be a solution for the same. The existing system in Indian Railways has decentralized train operations. There are huge challenges which include system dissonance at the interface, communication failure, SPAD due to wrong signal. The fixed infrastructure is monitored through IT systems like the track management system for track, the Supervisory control and data acquisition (SCADA) system for the overhead wires and the electronic interlocking systems for the signaling system. These systems presently work in isolation, however the benefits can be enormous if these communicate with each other and share relevant data authentically and proper coordination is there. Through the application of blockchain, the actors in the system can access data real time and transparently.

The condition of the infrastructure elements can also be captured in tamper-proof records. This data can further be utilized to ensure that the work of maintenance is carried out in an optimized manner thereby reducing the down time and improving the availability for train operations. Communicating through smart contracts with the track elements which include level crossings, signals, interlocking track circuits, releasing them and setting diversion points will provide efficiency in the system. No manual intervention will be required and would be done by the automated central system based on rules and without conflict. This can enhance the total track availability of the track thereby impacting the throughput of the track.

Presently train operations are controlled centrally, however these can be made more dynamic by decentralising the decision making of routing in a manner which is safe, reliable, efficient and auditable. The European train control system which Indian Railways plans to adopt is a step in this direction as it maintains "safe blocks" in which trains operate and takes corrective action in case the operator does not react timely. Most of the operations change every minute based on failure of equipment or other unforeseen circumstances. This decentralized system shall make it possible for the trains and track elements to work as first-level active participants and carry out several actions such as route locking, setting a changing point position, communicating with other trains, automatic speed control and braking, GPS drove Train localization. All these actions shall be network-driven and ensure compliance with the standard working rules. The reliability of which shall be derived from the matching of physical reality (trackside equipment) and the IT counterpart (Digital twin). This would enable the detection of trackside malfunctions and the prevention of accidents. Going by the cardinal principles of Railway Signaling which give utmost priority to the safety of human life, the validation of every decision on the go shall be done by strong authorities as multiple parties such as station masters. Or even a better alternative will be to involve the signaling rules and control procedures in the smart contract, a copy of which shall be carried by each train to be followed while making their own route-based decisions. In this case, a Blockchain API shall be superimposed over such Rail network enabling all the trains with conflicting movements for a particular route to act as the 'active nodes' dynamically and make a 'consensus-based-decision' as elaborated in the schematic.

Further, combination of these emerging technologies like blockchain, VC, RBC, the heartbeat signal from trains, ATO control algorithms, 5G using train-to-track/track-to-track communication links, can solve the problem of railway traffic conflict control along with the objective to minimize or nullify the overall individual train delay progressively. Blockchain can be used for all other transactions for example, for checking the driver's license and driver authorization, checking the compatibility of sub-systems and billing the infrastructure. As of now, the technology is in an experimental state where potential use cases are being developed and tested to affect the following areas.

5.2 Maintenance (Rolling Stock and Infrastructure)

Trains are complex equipment and have large number of subparts and assemblies. Maintenance is an important function to meet the objective of reliability, safety and dependability of the train. Maintenance are broadly of two types: Preventive-Scheduled and Reactive-Unscheduled Maintenance. It is seen that with higher preventive maintenance, need and cost for reactive maintenance gets reduced. Further, reactive maintenance has more negative impact because of unscheduled breakdowns and higher downtime cost. Scheduled maintenance are triggered by time period, usage or monitoring the condition of parts. Currently the rolling stock is maintained based on preventive maintenance and follow scheduled repairs based on time or usage. Another possibility is to have condition based monitoring wherein it is essential to monitor the performance of the parts in real time based on the condition of the equipment. This implies that we can predict when things will go wrong and pull back the train for repairs at the right time. Also these individual components or sub-assemblies are connected together and their performance can be observed in tandem. Remote monitoring of diesel locomotives by providing sensors in critical subassemblies has been in practice. The alerts are sent to a central control centre and corrective or preventive action initiated to prevent failure. These alerts can also be shared in real time with the manufactures and designers and this will help to close the loop and ensure better quality of the products as all defects and failures get logged and can be investigated using the data prior to the failure. It will be easier for the suppliers to link these to individual batches in their production and they would be able to zero in on the operators and the machines. The transaction costs of this traceability which is presently done manually will also be nominal.

This would also help in reduction in the warranty claims and effective spare parts management. The spares mostly rotables can be tracked and performance logged, thereby providing better control at a reduced cost. The onetime cost of the technology has to be traded off against the long-term benefit.

Since the components from many Original equipment manufacturers (OEMs) have to talk to each other, blockchain technology can be the solution for securing the network and linkages with complete traceability. Similar usage can be there in the smart coaches being developed by Indian Railways wherein different components for passenger interface, electrical and braking systems, water and infotainment etc. are monitored through sensors. Once this is done then all the data from all the coaches and locomotives can be shared and used for improving the maintenance thereby reducing the downtime and enhancing the utilisation of the asset. Further, there are track side equipment like hot box detectors, wheel impact load detectors etc. which externally monitor the condition of rolling stock. The data from these can also be integrated to enhance the overall output and understanding of the rolling stock. The movement of these stock is being planned based on the operational software which can share inputs with this system and this will ensure that the maintenance can be planned with least disturbance to the movement of trains. All these are important in the context multi operator regime when clear responsibility and traceability has to be ensured from a legal and contractual perspective. Smart contracts can be implemented to make rule based contracting. Further block chain supported by IOT and sensor technology shall provide the fundamentals for modern maintenance. IOT based condition monitoring and sensors which shall form dense, connected networks will transmit data to central data centres but alongside the strengths the threats to the system is equally precarious. Blockchain through its encrypted tamperproof system can provide a trustworthy system which is verifiable and the data trail is auditable.

5.3 Customer Service

The next set of service operations application for Blockchain is focussed around the customer experience and journey seamless and hassle free. Presently the customer buys tickets and orders for services like E-catering and blankets for which payments are made through different gateways and there is also the need to authenticate the identity of the traveller to ensure that only authorised passengers are travelling. This is even more important from the security perspective, Blockchain ledger can be useful in ensuring the above and facilitating instant payments with no need to authenticate further and also guaranteeing the identity. This can further be extended to intermodal transfers and movements which will further enhance the customer experience and provide seamless transport solutions. These would help journey planners to provide end-to end solutions across sea, road, air and rail. The same is the case for freight customers or in the parcel business where the need to carry the railway receipts as proof will be eliminated with use of blockchain technology. This will also be useful in delivering consignments to third parties and avoid documentations like Goods Receipt Note. Challenges of Short delivery, Incorrect deliveries can be addressed like many times the consignment would not get loaded in the train or would get over carried. Block chain will be a reliable and accurate medium to confirm the transaction. This shall help in building the brand image of the Railways as a safe and secure service provider and earn the loyalty of the customers.

There has been a focus on customer retention by rewarding travellers for their loyalty. It is being done both by the credit card companies and railways. For this, loyalty programs need to be accurate and user-friendly to ease the complexity involved in keeping an account of passenger's loyalty points and redemption as per the defined policies. In accuracy is a deterrent to future loyalty. Blockchain can be used to develop a platform to improve the interoperability between multiple schemes but also allows invoicing and payment transfer between partners using smart contracts. Redemption of reward points from one service to another such as automatic credit of points after a train journey, which can further be used to pay an online cab service shall add to a better travel experience.

For undertaking the journey in any reserved class on Indian railway, a prescribed proof of identity shall be available online. On the blockchain, identification of the passenger, change of booking or enquiry can be done using traveller IDs which eliminates the need to carry any hard copy of id proof during travel. This will enhance the travel experience of the passenger and the same can also be used for seamless inter-modal connectivity as most passengers use road or metro etc. for the first mile and last mile connectivity to the rail system.

Automatic billing can be done on Rail networks which provide access to other passenger and freight railways at regulated infrastructure charges. The passengers buy tickets manually and the payments to subcontractors and vendors are made through traditional methods. Blockchain technology enable all these payments to happen instantly on a secure network using Blockchain ledger. The apportionment of earnings between various players and also between various railway systems can be done easily without having system of manual checks which add to the inefficiency and delays. This will simplify the accounting as well as the statistical formulations. The traceability will also help in case of accidents where compensation is tenable as this can be handled easily and securely using the block chain technology.

On the freight side too blockchain can help connect the customer and the logistics provider in a better way. The access to real time loading information, train and truck movement during transfer and availability of crew can help reduce the cycle time and thereby bring down the logistics cost. The smoother flow of authenticated information across the various players can be a game changer and this can reduce the needs for the manual interfaces used presently to collect the information which is subject to noise and leads to sub-optimal decision making and planning.

6 Mapping the Railway Application to the Provenance Framework

From the literature, a Standard Provenance knowledge framework has been developed by [17] that has its application in showing how provenance knowledge can enhance assurances and reduce perceived risks via the application of blockchain (Fig. 4).

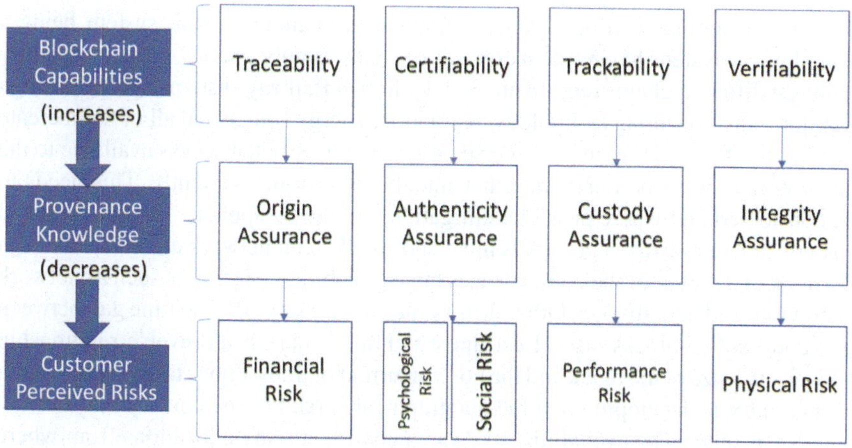

Fig. 4 Standard provenance knowledge framework

This model is very comprehensive and captures the contribution of Blockchain very succinctly. The risks perceived by the customer can broadly be classified into physical risk, financial risk, psychological risk, social risk and physical risk. These risks help in providing origin, authenticity, trust of custody and integrity to the supply chain. This eventually links to the blockchain capability of traceability, Certifiability, trackability and verifiability.

This standard model was explained by the authors to the focus group members. The focus group of seven Railway Experts through a brainstorming exercise have mapped it to the Railway Application Provenance Model. One of the author is a railway expert with 30 years of experience in Indian Railways. The total experience of the focus group is 140 years. This provides an understanding of the intrinsic reason of blockchain application.

It was observed that the blockchain capabilities that provide value to the chain are different. Whereas in the infrastructure and rolling stock vertical, the value is derived through trackability but in the case of Customer service it is authenticity of transactions and assurance that is required (Fig. 5).

Fig. 5 Railway application provenance model

Implementation of Blockchain are filled with challenges. Rail system being a complex ecosystem having convergence of many supply chains – interoperability amongst different chains require huge scale. Indian Railway system is going through a major transition along technology, regulation, energy source and all of them create their own set of uncertainties. Cybersecurity and unrestricted access available to the auditors may itself be misused, either intentionally or inadvertently. This needs to be considered and where possible, mitigations should be applied. The geographical span of Indian railways network is huge and Blockchain network alongside through the nook and corner of the country faces huge scalability constraints such as network bandwidth and data storage. Other factors such as size of block and time gap between two consecutive blocks also effect the scalability. There is an inverse relationship between the size of the block and the no. concurrent of transactions taking place. India as a country is developing and the geographical spread of the country and railway network is huge. The availability of 24×7 network at remote locations from where the data is generated can also be a constraint in the implementation, however, this is a steady slow process of technology acceptability and eventually the infrastructure constraints are overcome.

7 Conclusion

Transport is an important service and represents the essence of service operations management. Amongst the different modes of transportation railway system remains one of the most complex systems. In this book on 'Big data and service operations management' in this chapter, we have explored Blockchain-based application in the railway ecosystem from service operations perspective.

This is an era of Big data. The volume and variety of data generated is enormous and it is generated by human-users and machines. Any business or organization that can assimilate data to address the difficult questions about their operations can benefit from big data. the demand for big data transcend across all sectors and business. Those who work to understand their customers' business and their problems will be able to proactively identify big data solutions appropriate to their needs, and thus gain competitive advantage over their competitors. Blockchain which is radically changing the future of transactions in industries can help in smart contracts and as a digital currency. It can help in digital record keeping in operations and maintenance as well as for IPRs and Research. It can help in descriptive analytics, diagnostic, prescriptive, predictive and cognitive analytics.

Although the technology of Blockchain has been there for quite some time, but the advent of big data has made the application of Blockchain more significant. Blockchain can effectively store online data and perform monitoring and controlling accessibility of data across in an efficient manner. Various parties in a transaction store the transactional information in different ledgers. All of these parties can be granted access to one single network with the help of Blockchain.

Broadly considering three verticals along which the application can be categorized. These include infrastructure, rolling stock and customer service. All three orchestrate synchronously to provide the safe and reliable service. Thus we can understand what happened, why it happened and what will happen. Organizationally, this digital transformation which will have a cross functional impact will need multi-disciplinary teams for innovating, operationalizing and transforming. Each department like accounts, HR, operations, infrastructure, business development and procurement will be impacted and will need to modify their systems and processes to cope with the change. Thus IR will need to change how it educates its managers to understand, assimilate and champion the transformation. This would be a cultural transformation and would also need chief technology officers at all levels so as to help leaders think, act and react differently. Since new possibilities will arise they need to comprehend the same and think divergently. Teams will become collaborative and mobile and it will be easier to connect virtually across the geographical boundaries.

We can use simulation based decision making and the system can self-learn thereby reducing the human intervention and improving the quality of decision making. Each of these vertical application is a study in itself, infest with its own specific challenges. These can be specific directions for the study.

This is a conceptual paper of application of blockchain for Indian railways. The contribution of this paper is to provide a structure to understand blockchain application in Indian railways and provide a framework for blockchain applications in railway specific operations management. Future research can look at technology adoption models and being a complex ecosystem, the technology adoption and diffusion also will be fest with complications and difficulty. Models can be devised to address these issues.

The technology adoption will also include change management and human resource challenges which can be researched. This shall have socio-technical ramifications. This can also be a potential area of research. The interface and interaction of blockchain and supply chain is significant and has to prudently understood and analyzed before and during implementation. The supply chains of electronics, mechanical, electrical spare parts have to be properly aligned.

The limitation of this study is that it has limited number of members in the focus group and biases might arise. Further, the focus has been on one technology but AI, IOT along with Blockchain will provide a more comprehensive understanding and trajectory of technology change. Future studies can include the total impact of all these technologies jointly.

To Conclude, understanding the tradeoffs of security, privacy, transparency, functionality, auditability, risk, and scalability are delicate balances which have to be analyzed while consideration of the blockchain implementation in Indian railways.

Acknowledgement The infrastructural support provided by the FORE School of Management, New Delhi in completing this paper is gratefully acknowledged.

References

1. Charles, V. & Emrouznejad, A. (2019). Big Data for the greater good. In: *An introduction: Big data for the greater good.* Studies in Big Data. s.l.:Springer.
2. Charles, V., & Gherman, T. (2019). Big data analytics and ethnography: together for the greater good. In: *Studies in big data.* s.l.:Springer.
3. Choi, T. (2020). Creating all-win by blockchain technology in supply chains: Impacts of agents' risk attitudes towards cryptocurrency. *Journal of Operational Research Society.*
4. Choi, T., Feng, L. & Li, R. (2020a). Information disclosure structure in supply chains with rental service platforms in the blockchain technology era. *International Journal of Production Economics, 221*, 1–14.
5. Choi, T., Guo, S. & Luo, S (2020b). When blockchain meets social-media: Will the result benefit social media analytics for supply chain operations management? *Transportation Research Part E, 135*(101860), 1–21.
6. Choi, T., & Luo, S. (2019a). Data quality challenges for sustainable fashion supply chain operations in emerging markets: Roles of blockchain, government sponsors and environment taxes. *Transportation Research Part E: Logistics and Transportation Review, 131*, 139-152.
7. Choi, T., Wen, X., Sun, X. & Chung, S. (2019b). The mean-variance approach for global supply chain risk analysis with air logistics in the blockchain technology era. *Transportation Research Part E: Logistics and Transportation Review, 127*, 178–191.
8. Dutta, P., Choi, T. M., Somani, S., & Richa, B. (2020). Blockchain technology in supply chain operations: Applications, challenges and research opportunities. *Transportation Research Part E: Logistics and Transportation Review, 142*(102067), 1–33.
9. Emrouznejad, A. (2016). Big data optimization: Recent developments and challenges. In: the series of *Studies in big data.* Verlag: Springer.
10. Emrouznejad, A. & Marra, M. (2017). Big data: Who, what and where? Social cognitive and journals map of big data publication. In: *Big data optimization: Recent developments and challenges, studies in big data.* s.l.:Springer.
11. Es-Samaali, H., Outchakoucht, A., & Leroy, J. P. (2017). A blockchain-based access control for big data. *International Journal of Computer Networks and Communications Security, 5*(7), 137–147.
12. Gaofeng, H., et al. (2020a). Blockchain based federated learning for intelligent control in Heavy haul railway digital object identifier. *IEEE Access, 8*, 176830–176839.
13. Gaofeng, H., et al. (2020b). Blockchain-based federated learning for intelligent control in heavy haul railway. *IEEE Access, 8*, 176830 176839.
14. Hastig, G., & Sodhi, M. (2019). Blockchain for supply chain traceability: Business requirements and critical success factors. *Production Operations Management, 29*(4), 935–954.
15. Lee , I. (2017). Big data: Dimensions, evolution, impacts and challenges. *Business Horizons, 60*(3), 293–303.
16. Kuperberg, M., Kindler, D. & Jeschke, S. (2020). Are smart contracts and blockchains suitable for decentralized railway control? *Ledger 5.*
17. Montecchi, M., Plangger, K. & Etter, M. (2019). It's real, trust me! Establishing supply chain provenance using blockchain. *ScienceDirect,* 283–293.
18. Muniadi, G. (2020). Blockchain-enabled virtual coupling of automatic train operation fitted main line trains for railway traffic conflict control. *IET Intelligent Transport Systems, 14*(6), 611–619.
19. Peng, X., et al. (2020). Research on financial platform of railway freight supply chain based on blockchain. *Smart and Resilient Transport, 2*(2), 69–84.
20. Preece, J. D. & Easton, J. M. (2019a). A review of prospective applications of blockchain technology in the railway industry. *The International Journal of Railway Technology,* 1–21.
21. Preece, J. D., & Easton, J. M. (2019b). *Blockchain technology as a mechanism for digital railway ticketing,* 3599–3606.
22. Sharma, M. G., & Kumar, S. (2020). The implications of blockchain as a disruptive technology for construction industry. *IIM Kozikhode Society and Management Review, 9*, 177–188.

23. Tata, C. S. (2019). *Blockchain for a robust and efficient supply chain.* Tata Consultancy Services Limited, Mumbai.
24. Sivarajah, U., Kamal, M. M., Irani, Z., & Weerakkody, V. (n.d). Critical analysis of big data challenges and analytical methods. *Journal of Business Research, 70*, 263–286.

Mohita Gangwar Sharma is a Professor at FORE School of Management, New Delhi. An FPM (IIM-Lucknow), Masters (IIFT) and B.Tech. -IIT-BHU, she has more than 23 years of experience in the Industry (manufacturing and service) and Academics. She is a Chevening Rolls-Royce Science and Innovation Scholar from SAID Business School-, Oxford and N.T.S.E scholarship by NCERT. She has published in eminent academic journals: Transportation Research-D, IJLM, and Ivey case studies. She leads several executive education programs in the area of Supply Chain, Purchasing, Project Management and Quality Management. Her areas of research: Operations Management and Operations Strategy.

Sachinder M. Sharma is a Railway Mechanical Engineer with 30 years of experience in the transport sector, specifically Railways. His expertise is in rolling stock maintenance and operation, disaster management, fuel and alternate fuels, crew scheduling and in managing large teams. He is a fellow of the Institution of Mechanical Engineers, U.K and India. He has a MPA from the National University of Singapore, is a Chevening CRISP Scholar from the University of Oxford, U.K and is presently perusing his PHD from Indian Institute of Management, Kozhikode. He has various academic publications and has also been a faculty at the National Academy of Indian Railways.

Blockchain Interoperability Issues in Supply Chain: Exploration of Mass Adoption Procedures

Yaşanur Kayıkcı and Nachiappan Subramanian

Abstract Today's supply chains are becoming more data-driven with the impact of big data, but there are many challenges that need to be overcome in big data for better service operations management in supply chains. Blockchain has the great potential to improve big data services and applications with its decentralisation and security nature. However, blockchain interoperability is critical to realising more value creation in blockchain networks and achieving promising results for global supply chains that intersect with multiple business ecosystems and blockchain platforms. In addition, it is unclear how to address interoperability issues for mass adoption. In this chapter, a three-step approach is applied to analyse blockchain interoperability in supply chains for mass adoption. First, a literature review is conducted to explain blockchain technology and the widely used methodologies for blockchain interoperability in supply chains. Then, four real-case blockchain use cases in supply chains from different industry segments are analysed in terms of their technical capabilities addressing interoperability concerns. Finally, we discuss results of use case analysis based on the comments of interviewees. The analysis reveals that REST-APIs with a common interface and GS1 standards are very useful to integrate with blockchain applications in supply chains for mass adoption.

Keywords Blockchain interoperability · Supply chain · Use case · API · Blockchain ecosystem

---

Y. Kayıkcı (✉)
Department of Industrial Engineering, Turkish-German University, Istanbul, Turkey
e-mail: yasanur@tau.edu.tr

N. Subramanian
Science Policy Research Unit, University of Sussex Business School, Brighton, UK
e-mail: N.Subramanian@sussex.ac.uk

A. Emrouznejad and V. Charles (eds.), *Big Data and Blockchain for Service Operations Management*, Studies in Big Data 98,
https://doi.org/10.1007/978-3-030-87304-2_13

1 Introduction

Big data is dramatically revolutionising today's supply chains [15]. As big data becomes a larger and more user-friendly asset, supply chains are structured to be more data-driven for better service operations management [6, 7]. Vast amounts of data are used in supply chains, they can be surrounded by structured/unstructured data, and decentralised and connected data streams [14]. As a decentralised, ledger-based technology, blockchain technology plays a critical role in managing such big data to reduce risks, increase efficiency and visibility, and maintain transparency across end-to-end supply chains [23].

Blockchain is a digital data management infrastructure for big data that is both stable and flexible [8]. Many public and private blockchain applications with varying features have been developed since Bitcoin was launched in 2009. As companies develop blockchain solutions, they are increasingly concerned about the proliferation of blockchains, the linking of blockchain applications developed by several providers, and legacy systems, the fear being of locking up solutions too early [26]. For this reason, blockchain technology must develop standards and multi-channel information exchange protocol [10]. Blockchain technology, like any other evolving technology or platform, has several technical components and factors in order to operate efficiently [41]. In the present context, there are several blockchain applications that address supply chain challenges and utilise existing blockchain platforms, such as Ethereum, R3 Corda, Hyperledger Fabric. The key concern in having blockchain-as-a-service (BaaS) architecture for mass adoption is the access, use, transfer and interpreting of data cohesively between different versions of platforms in a heterogeneous blockchain landscape [9, 25, 35]. Since a single blockchain network cannot meet all the needs for any given business transaction, multiple networks will need to collaborate to offer unique value. This is often referred to as interoperability.

The exchange of information for different blockchain transactions, or the use of information by different blockchain providers, necessitates an important interoperable architecture [13]. As the number of blockchain applications in the supply chain rapidly increases, companies prefer blockchain platforms that they can use without limitation in external collaboration options in the future. Therefore, interoperability is a critical concern for decision-makers interested in building blockchain solutions [2, 38]. Interoperability is an important element in the maximisation of efficiency and effectiveness when working with various public and private blockchain systems as well as non-blockchain systems together.

This chapter provides an insight into blockchain interoperability in supply chains for mass adoption. A three-step approach is applied through conducting a literature review of blockchain technology and commonly used methodologies for blockchain interoperability, analysing four blockchain real-life use case applications in supply chains that address interoperability concerns and mass adoption, and discussing results of the analysis based on the comments of interviewees. The rest of the chapter is structured as follows: Sect. 2 gives a broad overview of blockchain technology

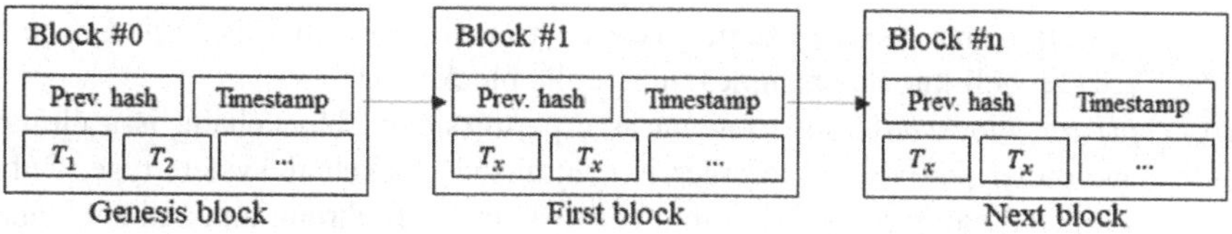

Fig. 1 Blockchain structure. *Source* Authors

and its application in supply chains. Then in Sect. 3, the context of blockchain interoperability is explained in terms of data interoperability standards and cross-chain interoperability. Section 4 provides insights into blockchain interoperability and mass adoption using four blockchain use cases in the supply chain. The chapter ends with conclusions and future research directions in Sect. 5.

2 Scope of Blockchain Technology

2.1 *Blockchain Definition*

Blockchain technology is a distributed ledger technology. It is simply a digital database of transactions used by an extended network to record, share and synchronise without having central authority [34]. Each transaction included in blockchain gets encrypted with its own key pair and linked to the previous and next transaction chronologically. Groups of transactions are blocked together, named the block; this contains transactions, data and references to the previous blockchain (creating the chain), shown in Fig. 1 [30]. Every block with its own hash (digital fingerprint) is added to the next one, together with the timestamp, forming an unalterable chain. Blocks to be added to the blockchain must be achieved through consensus.

2.2 *Blockchain Types*

Not all blockchains are the same. There are four different types of blockchain systems [34]: public blockchain, private blockchain, consortium blockchain, and hybrid blockchain. Each type of blockchain meets specific objectives and requirements:

(i) *public blockchain*, also known as a permissionless blockchain, is the most used blockchain system type. Nobody requires permission to join, interact with, or forge consensus. This blockchain is an open, decentralised, secured, and immutable ledger system. However, the consensus efficiency is very low and the transaction speed is very slow. A transaction is recorded anonymously on the blockchain, but is transparent to anyone. Therefore, these are often ideal for operating and managing digital assets/currencies. Bitcoin, Ethereum, Litecoin are the well-known examples of a public blockchain.

(ii) *private blockchain*, also known as a permissioned blockchain, is a closed and either partially or completely centralised blockchain system type. Only validated participants can join a network and only through an authentic and verified invitation. A single trusted central organisation has authority over the network and it establishes rules dictating network participants who can see, read, write or validate transactions on the blockchain. Furthermore, this organisation has the right to override, edit or delete transactions on the blockchain. This type of blockchain is highly efficient as verification is done by the owner of blockchain network. However, it is difficult to align different organisations to the same blockchain. Hyperledger Fabric, Ripple, MultiChain are typical examples of a private blockchain.

(iii) *consortium blockchain*, also known as a federated blockchain, is a partially decentralised, semi-private blockchain system type. This is governed by a group of organisations (consortium), rather than a single organisation, that shares the responsibilities of maintaining a blockchain. This blockchain is best suited for organisations that need both public and private blockchains. Only a few pre-selected organisations or nodes in the network have the right to authorise transactions and supervise the consensus process. Therefore, there is no single authority in control. IBM Food Trust is a typical application of consortium blockchain. Quorum, R3 Corda are well-known examples of a consortium blockchain

(iv) *hybrid blockchain*, which means controlled access and freedom at the same time, is a combination of the public and private blockchain. This blockchain combines the features of both types, although is not open to everyone, and still offers blockchain features such as integrity, transparency, and security. The system architecture is fully customisable; the participants can decide who can participate or which transactions are made public. This provides the best solution from both public and private and ensures a company to work with its stakeholders in the best possible way. Dragonchain, XinFin, Libra are typical examples of a hybrid blockchain.

2.3 Evolution of Blockchain Ecosystem

Blockchain technology has evolved over the years and its evolution can be divided into three eras, as shown in Fig. 2:

(i) *Blockchain 1.0 era*: The first-generation blockchains started with the emergence of Bitcoin in 2009. This was the first blockchain application and dominated for the first six years [26]. The public blockchain applications enabling digital currency transactions emerged in this era; however, they had multiple drawbacks, including efficiency and scalability. The blockchain ecosystems

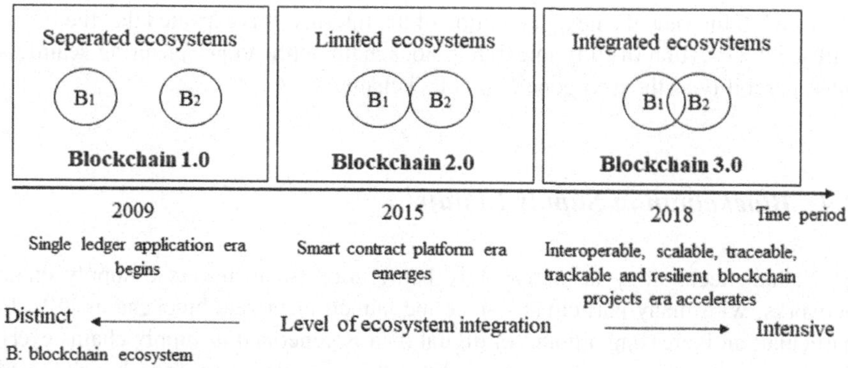

Fig. 2 The evolution of blockchain ecosystem. *Source* Authors

were separated in this era, meaning that there was no integration between different blockchain systems, so no interoperability.

(ii) *Blockchain 2.0 era*: The second-generation blockchains began in 2015 with the launch of Ethereum, a decentralised platform that allowed companies to use smart contracts to build a set of applications in supply chains [26]. The private blockchain emerged in this era, but they did not meet every use case requirement. There was only limited integration between blockchain ecosystems, so there was limited interoperability based on smart contracts.

(iii) *Blockchain 3.0 era*: The third-generation blockchain had a less identifiable launch, but the term is used to describe numerous applications including improving scalability, interoperability, security, privacy, resilience and sustainability [1, 26]. The Blockchain 3.0 applications aimed at seamless integration across multiple public blockchains (Bitcoin and Ethereum), multiple private blockchains (Hyperledger Fabric and R3 Corda), public and private blockchains (Ethereum and Hyperledger) and blockchains with legacy systems (Ripple and SWIFT).

Organisations are willing to implement blockchain technology in their ecosystems. However, the technical complexities in building, operating, and maintaining blockchain infrastructure result in creating isolated systems and not allowing full integration between different blockchain ecosystems. These constitute obstacles to the mass adoption of blockchain technology. BaaS vendors develop and deliver solutions to support organisations to overcome interoperability barriers in using this technology. For that reason, Blockchain 3.0 has further expanded to enable the multichain future, especially in 2021. Cross-chain interoperability solutions are being developed with new decentralised protocols such as Polkadot, Cardano, Kusama and more. Also, blockchain markets and use cases like DeFi (Decentralised Finance), NFTs (non-fungible tokens), and decentralised applications have grown extensively. In addition, the development of interoperable, seamlessly integrated, automated, decentralised blockchain technology also adds incredible value to the capabilities of the Web 3.0 Internet, the next generation of the Internet. It is expected that the future will see the creation of fully integrated blockchain applications, ensuring seamless interoperability as the next generation blockchain.

2.4 Blockchain in Supply Chains

Blockchain technology is increasingly being adopted in dispersed supply chain networks, with many participants since the launch of private blockchains [12]. In particular, an increasing amount of digital data is generated in supply chains every day, as more and more customers tend to buy online [8], therefore, big data and blockchains need to work together to create more opportunities and better service operations. Blockchain technology is expected to increase transparency and accountability for the streams of goods, commodities, money and information between parties to enable more integrated, flexible, reliable, secure and efficient value chains, while providing an immutable, auditable record of activities along the end-to-end transaction journey [23, 34]. Blockchain enabled supply chains can have the potential to simplify processes, identify counterfeit products, facilitate proof of provenance, enhance track and traceability (e.g., ensuring refrigerated goods are kept within a consistent temperature range), reduce paperwork processing, and operate the Internet of Things (IoT), while reducing costs and minimising risk [5, 36, 39]. Indeed, the inherently "anti-authoritarian" nature of blockchain technology contributes to sustainability in a supply chain network and across different supply chain networks [23].

Figure 3 shows a blockchain ecosystem with participants in the food industry where all network partners, including farmers, cooperatives, manufacturers, distributors, transporters, retailers, and customers as well as interrelated institutions such as banks, insurance services, certification bodies and governmental authorities (e.g., customs, tax), form a consortium in their respective ecosystem running on the

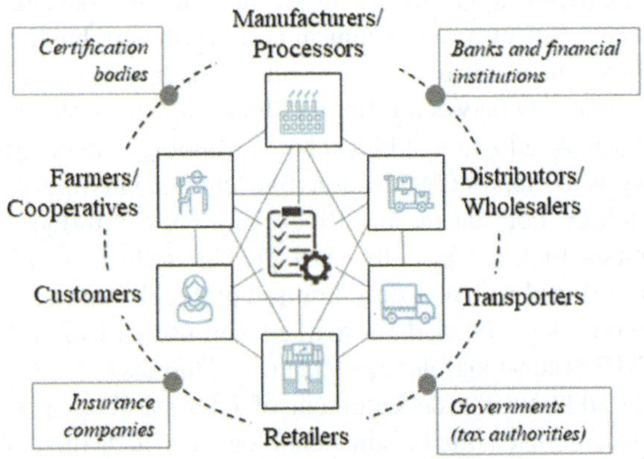

Fig. 3 A blockchain ecosystem in supply chain. *Source* Authors

blockchain platform. Here, technical alignment and collaboration within supply chain partners is essential. In order to reveal the real blockchain potential in supply chains, seamless data exchange within and between ecosystems is required [32]. The data exchange within an ecosystem refers to the intra-ecosystem relationship and seamless interoperability between the same network partners along the vertical supply chain, while data exchange between ecosystems denotes the inter-ecosystem relationship between different network partners along the horizontal supply chain. The existing blockchain use cases in supply chains from manufacturing, healthcare, retail, transportation, and food to other industries, show that most supply chain consortia operate on private, consortium or hybrid blockchain systems [34].

3 Blockchain Interoperability

Blockchain adoption is on the rise across supply chains of different industries. However, the lack of interoperability standards is a major obstacle to mass adoption [5, 31]. More and more commercial enterprises build and implement private, consortium or hybrid blockchains [41], which create often closed ecosystems. They use multiple data standards and protocols that isolate blockchains in their respective environments. Due to the nature of the business, blockchain networks require integration between both disparate blockchain and non-blockchain ecosystems. In a large-scale multi-chain environment, one of the important issues is to integrate different ecosystems together to sustain cross-chain information sharing and seamless secure data exchange between those ecosystems [2, 27]. This issue refers to interoperability. Figure 4 illustrates the blockchain interoperability for three cross-sectorial ecosystems in food supply chains, where two are the blockchain ecosystem and the third is the non-blockchain ecosystem. In an ecosystem, both *inter-blockchain interoperability* (in order to integrate blockchain ledger and legacy systems of partners) and *intra-blockchain interoperability* (for the purpose of cross-chain interoperability) is required *for blockchain-to-blockchain* and *blockchain-to-non-blockchain* seamless communication.

Interoperability simply refers to the capability of computer systems to exchange and use information, as well as to transfer an asset between two or more systems while keeping the status and uniqueness of an asset consistent [40]. Interoperability standardisation can create commonly agreed blockchain terminology. Implementation of common standards and protocols in blockchains can provide scalable interoperability and integration with different blockchains and non-blockchain systems. Resolving collaboration and cross-chain interoperability problems between public, private, consortium and hybrid blockchains can open the way to a hyper-connected world [42]. The general definition of blockchain interoperability provided by the US National Institute of Standards and Technology (NIST) is:

> An interoperable blockchain architecture is a composition of distinguishable blockchain systems, each representing a distributed data ledger, where transaction execution may span multiple blockchain systems, and where data recorded in one blockchain is reachable and

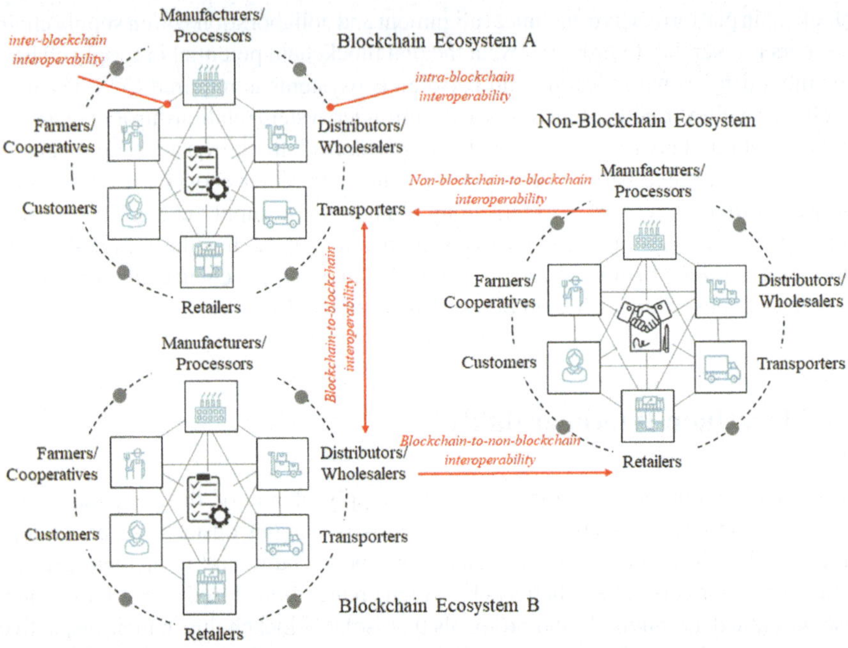

Fig. 4 Blockchain interoperability for different ecosystems in supply chains. *Source* Authors

verifiable by another possibly foreign transaction in a semantically compatible manner. [18, p. 4]

Traceability is a challenge of a multiple stakeholder and multiple networks. The seamless data exchange between different blockchain systems is crucial to the success of every business. In this sense, it is necessary for blockchains to be standardised to be future-proof [2]. This would allow parties to speak the same language, so that they can interact and transact with other existing or upcoming blockchain networks, as well as incorporate and share common capabilities and feature sets related to consensus models, transactions, and contract functionalities [20]. Blockchain interoperability allows blockchain systems to communicate with each other without the help of intermediaries [1, 10].

3.1 Data Interoperability Standards

Data stored or referenced by blockchain supply networks can be configured for shared communications and interoperability through the use of standards and protocols. Different data standards are developed to solve the interoperability problems in the blockchain world. The most commonly used open supply chain standards for blockchains are the GS1 global traceability standard [17] and ISO (the International

Organization for Standardization) open standards for blockchain and distributed ledger technologies [19]. These impact parties in supply chains for sharing business value in a standardised way [24] and enable interoperability between them by creating a common set of rules for identification, data capture, data sharing, and data usage [17]. In addition to these two standards, UN/CEFACT (United Nations Centre for Trade Facilitation and Electronic Business) launched a project to develop a strategy for the use of blockchain technology and created the UN/CEFACT standard with Supply Chain Reference Data Model (SCRDM) for trade facilitation [37]. Also, some associations are developing blockchain industry standards for supply chains and logistics. The International Port Community Systems Association (IPCSA)[1] started a blockchain bill of lading initiative with the cooperation of Alibaba and LOGINK; this aims to standardise blockchain applications for logistics and e-commerce to facilitate global port system operations and eliminate paperwork. The Blockchain in Transport Alliance (BiTA)[2] established a common framework and industry standards for DLT and blockchain applications/solutions in the transportation, logistics, supply chain and freight industry. Lin et al. [28] developed a blockchain-based traceability system for food safety using joint GS1 and ISO interoperability standards called EPCIS (Electronic Product Code Information Services). This system aims to enable a management architecture for standard data exchange and item-level tracking from blockchain and non-blockchain ecosystems where the traceability system can overcome potential data problems. According to [38], the developed ISO/T3-307 standard might provide better accessibility and alleviate interoperability issues by representing a tool to disseminate innovation and harmonise different protocols. Kan et al. [21] proposed an inter-blockchain communication model for multi blockchain architecture.

GS1 standards enable blockchain consortia to seamlessly interoperate; the network of supply chain stakeholders depends on industry's commitment to standards. The GS1 Standards System is an integrated global standards package that provides supply chain visibility by accurately identifying, capturing and sharing information about products, parties, locations, assets and services. In this sense, the GS1 standards support the core service concepts of identity, capture, and share to encourage organisations to improve traceability for the benefit of all participants in the global supply chains [17, 20].

- The *GS1 identity keys* are used in six areas [17]:
  - (i) Company: Global GS1 Company Prefix; Global Location Number (GLN);
  - (ii) Product: Global Trade Item Number (GTIN); Serialized Global Trade Item Number (EPC*/SGT);
  - (iii) Location: Global Location Number (GLN);
  - (iv) Logistics: Serial Shipping Container Code (SSCC); Global Shipment Identification Number (GSIN);

[1] https://ipcsa.international/initiatives/logistics-visibility-task-force/.

[2] https://www.bita.studio/.

(v) Assets: Global Individual Asset Identifier (GIAI); Global Returnable Asset Identifier (GRAI);
(vi) Services and other: Global Service Relation Number (GSRN); Global Document Type Identifier (GDTI).

- The *GS1 capture keys* are categorised in two classes [16]:

(i) Barcode, which contains UPC/EAN, Code 39, Code 128, Code 16 K, Code 49, PDF 417, Data Bar, DataMatrix, QR Code, Aztec Code and MaxiCode;
(ii) Electronic product code, which contains only EPC (UHF, HF Gen2)/Radio Frequency Identification (RFID).

- The *GS1 share keys* are divided into three data exchange types [17, 20]:

(i) Visibility event data from physical/digital object: EPCIS and Core Business Vocabulary (CBV);
(ii) Transactional data of objects: GS1 EDI (GS1 XML, GS1 EANCom, eCom);
(iii) Master data: Global Data Synchronization Network (GDSN), GS1 source, GS1 smart search, GLN service.

3.2 Cross-Chain Interoperability

In practice, blockchain solutions are formed around existing smaller and even closed ecosystems. Therefore, they often remain disconnected from each other, like islands with their own communities, unable to exchange information or values with the outside world. However, the global supply chains intersect multiple types of ecosystems and require blockchain to blockchain as well as blockchain to non-blockchain integrations [33]. Cross-chain interoperability not only increases the scalability of the blockchain, but also opens up the channel of value circulation between different chains. That means that blockchains can talk to each other and one can read the other's state. Two types of data are used for interoperability [2, 4]: digital asset exchange and arbitrary data exchange. Digital asset exchange provides the ability to transfer and exchange anything representing value (e.g., cryptocurrencies, non-fungible tokens) originating from different blockchains, whereas arbitrary data exchange provides the ability to use data (e.g., making payment to a supplier) on one blockchain platform to affect state changes on another. This refers to the capacity to make advanced blockchain to blockchain Application Programming Interface (API) calls by extending to the smart contract code on a blockchain that directly verifies the consensus certainty of events on other blockchains. The most use cases in supply chains use arbitrary data exchange [40].

In order to achieve blockchain interoperability, some approaches were developed in the literature from a technical perspective. Interoperability approaches across public blockchains are well-known and widely used. They were categorised under three classes [40]:

(i) *Oracle*: this approach uses a tool known as an oracle to get data and alike to send data; oracle employs smart contracts to add actual data (e.g., temperatures, humidity, prices) onto blockchains for business process automation [3]. This method can be easily implemented, especially in a non-blockchain environment [29]. In fact, oracles do not create a real blockchain to blockchain interoperability; they combine non-blockchain systems to make a blockchain interoperable with those systems [41].

(ii) *API gateway*: this organises several APIs and essentially acts as an intermediary to communicate between two or more blockchain platforms/applications; however, it still requires cross-authentication between nodes. An API is a piece of code that manages the access point to a server, and rule developers must follow this to interact with a database, library, software tool, or programming language [40]. Creating and implementing an API gateway is easier than other interoperability approaches [41], also, an API gateway centralises trust in the API operator.

(iii) *cross authentication*: this provides cross-chain interoperability between both different blockchain to blockchain and blockchain to non-blockchain systems. Cross authentication covers three methods as follows [4]:

- *Notaries* (both digital asset and arbitrary data exchange): the simplest way to facilitate most cross-chain operations is to use an intermediary trusted entity (a third-party) or group called a notary scheme/a set of notaries (also known as one-way pegs or two-way pegs) that confirms the status of an interactive blockchain to enable operations [4]. The notaries come to an agreement through a consensus algorithm. Notaries follow a centralised architecture to ensure cross-chain interoperability [22]. The disadvantage of this system is the third-party dependency; the system needs the active participation of a trusted and centralised entity [11]. Therefore, notaries provide a low level of decentralisation. This method is also proven to a limited extent in few solutions [25]. Liquid using federated pegged sidechain is a primary solution in this category [20].
- *Relays* (both digital asset and arbitrary data exchange): a relay scheme, also known as a sidechain, secondary chain or satellite chain, is a mechanism for two existing blockchains to interoperate [2], which is used in many blockchain interoperability solutions. This method provides a high level of decentralisation [33]. Although many interoperability solutions expect to expand functionality, this method has not been well proven. The most prominent efforts in this category are Cosmos, Polkadot, and ChainLink [20].
- *Hash-locking* (only digital assets exchange): a hashed time lock, also known as an atomic swap, is the most practical technical approach to blockchain interoperability and provides a high level of decentralisation. This method is proven technically in several solutions; however, it is the most limiting solution in terms of functionality. A hash time locked contract (HLTC) is a time-bound conditional blockchain-based payment that uses

hashlocks and timelocks to require the receiver of a payment to either acknowledge receiving the payment prior to a deadline or forfeit the ability to claim the payment, therefore returning it to the payer. HTLCs allow cross-chain atomic swaps and fully funded bi-directional payment channels between assets on certain types of blockchain. Interledger Protocol and ARK Core Series are the most prominent solutions of hash-locking.

The industrial solutions (e.g., Polkadot) that enable cross-chain operations are developed based on the aforementioned interoperability methods and act as a bridge to transfer digital assets and/or arbitrary data from one blockchain ecosystem to another. The chains may have different governance models, protocols, rules, and communities, but the bridge provides a cohesive way for both parties to interoperate securely. There are different designs for the bridges depending on the selected methodology. As a summary, they can mainly be divided in two categories: "trusted" or more centralised bridges (Notaries) based on trust or federation, and "non-trusted" or more decentralised bridges (Relays and Hash-locking) without having a third-party. The trusted bridge depends on some form of central authority that establishes trust for users to operate within a particular third-party app or service. The non-trusted bridge does not need a legal entity or authority to be trusted, but rather relies on the mathematical logic embedded into the code, enforced by the technology and/or incentive mechanism behind the system. Since 2021, many industrial solutions based on such interoperability methods have been developed, notably in line with the foundation of Web 3.0 Internet technology. In particular, the non-trusted types of interoperability are often preferred in blockchain ecosystems. In the next section, we analyse these interoperability methods with different use cases.

4 Blockchain Mass Adoption

4.1 Blockchain Use Cases in Supply Chains

In this section, in light of the interoperability methods explained previously, we want to analyse some of the real-life blockchain use cases to understand their potential for blockchain interoperability and mass adoption. A total of four blockchain use cases from different industry segments were selected, and the procedures are documented through multiple interviews.

Data collection: the selection process of use cases was made due to market size, recognition, and presence in the global supply chain ecosystem. First, the ten most appropriate use cases were identified based on the requirements of this chapter. We contacted them via email and LinkedIn to invite them for an interview: four of them agreed to participate in the interview. The selected use cases were from the network of the first author. The interviews were carried out online on an agreed day, and the duration of each interview was less than an hour. The interviews were

Table 1 Data collection

Use case company	The person interviewed	Industry area	The duration of interview (min)	Secondary data used
UC-1	Founder (Interviewee 1)	Food and agriculture	55	News articles, public interviews of founder
UC-2	Managing director (Interviewee 2)	Recycling	40	News articles, public videos and interviews
UC-3	Business Development representative (Interviewee 3)	Logistics/Shipping	45	YouTube videos, news articles, press releases
UC-4	Chief technology officer (Interviewee 4)	Textile, Garment and Accessories	50	LinkedIn article, public podcasts, press releases

Source Authors

recorded with the permission of the participants. The interview questions were open-ended to understand interoperability issues as well as mass adoption in terms of use cases; a checklist was used as a guide to ensure content comparability. We mainly talked with companies about the technical aspects of their solutions, such as the blockchain framework used, the features and functionality of blockchain ecosystems, the interoperability model, data standardisation, and mass adoption capabilities. After the interviews were completed, the recorded videos were coded and summarised; in this way, the primary dataset based on each use case interview was prepared. In addition, secondary data about each use case from different sources, such as news articles, press releases, company presentation and so on, were collected to use in this section. Table 1 shows the details of the data collection.

Data analysis: the data collected during the interviews and from other sources were analysed for each use case. The authors read each text several times to examine the whole context according to the business environment, the purpose of blockchain use, interoperability model and data standards. Table 2 denotes the analysis of blockchain use cases. The details of use cases are explained as follows:

UC-1 is a blockchain platform powered by the Corda framework to provide end-to-end traceability, cold chain monitoring (temperature, humidity) and organic certification for food products in the food supply chain. The blockchain platform is based on a consortium blockchain and uses most food related data standards for GS1, EPCIS & CBV, and GTS and APIs and relays approaches as interoperability methods.

UC-2 is a blockchain platform powered by a Hyperledger Sawtooth framework to provide origin tracking (source) for vehicle batteries and lithium-ion batteries in the electronics recycling industry involved in the battery recycling supply chain. It also tracks recycling processes to generate a battery passport. A recycler can use the blockchain platform to easily separate strings for use from one battery to another without losing its originating data. This platform is based on a consortium blockchain

Table 2 Analysis of blockchain use cases in supply chains

#	Business industry	Blockchain framework	Blockchain functionality	Inter-operability model	Data standards	Mass adoption
UC-1	Food	Corda (consortium blockchain)	Food traceability; cold chain monitoring; organic certification	Relays; APIs (CorDapp)	GS1, EPCIS & CBV, GTS 2.0	Low
UC-2	Recycling	Hyperledger Sawtooth (consortium blockchain)	Origin tracking; battery passport	Oracle; HTLC; API Gateway	ISO, GS1	Medium
UC-3	Logistics/Shipping	Hyperledger Fabric (private blockchain)	Trade facilitation; workflow automation; vehicle monitoring and tracking; freight payments	Oracle; HTLC; API Gateway	UN/CEFAC, GS1	Medium
UC-4	Textile, Garment and Accessories	Ethereum/Quorum (private blockchain)	Origin and ownership tracking; counterfeit detecting; original certification	Oracle; HTLC; APIs	GS1	Low

Source Authors

and uses ISO and GS1 standards as data standards, and oracle, HTLC and REST APIs as interoperability methods that simplify the creation and transmission of transactions and batches to Sawtooth network nodes. This blockchain platform aims to bring relevant companies in battery recycling ecosystems together to increase higher efficiency, engagement and productivity in battery recycling operations. In addition, the platform provides a consensus mechanism built into the battery recycling supply chain network.

UC-3 is a blockchain platform powered by a Hyperledger Fabric framework to coordinate the business network for all multimodal shipment activities among shippers, transport providers, ports and terminals, ocean carriers, government authorities, customs brokers, and more. The information shared by each participant on the platform can be tracked, recorded and managed throughout the shipment's journey. This platform is used to facilitate trade, automatise workflow, track logistics operations (vehicle monitoring and tracking) and payment processes (automated freight payments, shipping cash flows). This blockchain platform operates in a private blockchain ecosystem and uses oracle, HTLC and API gateway approaches for interoperability, where REST APIs integrate into the in-house systems (e.g., ERP, TMS) of participants. UN/CEFACT and GS1 data standards are used for data exchange.

UC-4 is a platform powered by an Ethereum/Quorum framework to enable origin tracking, counterfeit detecting and original certification of luxury components and products for textiles, garments and accessories. The blockchain platform is based on a private blockchain, so that the participants from processors to end-users can track real-time information on origin and ownership, authenticity, maintenance and equipment. The platform uses oracle, HTLCs and APIs as interoperability approaches, also GS1 standards as data standards.

4.2 Discussion

In this section, the results obtained from the interviews are discussed by quoting the interviewees. The most used blockchain platforms operate on open-source BaaS frameworks; these have different technical competence and interoperability capabilities. Therefore, the use cases of the platforms examined in this section have different infrastructures and implementations. All interviewees acknowledged "interoperability is crucial to business success", but although "interoperability is not a blockchain-specific concept, it requires coordinated standards to allow data exchange in a secure and validated manner" (interviewee 1). Particularly, interviewee 2 and interviewee 3 emphasised "the interoperability capabilities provide seamless, user-friendly experiences for participants across different platforms in the wider blockchain ecosystems". Furthermore, interviewee 4 added that "incompatible technologies, lack of transparency and possible confusion cause friction and lead to inefficiency for their business".

In addition, interviewees highlighted the importance of mass adoption to keep their business going. This also indicated that "the rate of mass adoption depends on the level of interoperability that aligns with the capabilities of their blockchain platforms" (interviewee 2 and interviewee 3). It can be summarised that non-interoperable blockchain ecosystems will be siloed and disconnected from each other, operating side by side without communication and transaction, while interoperable blockchain ecosystems ensure multiple participants from separate blockchain networks to interact without spending resources to translate received information and experience downtime.

All interviewees believed that using APIs with other cross-chain interoperability methods can enable blockchain use cases to have data transparency and end-to-end visibility throughout all stages of the supply chain lifecycle, from suppliers to retailers, and facilitate communications among them. It is acknowledged that "interoperable blockchain systems allow information to be easily processed and sent a response accordingly" (interviewee 1 and interviewee 3), also "interoperability is conceptually simple and can be solved relatively easily with APIs" (Interviewee 2). However, interoperability across different blockchains is more extensive and complicated, as blockchain frameworks (e.g., Ethereum and Hyperledger or Ethereum and Corda) use different hash algorithms and consensus protocols; even these may differ when new versions arrive. On the other hand, the use cases agreed that using a one-size-fits-all standard for interoperability is not suitable for blockchain platforms. Each use case has different requirements to meet and the use of a single standard cannot respond to this. On the other hand, setting a single standard would cause important time delays and hinder the critical innovations required. Therefore, it is believed that "establishing open standards can allow all parties in a supply chain to contribute naturally and create a single, state-of-the-art public blockchain" (all interviewees). Nevertheless, interviewee 3 contributed that "the standards need to be as simple as possible to integrate the solutions to each other and to release new blockchain based business models".

In addition, interviewees believed that "their participants can leverage interoperability to operate with greater confidence and less risk of obsolescence, thereby improving mass adoption". The analysis of blockchain use cases showed that the rate of mass adoption ranged between low and medium, depending on the current level of interoperability. As a final remark, interviewees noted "the implementation of REST-APIs with a common interface and GS1 industry standards contributed to interoperability for mass adoption in the supply chain".

The use case analysis showed that different interoperability methodologies are used, such as relays, oracles and HTLCs. This is mostly in line with the capabilities of the respective blockchain frameworks (ledger technologies) to exchange value or information based on existing technologies. This enabled different stakeholders in blockchain ecosystems to access blockchain data with different read access levels. However, under these circumstances, no blockchain use case will achieve full mass adoption. More decentralised protocols must be used to communicate with the outside world. In particular, the siloed nature of today's blockchain networks cannot be aligned with the principle of decentralisation and reflects the partition of the current centralised Web 2.0 Internet. This also means that blockchain technology is not yet mature and there is still room for improvement in this technology in terms of interoperability. As a result, we can say that the next generation blockchains with the use of a trusted, secure and decentralised Web 3.0 Internet will be more capable and efficient in providing seamless interoperability.

5 Conclusions

There is an increasing need for advanced technologies to overcome the hype and mature into useable technologies adopted on a large scale for economic benefits, while the growth of big data increases the need to securely manage, store and exchange data across supply chains. Blockchain is one of those technologies used to scale up. However, it still has many technical capability shortcomings due to its immaturity. One important technical challenge is the lack of interoperability between different blockchain systems, often seen as the biggest deterrent to mass adoption. The literature review and use case analysis reveal that interoperability is a critical element for the existence and sustainability of blockchain-based platforms and business models. In the short term, this could mean transferring value from one organisation's legacy system to another organisation's blockchain ledger, while in the long term, it could mean trading between different types of blockchain as well as non-blockchain and other legacy systems. There are some industrial solutions that use interoperability methodologies to solve interoperability challenges and they are still in development. Few bridges can provide a good level of interoperability, but there is still no solution for full interoperability that can integrate every kind of blockchain ecosystem. It can be said that the sooner the blockchain matures, the higher the rate of mass adoption will be.

Today's supply chains are heavily influenced by big data. Blockchain technology is the ideal solution for the many service operation challenges faced by data-driven supply chains. The respective stakeholders join a blockchain platform to seamlessly exchange massive amounts of data throughout supply chains. Data exchange can take place in the vertical integration of a blockchain ecosystem. However, no blockchain network can exist without the horizontal integration of other ecosystems, otherwise such isolated blockchains will sooner or later disappear from the market. Therefore, the blockchain platforms should be designed to be as interoperable as possible to transfer a digital asset or arbitrary data seemingly from one blockchain ecosystem to another; this is because the future of blockchain will evolve on multi-party and multi-chain due to current business challenges. In the future, more research is needed on blockchain interoperability highlighting the multi-chain aspect. This could include the maturity assessment of blockchain interoperability in multi-party and multi-chain ecosystems, making new implementation decisions on existing blockchain solutions in terms of interoperability for the development of new multi-chain blockchain use cases, or checking the interoperability capabilities of different ecosystems to decide on multi-chain collaboration.

This chapter has some limitations. First, the presented case analysis covered only four blockchain use cases. More use cases could be included in the analysis to understand the relationship between blockchain interoperability and mass adoption. Second, we analysed four blockchain use cases from different industry segments, and their ecosystems had no relationship with each other. This analysis could also be done for blockchain use cases belonging to different ecosystems, such as central complementary, supplementary or competitive ecosystems that are interrelated.

Acknowledgement The second author would like to thank the UK Research and Innovation funding service for awarding the project "REMP: Remanufacturing e-marketplace for circular digital supply chain" under the competition Innovate UK-Manufacturing made smarter: digital supply chain, feasibility studies.

References

1. Ackermann, J., & Meier, M. (2018). *Blockchain 3.0: The next generation of blockchain systems*. Advanced Seminar Blockchain Technologies, Summer Term 2018, Technical University Munich.
2. Belchior, R., Vasconcelos, A., Guerreiro, S., et al. (2021). *A survey on blockchain interoperability: Past, present, and future trends*. Retrieved April 15, 2021, from https://arxiv.org/pdf/2005.14282.pdf.
3. Beniiche, A. (2020). *A study of blockchain oracles*. Retrieved April 24, 2021, from http://arxiv.org/abs/2004.07140.
4. Buterin, V. (2016). *Chain interoperability*. Retrieved April 11, 2021, from https://allquantor.at/blockchainbib/pdf/vitalik2016chain.pdf.
5. Casino, F., Dasaklis, T. K., & Patsakis, C. (2019). A systematic literature review of blockchain-based applications: Current status, classification and open issues. *Telemat Inform, 36*, 55–81.
6. Charles, V., & Emrouznejad, A. (2019). Big data for the greater good: An introduction. In A. Emrouznejad & V. Charles (Eds.), *Big data for the greater good*. Studies in Big Data (Vol. 42). Springer. https://doi.org/10.1007/978-3-319-93061-9_1
7. Charles, V., & Gherman, T. (2019). Big data analytics and ethnography: Together for the greater good. In A. Emrouznejad & V. Charles (Eds.), *Big data for the greater good*. Studies in Big Data (Vol. 42). Springer. https://doi.org/10.1007/978-3-319-93061-9_2
8. Deepa, N., Pham, Q.-V., Nguyen, D. C., et al. (2020). A survey on blockchain for big data: Approaches, opportunities, and future directions. arXiv preprint arXiv:2009.00858.
9. Deng, L., Chen, H., Zeng, J., et al. (2018). Research on cross-chain technology based on sidechain and hash-locking. In S. Liu, B. Tekinerdogan, M. Aoyama, et al. (Eds.), *Edge computing – EDGE 2018*. Lecture Notes in Computer Science (Vol. 10973). Springer.
10. Dhuddu, R., & Mahankali, S. (2021). Consortium approach-operational and architectural considerations. In *Blockchain in e-governance: Driving the next frontier in G2C services*. BPB Publications.
11. Dilley, J., Poelstra, A., Wilkins, J., et al. (2016). Strong federations: An interoperable blockchain solution to centralized third party risks. arXiv preprint arXiv:1612.05491
12. Dujak, D., Sajter, D. (2019). Blockchain applications in supply chain. In A. Kawa & A. Maryniak (Eds.), *SMART supply network, EcoProduction* (pp. 21–46). Springer.
13. Dutta, P., Choi, T.-M., Somani, S., et al. (2020). Blockchain technology in supply chain operations: Applications, challenges and research opportunities. *Transportation Research Part E: Logistics and Transportation Review, 142*, 102067.
14. Emrouznejad, A., & Marra, M. (2017). Big data: Who, what and where? Social, cognitive and journals map of big data publication. In A. Emrouznejad (Ed.), *Big data optimization: Recent developments and challenges*. Studies in Big Data (Vol. 17). Springer. Retrieved June 08, 2021, from https://link.springer.com/chapter/10.1007%2F978-3-319-30265-2_1.

15. Emrouznejad, A. (2016). *Big data optimization: Recent developments and challenges*. Studies in Big Data. Springer-Verlag. ISBN: 978-3-319-30263-8.
16. GS1. (2016). GS1 RFID/Barcode interoperability guideline. GS1.
17. GS1. (2017). GS1 global traceability standard. Release 2.0. Ratified 2017 (GS1's framework for the design of interoperable traceability systems for supply chains). GS1.
18. Hardjono, T., Lipton, A., & Pentland, A. (2019). Toward an interoperability architecture for blockchain autonomous systems. *IEEE Transactions on Engineering Management, 1–12*. https://doi.org/10.1109/tem.2019.2920154
19. ISO. (2016). ISO/TC 307 Blockchain and distributed ledger technologies. *ISO*. Retrieved April 11, 2021, from https://www.iso.org/committee/6266604.html.
20. Jabbar, S., Lloyd, H., Hammoudeh, M., et al. (2020). Blockchain-enabled supply chain: Analysis, challenges, and future directions. *Multimedia Systems*. https://doi.org/10.1007/s00530-020-00687-0
21. Kan, L., Wei, Y., Hafiz Muhammad, A., et al. (2018). A multiple blockchains architecture on inter-blockchain communication. In *2018 IEEE International Conference on Software Quality, Reliability and Security Companion (QRS-C)*. https://doi.org/10.1109/qrs-c.2018.00037
22. Kannengießer, N., Pfister, M., Greulich, M., et al. (2020). Bridges between islands: Cross-chain technology for distributed ledger technology. In *Proceeding of the 53rd Hawaii International Conference on System Sciences* (pp. 5298–5307)
23. Kayikci, Y. (2021). Blockchain driven supply chain management: The application potential of blockchain technology in supply chain and logistics. In T. Paksoy, C.G. Kochan & S.S. Ali (Eds.), *Logistics 4.0: Digital transformation of supply chain management* (pp. 146–155). CRC Press. https://doi.org/10.1201/9780429327636-16
24. Keogh, J. G., Rejeb, A., Khan, N., et al. (2020). Blockchain and GS1 standards in the food chain: A review of the possibilities and challenges. In D. Detwiler (Ed.), *Building the future of food safety technology* (pp. 171–204). Elsevier. https://doi.org/10.1016/B978-0-12-818956-6.00017-8
25. Koens, T., & Poll, E. (2019). Assessing interoperability solutions for distributed ledgers. *Pervasive and Mobile Computing, 59*. https://doi.org/10.1016/j.pmcj.2019.101079
26. Lacity, M., Steelman, Z., & Cronan, P. (2019). *Towards blockchain 3.0 interoperability: Business and technical considerations* (White Paper, Nr. BC CoE 2019-01, Blockchain Center of Excellence).
27. Li, D., Yu, J., Gao, X., et al. (2020). Research on multidomain authentication of IoT based on cross-chain technology. *Secur Commun Netw, 2020*, 6679022.
28. Lin, Q., Wang, H., Pei, X., et al. (2019). Food safety traceability system based on blockchain and EPCIS. *IEEE Access, 7*, 20698–20707.
29. Mammadzada, K., Iqbal, M., Milani, F., et al. (2020). Blockchain oracles: A framework for blockchain-based applications. In A. Asatiani, et al. (Eds.), *Business process management: Blockchain and robotic process automation forum. BPM 2020*. Lecture Notes in Business Information Processing (Vol. 393). Springer.
30. Nakamoto, S. (2008). *Bitcoin: A peer-to-peer electronic cash system* (White Paper). Retrieved April 15, 2021, from https://bitcoin.org/bitcoin.pdf.
31. O'Leary, D. E. (2019). Some issues in blockchain for accounting and the supply chain, with an application of distributed databases to virtual organizations. *International Journal of Intelligent Systems in Accounting, Finance & Management, 26*(3), 137–149.
32. Pan, S., Trentesaux, D., McFarlane, D., et al. (2021). Digital interoperability and transformation in logistics and supply chain management: Editorial. *Computers in Industry, 129*, 103462.
33. Pillai, B., Biswas, K., & Muthukkumarasamy, V. (2020). Cross-chain interoperability among blockchain-based systems using transactions. *Knowledge Engineering Review, 35*.
34. Subramanian, N., Chaudhuri, A., & Kayıkcı, Y. (2020). *Blockchain and supply chain logistics*. Palgrave Pivot. https://doi.org/10.1007/978-3-030-47531-4
35. Syahputra, H., & Weigand, H. (2019). The development of smart contracts for heterogeneous blockchains. In K. Popplewell, K.D. Thoben, T. Knothe, et al. (Eds.), *Enterprise interoperability. VIII. Proceedings of the I-ESA Conferences* (Vol. 9). Springer. https://doi.org/10.1007/978-3-030-13693-2_19

36. Tijan, E., Aksentijević, S., Ivanić, K., et al. (2019). Blockchain technology implementation in logistics. *Sustde, 11*(4), 1185.
37. UN/CEFACT. (2019). *The technical applications of Blockchain to United Nations Centre for Trade Facilitation and Electronic Business (UN/CEFACT) deliverables* (White Paper). Retrieved April 18, 2021, from https://unece.org/DAM/cefact/cf_plenary/2019_plenary/ECE_TRADE_C_CEFACT_2019_08E.pdf.
38. Valle, F. D., & Oliver, M. (2020). Blockchain enablers for supply chains: How to boost implementation in industry. *IEEE Access, 8*, 209699–209716.
39. Wang, Y., Chen, C. H., & Zghari-Sales, A. (2020). Designing a blockchain enabled supply chain. *International Journal of Production Research, 1–26*. https://doi.org/10.1080/00207543.2020.1824086
40. WEF. (2020). *Inclusive deployment of blockchain for supply chains: Part 6: A framework for blockchain interoperability* (White Paper, World Economic Forum and Deloitte).
41. Wilkie, A., & Smith, S. S. (2021). Blockchain: Speed, efficiency, decreased costs, and technical challenges. In: H. K. Baker, E. Nikbakht, & S.S. Smith (Eds.), *The emerald handbook of blockchain for business* (pp. 157–170). Emerald. https://doi.org/10.1108/978-1-83982-198-120211014
42. Yaga, D., Mell, P., Roby, N., et al. (2018). *Blockchain technology overview*. NIST Draft NISTIR (Vol. 8202). Retrieved April 23, 2021 from https://nvlpubs.nist.gov/nistpubs/ir/2018/NIST.IR.8202.pdf.

Dr. Yaşanur Kayıkcı is a Consultant and Associate Professor of Industrial Engineering for the working area of Logistics and Technology Management. She is currently serving as a Head of Industrial Engineering Program at the Turkish-German University in Istanbul, Turkey. Her research interests include information science, new technologies and innovations (AI, Industry 4.0, Internet of Things, blockchain, and big data), logistics information systems and technology management, digital supply chain, platform models and strategies, industrial digital transformation, collaborative business models for on-demand transport services and smart circular economy.

Nachiappan Subramanian (Nachi) is the Professor of Operations and Logistics Management & Supply Chains at the University of Sussex UK. Nachi has 22 years of teaching experience in the UK, China, Australia and India. Nachi has published 106 peer reviewed journal articles in technology interventions in supply chain, sustainable supply chains and risk and resilience. He is the Senior Associate Editor of the International Journal of Physical Distribution and Logistics Management and the Associate Editor for the International Journal of Logistics Management. He is the Fellow of Chartered Institute of Logistics and Transport UK and a Senior Fellow of Higher Education Academy, UK. Nachi authored a book on the title "Blockchain and Supply Chain Logistics: Evolutionary Case Studies (ISBN 9783030475307)" along with Chaudhuri A, Kayıkcı Y published by Palgrave McMillan.

Blockchain Technology Enablers in Physical Distribution and Logistics Management

Rohit Sharma and Anjali Shishodia

Abstract Blockchain technology has disrupted the present business environment. With the ever increasing supply chain disruptions, real time information sharing plays a critical role in achieving the organization goals. As majority of the operations are data intensive, a lot of (*big*) data is generated which can yield fruitful results. Through integration of modern information and communication technology tools and disruptive technology platforms, this real-time data can be used for leveraging competitive advantage and enhancing the operational efficiency. However, blockchain being a nascent technology, its adoption in logistics and physical distribution, and service operations sector is relatively low. Some of the reasons which can be attributed to it are lack of awareness of blockchain enablers and lack of comprehensive benefits about blockchain technology. The present study aims at finding out the blockchain technology enablers through a pertinent literature review. These enablers were then validated by experts for highlighting their prioritization using analytic hierarchy process (AHP). The findings from the study highlight that traceability and transparency were the factors given the utmost priority. The findings also highlight how the adoption of blockchain technology will enhance the operational efficiency in case of service operations management. Further, implications for research and practice are discussed which are followed by limitations and future research directions.

Keywords Blockchain · Logistics and physical distribution · Supply chain management · Service operations management

R. Sharma (✉)
Jaipuria Institute of Management, Noida, India
e-mail: Rohit.sharma@jaipuria.ac.in

A. Shishodia
LM Thapar School of Management, Dera Bassi, Mohali, India
e-mail: Anjali.shishodia@thapar.edu

A. Emrouznejad and V. Charles (eds.), *Big Data and Blockchain for Service Operations Management*, Studies in Big Data 98,
https://doi.org/10.1007/978-3-030-87304-2_14

1 Introduction

Majority of the logistics and physical delivery (LPD) organizations worldwide face the problems of meeting ever-rising customer demands, effective transportation cost controls, better supplier/partner relationships, inventory accuracy and location issues, and effective planning and risk management. Adding to it, there are operational issues that require streamlining in order to increase the organizational efficiency. To sum it up, the current operations require cost efficiency, reliability, speed, trust among stakeholders, and effective logistics management systems [12]. It has been highlighted that lack of decentralized and real-time information sharing solutions leads to visibility and trust issues among the various stakeholders in LPD sector. Enhancing the information and communication technologies' (ICT) applications in the LPD sector can automate the processes and lead to reductions in operational costs and elimination of non-value added activities [1]. More and more supply chain management operations (SCMO) are undergoing the paradigm shift from conventional SCMO practices to digital SCMO practices. As one of the emerging technologies in SCMO, blockchain (BC) holds a lot of potential in disrupting the current status quo in LPD sector [37]. As majority of the LPD operations are data intensive (due to data volume, variety, and velocity) [19], this has led to a paradigm shift in the operations which is indicated by the emergence of big data business analytics for gaining a competitive advantage [49]. Consequently, SCM organizations world over are now utilizing big data analytics for improving demand forecast and planning. Studies from the recent past (see [7, 38]) define the amalgamation of big data analytics in SCM as supply chain analytics (SCA). In the case of LPD operations, when the focus is on devising a logistics and SCM strategy, SCA finds its applications in insourcing decisions, supply chain network design, and new product design and development. SCA assists practitioners in adapting to the business environment and helps in decision-making for enhancing SC flexibility and profitability [49]. Therefore, it is anticipated that BC will leverage SCA applications in LPD sector by linking the operational performance to the business drivers on a real-time basis. The real-time information sharing capabilities of BC will further improve facets such as service technology levels, service design, and enhance operational efficiency in case of service operations management [25].

BC is a peer-to-peer transaction platform that is decentralized, distributed, and operates without the need of a trusted intermediary [16]. As reported by Wang et al. [50], despite garnering a lot of interest from the academia and practitioners, the adoption of BC in the LPD is still in nascent stages. Studies have been conducted pertaining to BC adoption in various fields such as agriculture, healthcare, SCM, transportation, governance, and other sectors [6, 20, 21, 26, 48]. There is a lack of clarity and uncertainty as how the LPD sector will adopt BC as [5] suggests that it will be a complex undertaking. There is limited literature highlighting the adoption of BC in LPD sector and the comprehensive benefits that follow its adoption. Against this backdrop, the present study aims to highlight the potential enablers of BC adoption in the LPD sector. Using analytic hierarchy process (AHP), opinions of practitioners from Indian LPD sector were sought for prioritization of the critical enablers of BC adoption in developing economies such as India.

Following this introduction, rest of this chapter is organized as: Sect. 2 highlights the systematic literature review and the critical enablers of BC adoption. Section 3 discusses the research methodology. The results are presented in Sect. 4 and it is followed by discussions, implications, and scope for future work are presented in Sect. 5.

2 Literature Review

A systematic literature review (SLR) was carried out for identifying the reasons for adoption of BC in LPD sector. Articles published in electronic databases such as Scopus, ScienceDirect to name a few, were included in the literature review. Keywords for the literature search included the usage of the keyword individually and a combination of keywords: blockchain, logistics, warehousing, physical distribution, and supply chain management. The flowchart for the SLR is highlighted in Fig. 1.

The literature review for the present study is segregated into two parts viz. challenges in the LPD sector, BC and its enablers.

2.1 Challenges in the LPD Sector

Many challenges impede the operational processes in the LPD sector. The usage of obsolete technology leads to lack of standardization and the inability to share real-time operational information which further hinders effective resource management in the LPD sector [31]. There are also challenges associated with unsatisfactory

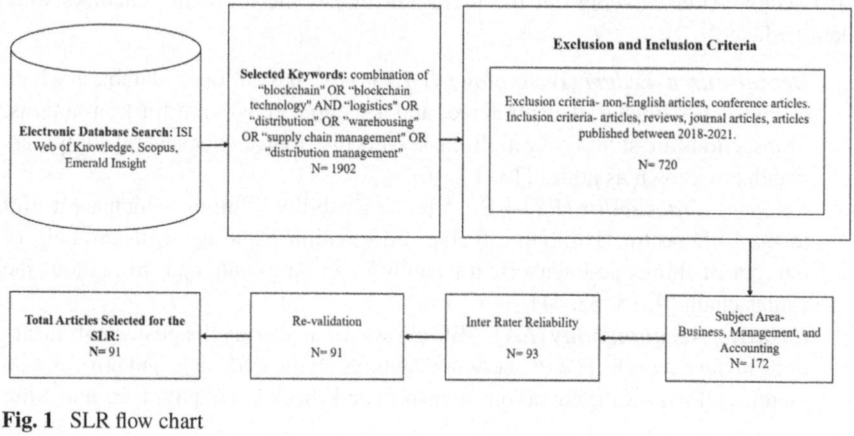

Fig. 1 SLR flow chart

and extensive trade documentation due to slow data movement [53]. Studies have highlighted that due to lack of accountability, dishonest behaviour and cargo thefts give rise to lack of trust among various stakeholders in the LPD sector [15, 45]. There are also issues aplenty related to payment delays and long payment cycles that render the LPD system ineffective thereby decreasing speed, agility, and operational flexibility [39]. Other issues include lack of effective control and freight monitoring which hamper the process efficiency in the LPD sector [3].

It is anticipated that BC adoption can lead to many advantages in the LPD sector but very little is known on the disruptions that BC can cause therefore, the research demands urgent investigation. The following section discusses about the BC enablers and the potential benefits of BC adoption in the LPD sector.

2.2 Blockchain Technology

BC is a unique platform which can validate and manage transactions between different stakeholders without the need of a trusted intermediary. It is a distributed ledger that can automate contracts and execute compliance checks running on many nodes at a time which are distributed geographically across different organizations and stakeholders [42]. BC has its unique ability in validating, recording, distributing the records and transactions in immutable, encrypted ledgers [50]. BC maintains immutable information related to products and the stakeholders throughout the supply chain (i.e. the complete product cycle) and governs the information and financial flows [36]. Studies have highlighted that BC serves as a critical enabler for transforming the business practices thereby offering new value creation [32]. Moreover, BC has enhanced cybersecurity mechanisms that make it resilient to hacking and other cyber threats [30].

The features of BC are highlighted in Table 1.

It is anticipated that BC will bring about the much-needed transformation in the LPD sector. After carrying out the literature review, the following enablers were identified.

a. ***Decentralized Ledger Technology (DLT)***—BC has an open database which enables a distributed, decentralized, and a secured ecosystem for transactions. Transactions are stored over multiple servers distributed over several computers which are known as nodes [13, 14, 16].
b. ***Ensuring Traceability (ET)***—BC offers traceability solutions which are useful in the LPD sector. Through effective information gathering with the help of internet of things and sensors, traceability can be maintained throughout the values chain [9, 18, 22, 41].
c. ***Ensuring Accountability (EA)***—BC allows for analyzing the posterior transactions in the network. The BC network captures all the trade data and information thereby, allows the organizations to inspect and check bill of payments and other

Table 1 BC features

BC features	Description
Smart contracts	Computer protocol that controls digital assets. Formulates stakeholders' rights and obligations. Triggers the transactions on meeting the conditions
Consensus algorithms	Set of rules and procedures for maintaining coherency between multiple participating peers. Examples include Proof of Work, Proof of Authority, Proof of Stake to name a few
Hashing systems	Unique fingerprints assigned to transaction blocks
Transactions	Records of code and variables (or) results of function calls
Ledger	Encrypted transactions that can be shared with the stakeholders in the network
Communication channel	Constructs that implement a prescribes interface (e.g. Smart contracts)
Validators	Validators run/endorse a transaction. Actions include deploy, invoke, and query ledgers
Nodes/Peers	Stakeholders enrolled on the BC network
Block	Running history of the transactions

trade invoices, hence increasing accountability and efficiency of stakeholders [11, 52].

d. ***Ensuring Transparency (ETr)***—BC has an inherent mechanism that coordinates different stakeholders who do not trust each other. This aids in providing process transparency. The consensus mechanism in a BC network leads to higher authenticity thereby enhancing the transparency of transactions [8, 41, 46].
e. ***Reduces Transaction Costs (RTC)***—BC removes the need of a trusted intermediary thereby helping in reducing the associated transaction costs [2, 16].
f. ***Speed of Transactions (ST)***—based on algorithms and high-speed data networks, BC enhances the rate of transactions [10, 23].
g. ***Record Auditability (RA)***—Facilitating auditability of transaction records is a unique feature of the BC [28, 54].
h. ***Cybersecurity (CS)***—The advanced cryptographic algorithms in BC have the capability to fend off cyber-attacks and threats (Demirkan et al., 2020; Parizi et al., 2020; Riesco et al., 2020).
i. ***Risk Management (RM)***—BC has the unique capabilities to provide decentralized and tamper-proof ledgers thereby enhancing the risk management practices [43, 55].
j. ***Data Privacy and Anonymity (DPA)***—BC has the unique property of privacy preserving cryptographic signature framework. It allows enhanced information security, cybersecurity, and protects the network from cyber threats. BC network utilizes a unique cryptographic private key mechanism which ensures anonymity and data privacy. Ring signatures are used for anonymization [4, 27].

In total, fifteen critical enablers for the adoption of BC in LPD sector were identified. These were then validated by academicians and practitioners having more than twenty years of experience in the related fields. The five enablers which were not included were either dropped or clubbed with the ten identified enablers. The research methodology adopted for this study is discussed in the below sections.

3 Research Methodology

Figure 2 highlights the research framework used for this study. The current research comprised two stages. In stage 1, a literature review was carried out for finding out the critical enablers for adoption of BC in LPD. In the second stage, the identified enablers were evaluated by experts' opinion using AHP. Finally, managerial and practical implications are presented. These will help the managers and policymakers in formulating the decision strategies for enhancing the operational efficiency in LPD using BC.

The AHP methodology is presented in the below subsection (see Fig. 3).

3.1 AHP Methodology

Owing to the multiple nature of the enablers in the BC adoption in LPD, it can be considered as a multi-criterion decision-making (MCDM) problem. MCDM technique is gaining a lot of popularity among practitioners as they offer solutions to intriguing problems by offering solutions and trade-offs. In the present research study, AHP is used for prioritizing the identified enablers for BC adoption in the LPD sector. AHP is the most widely and commonly used MCDM techniques in the

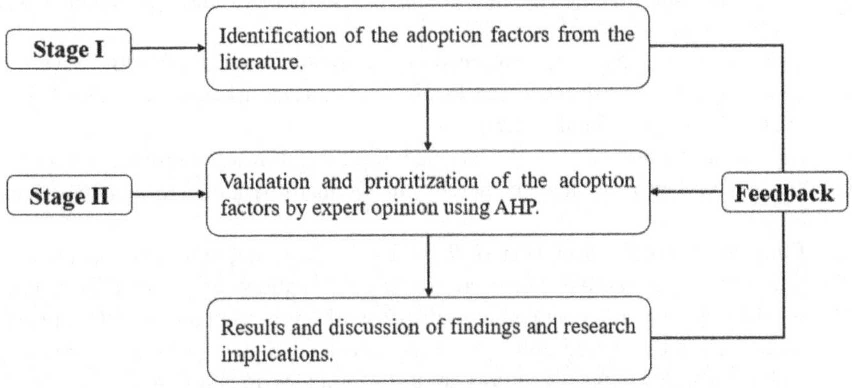

Fig. 2 Research methodology

Fig. 3 AHP flowchart

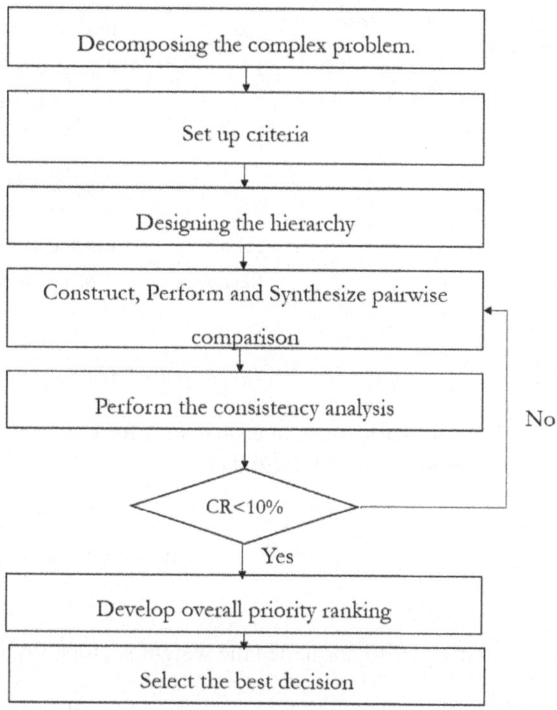

literature [29]. The AHP technique [35] was used to prioritize the factors obtained from the literature review for aiding the adoption of BC. The AHP technique provides an extensive means to analyse complex problems and makes use of hierarchical decision models. It is an MCDM tool which supports in modelling complex problems where priorities are to be synthesized. The AHP is an innovative methodology by which, one can relate intangibles to tangibles, the subjective to the objective, and link both to their purposes. AHP offers a unique way for integrating complexity; setting the right targets; establishing priorities and determining the overall value of each alternative solution. The advantage of using AHP over other MCDM techniques is that it holds the capability for providing numerical weights to different alternatives in an environment where the subjective judgment of qualitative and quantitative options makes up a critical part of the decision-making process. Other advantages of AHP include its ability to work out sensitivity analysis, essential justification of the alternatives, universal acceptance, and easy application in some areas [47].

Following are the detailed steps involved in using the AHP technique:

1. Established a hierarchical structure by decomposing the complex problem and then illustrated the hierarchical structure.
2. Developed a pair wise comparison matrix for indicating the relative importance of alternatives and suggested numerical rating as per the below Table 2. Afterwards, a judgment matrix shown by Eq. (1) can be procured.

Table 2 Numerical ratings in the AHP method

Scale	Explanation (when A factor compared to B)
1	A is equally important as B
3	A is moderately more important than B
5	A is strongly more important than B
7	A is demonstrated more important than B
9	A is extremely more important than B
2,4,6,8	Intermediate values
Reciprocals	When B is compared to A

$$p = (p_{ij})_{n*n}$$

3. The geometric mean of each matrix for weight vectors can be determined using the formula shown in Eq. (1).

$$W = \sqrt[n]{\prod_{i=1}^{n} pij} \quad (1)$$

In order to normalize the weight vectors, Eq. (2) is used.

$$W_i = \frac{W_i}{\sum_{i=1}^{n} W_i} \quad (2)$$

4. The relative weights and largest eigenvalue λ_{max} are then calculated according to the following Eq. (3):

$$PW = \lambda_{max} \quad (3)$$

Using Eq. (4) the consistency ratio is calculated whereas CI can be computed by Eq. (5). RI denotes the random consistency index whose value can be extracted from Table 3.

$$CR = CI/RI \quad (4)$$

$$CI = \frac{(\lambda_{max} - n)}{n - 1} \quad (5)$$

Table 3 Random consistency index RI

n	1,2	3	4	5	6	7	8	9	10
RI	0	0.52	0.89	1.12	1.12	1.36	1.41	1.46	1.49

When $\lambda_{max} = n$, matrix P is a completely consistent matrix. If CR ≤ 0.1 matrix P is considered as being acceptable. Otherwise, the comparison matrix needs to be adjusted. Similarly, the weights of the secondary criteria are calculated.

Finally, the weights in every hierarchy are summarized and the total consistency ratio is calculated by Eq. (6), where w_i represents the weights of the primary criteria, C_{1i} denotes the corresponding consistency ratio of the secondary criteria, and R_{1i} stands for the random consistency index of the corresponding secondary criteria which can also be illustrated in Table 2. If $CR_{total} \leq 0.1$, the matrix is considered as being acceptable.

$$CR_{total} = \frac{\sum_{i=1}^{n} w_i C_{1i}}{\sum_{i=1}^{n} w_i C_{1i}} \quad (6)$$

The expert panel consisted of LPD practitioners (domain areas: logistics and supply chain, warehouse pledging and banking services, computer science, and BC), academicians (with relevant experience in logistics and supply chain management). The total number of experts was seven and satisfied the group size requirements for executing exploratory studies as suggested by Murry and Hammons [33]. Respondent details and demographics are provided in Table 4.

Table 4 Respondent profiles

Designation	Sector/Industry	Educational qualifications	Work experience	Domain knowledge
Professor	Education; Teaching and Consultancy	Doctorate	20+ years	BC, Logistics and SCM
Professor	Education; Teaching and Consultancy	Doctorate	20+ years	BC, Logistics and SCM
General Manager	Warehousing	Postgraduate	15+ years	Logistics and Warehousing
Senior IT Professional	IT Consultancy and BC	Postgraduate	5–10 years	BC
Vice President	Warehousing	Postgraduate	15+ years	Logistics and Warehousing
Vice President (Ops.)	Food Processing	Postgraduate	20+ years	Warehousing Pledging and Finance
Chief Supply Chain Officer	Food Processing	Postgraduate	20+ years	Logistics and Supply chain

3.2 Results

The response from the experts on the importance attached to the factors for BC adoption in LPD was collected on a 9-point Likert scale (1 = equal importance and 9 = extreme importance). The AHP technique was discussed in Sect. 3.1 and applied to prioritize the enablers. The results are presented in Table 5. Based on the prioritization of the enablers, it can be seen that the enabler BC ensures Traceability (0.257) is perceived to be the most important enabler for adoption of BC in LPD, followed by Ensuring Transparency (0.2300), Ensuring Accountability (0.1310), Speed of Transactions (0.0890), and so on. The AHP model is highlighted in Fig. 4.

Table 5 Prioritization of factors using AHP

Factors	Percentage (%)	Rank
Ensuring traceability	25.70	1
Ensuring transparency	23.00	2
Ensuring accountability	13.10	3
Speed of transactions	8.90	4
Record auditability	7.80	5
Reduces transaction costs	7.10	6
Risk management	5.20	7
Cybersecurity	4.30	8
Data privacy and anonymity	3.40	9
Decentralized ledger technology	1.50	12

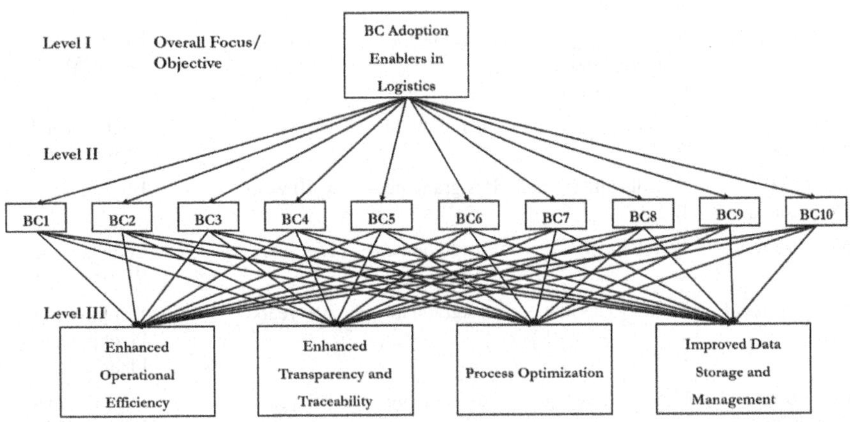

Fig. 4 AHP model

4 Discussion and Implications

The present study highlights ten critical enablers for BC adoption in the LPD sector. After the identification of these enablers, they were validated by experts using AHP. It was found that ensuring transparency and traceability were the top priority enablers. Along with other disruptive technologies, BC is also in contention of becoming one of the key enabling technologies in the LPD sector. BC adoption can become a key tool in the business transformation towards digitalization of processes in LPD [34]. The successful adoption of BC can bring in the much needed transparency in the LPD sector and enhance the trust among the LPD stakeholders. Overall, BC adoption can enhance the organizational efficiency in the LPD sector.

4.1 Managerial Implications

The present study makes significant managerial contributions. It is anticipated that BC will leverage the six Vs (namely Volume, Variety, Velocity, Veracity, Variability, and Value) [40] of big data in context of LPD sector. In terms of data volume, a lot of data are generated across various phases of LPD operations therefore, it is critical for the LPD practitioners to focus on real-time data for reducing the lead-times. Industries should invest in creating and providing the requisite infrastructure (such as sensor networks, high speed computers, etc.) for reducing the disparity in data flows to reduce data variability and complexity. Such steps will not only enhance data velocity but also help in enhancing the data storage and mining across various LPD verticals.

The results emphasized on utilizing BC adoption to reduce time delays or bottlenecks in material handling and real time tracking of services. This will not only increase the speed of transactions during business transactions (particularly in international logistics and distribution) but also helps in achieving operational efficiency. The higher transaction costs can be optimized by using smart contracts which further help in considerable savings in comparison with the traditional settings. Furthermore, adoption of blockchain will help in reducing the fraudulent documentation (pertaining to data management and storage), trade documentation, cargo tracking, and certifications. This will help in enhancing the transaction efficiency and developing trust among different organizations and stakeholders involved in the LPD ecosystem. Application of BC technology and big data in service operations management domain (such as LPD) will lead to value generation and thereby improve firm's competitiveness [17]. It is also expected that BC adoption will reduce the information distortion in the processes which will further reduce the probable bullwhip effects [40].

Overall, the adoption of BC and big data is service operations management will lead to creation of new business models, and optimization of business processes.

4.2 Policy Implications

Policymakers can consider this chapter as a tool for drawing/framing recommendations/regulations in the LPD sector. At the very outset, policymakers should assess and evaluate the organizational readiness for BC adoption as it will require addition of new skillset and knowledge pertaining to BC [44]. As BC adoption will provide traceability solutions throughout the SCM processes, it's adoption will enhance operational efficiency of the LPD sector. Enhanced traceability and transparency will lead to effective management and distribution of shipment. Regulatory agencies should also focus on the provision of a legal framework for facilitating BC adoption. Regulatory agencies and governing bodies can provide a future roadmap and financial incentives to reduce the perceived high costs of investing in BC. The governing bodies can implement policies to coordinate regulations and consolidate the legal systems to encourage LPD organizations to adopt BC. The smart contracts based on BC will provide data security and secure transactions. The regulators need to be aware of the potentials and opportunities that BC can create rather than focusing on its technical incompetence [24]. On the sustainability front, BC adoption will lead to complete product traceability and thereby help in effective waste management.

5 Conclusion and Future Research

In the present study, the researchers have identified ten critical enablers for adoption of BC in the LPD sector. After the identification of the enablers, they were validated by experts and prioritized using AHP. It can be said that the combination of BC and big data can provide the requisite traceability in the LPD and service operations sector. In a BC network, as there are many stakeholders complete security may be an issue but BC has provisions for real-time fraud prevention therefore, LPD organizations can contribute effectively towards social responsibility [51]. As widespread deployment of BC in the LPD sector is still in nascent stages, a large number of challenges and issues need to be carefully tackled before BCs widespread adoption. Some of the prominent issues and challenges that need to be overcome for successful adoption of BC are:

- Technology scalability and performance issues related to BC as it is still in nascent stages.
- Technology complexity pertaining to BC technical expertise.
- BC interoperability with existing systems.
- Perceived high implementation cost of BC.

The key takeaways from this study include:

- BC can play a crucial role in enhancing trust, improving security, enhancing traceability and transparency in the current LPD ecosystem.

- BC can eliminate frauds related to documentation and improve data storage and data management practices.
- BC can enhance transaction efficiency and enhance trust among various LPD stakeholders.
- BC can reduce the turnaround time and enhance the supply chain visibility in the LPD sector.
- Extensive automation in the LPD sector can minimize carbon emissions through effective resource utilization.

The present research has a few limitations. We adopted an AHP approach, there are issues related with individual's judgement which is always not attainable due to pairwise comparisons. Moreover, there is a chance that results may be biased if the criteria are independent of the results obtained.

Future research studies can focus on utilizing survey and qualitative methods for further exploring the identified enablers. Given the potential BC has across various dimensions in the LPD sector, future work should focus on highlighting the advantages of BC on the sustainability frontier such as sustainability performance of BC. Further, models or frameworks should be developed that highlight advance the body of knowledge on the amalgamation of BC and big data.

References

1. Acciaro, M., Renken, K., & El Khadiri, N. (2020). Technological change and logistics development in european ports. *European Port Cities in Transition*, 73–88.
2. Ahluwalia, S., Mahto, R. V., & Guerrero, M. (2020). Blockchain technology and startup financing: A transaction cost economics perspective. *Technological Forecasting and Social Change, 151*, 119854.
3. Ahmad, R. W., Hasan, H., Jayaraman, R., Salah, K., & Omar, M. (2021). Blockchain applications and architectures for port operations and logistics management. *Research in Transportation Business & Management*, 100620.
4. Andola, N., Yadav, V. K., Venkatesan, S., & Verma, S. (2021). Anonymity on blockchain based e-cash protocols—A survey. *Computer Science Review, 40*, 100394.
5. Andoni, M., Robu, V., Flynn, D., Abram, S., Geach, D., Jenkins, D., McCallum, P., & Peacock, A. (2019). Blockchain technology in the energy sector: A systematic review of challenges and opportunities. *Renewable and Sustainable Energy Reviews, 100*, 143–174.
6. Ar, I. M., Erol, I., Peker, I., Ozdemir, A. I., Medeni, T. D., & Medeni, I. T. (2020). Evaluating the feasibility of blockchain in logistics operations: A decision framework. *Expert Systems with Applications, 158*, 113543.
7. Asmussen, C. B., & Møller, C. (2020). Enabling supply chain analytics for enterprise information systems: A topic modelling literature review and future research agenda. *Enterprise Information Systems, 14*(5), 563–610.
8. Bai, C., & Sarkis, J. (2020). A supply chain transparency and sustainability technology appraisal model for blockchain technology. *International Journal of Production Research, 58*(7), 2142–2162.
9. Behnke, K., & Janssen, M. (2020). Boundary conditions for traceability in food supply chains using blockchain technology. *International Journal of Information Management, 52*, 101969.

10. Cai, W., Jiang, W., Xie, K., Zhu, Y., Liu, Y., & Shen, T. (2020). Dynamic reputation–based consensus mechanism: Real-time transactions for energy blockchain. *International Journal of Distributed Sensor Networks, 16*(3), 1550147720907335.
11. Cao, H., Li, R., Tian, W., Xu, Z., & Xiao, W. (2020). Blockchain-based accountability for multi-party oblivious RAM. *Journal of Parallel and Distributed Computing, 137*, 224–237.
12. Carlan, V., Coppens, F., Sys, C., Vanelslander, T., & Van Gastel, G. (2020). Blockchain technology as key contributor to the integration of maritime supply chain? In *Maritime supply chains* (pp. 229–259). Elsevier.
13. Carvalho, A., Merhout, J. W., Kadiyala, Y., & Bentley II, J. (2021). When good blocks go bad: Managing unwanted blockchain data. *International Journal of Information Management, 57*, 102263.
14. Centobelli, P., Cerchione, R., Esposito, E., & Oropallo, E. (2021). Surfing blockchain wave, or drowning? Shaping the future of distributed ledgers and decentralized technologies. *Technological Forecasting and Social Change, 165*, 120463.
15. Chang, Y., Iakovou, E., & Shi, W. (2020). Blockchain in global supply chains and cross border trade: A critical synthesis of the state-of-the-art, challenges and opportunities. *International Journal of Production Research, 58*(7), 2082–2099.
16. Chen, Y., & Bellavitis, C. (2020). Blockchain disruption and decentralized finance: The rise of decentralized business models. *Journal of Business Venturing Insights, 13*, e00151.
17. Dolgui, A., Ivanov, D., Potryasaev, S., Sokolov, B., Ivanova, M., & Werner, F. (2020). Blockchain-oriented dynamic modelling of smart contract design and execution in the supply chain. *International Journal of Production Research, 58*(7), 2184–2199.
18. Feng, H., Wang, X., Duan, Y., Zhang, J., & Zhang, X. (2020). Applying blockchain technology to improve agri-food traceability: A review of development methods, benefits and challenges. *Journal of Cleaner Production, 260*, 121031.
19. Feng, Q., & Shanthikumar, J. G. (2018). How research in production and operations management may evolve in the era of big data. *Production and Operations Management, 27*(9), 1670–1684.
20. Grover, B. A., Chaudhary, B., Rajput, N. K., & Dukiya, O. (2021). Blockchain and Governance: Theory, Applications and Challenges. *Blockchain for Business: How It Works and Creates Value*, 113–139.
21. Hasselgren, A., Kralevska, K., Gligoroski, D., Pedersen, S. A., & Faxvaag, A. (2020). Blockchain in healthcare and health sciences—A scoping review. *International Journal of Medical Informatics, 134*, 104040.
22. Hastig, G. M., & Sodhi, M. S. (2020). Blockchain for supply chain traceability: Business requirements and critical success factors. *Production and Operations Management, 29*(4), 935–954.
23. Jabbar, A., & Dani, S. (2020). Investigating the link between transaction and computational costs in a blockchain environment. *International Journal of Production Research, 58*(11), 3423–3436.
24. Jamison, M. A., & Tariq, P. (2018). Five things regulators should know about blockchain (and three myths to forget). *The Electricity Journal, 31*(9), 20–23.
25. Johnston, R. (2005). Service operations management: From the roots up. *International Journal of Operations & Production Management*.
26. Kamble, S. S., Gunasekaran, A., & Sharma, R. (2020). Modeling the blockchain enabled traceability in agriculture supply chain. *International Journal of Information Management, 52*, 101967.
27. Li, X., Mei, Y., Gong, J., Xiang, F., & Sun, Z. (2020). A blockchain privacy protection scheme based on ring signature. *IEEE Access, 8*, 76765–76772.
28. Maesa, D. D. F., Mori, P., & Ricci, L. (2019). A blockchain based approach for the definition of auditable Access Control systems. *Computers & Security, 84*, 93–119.
29. Mardani, A., Jusoh, A., Nor, K., Khalifah, Z., Zakwan, N., & Valipour, A. (2015). Multiple criteria decision-making techniques and their applications—A review of the literature from 2000 to 2014. *Economic Research-Ekonomska Istraživanja, 28*(1), 516–571.

30. Min, H. (2019). Blockchain technology for enhancing supply chain resilience. *Business Horizons, 62*(1), 35–45.
31. Moons, K., Waeyenbergh, G., & Pintelon, L. (2019). Measuring the logistics performance of internal hospital supply chains—A literature study. *Omega, 82*, 205–217.
32. Morkunas, V. J., Paschen, J., & Boon, E. (2019). How blockchain technologies impact your business model. *Business Horizons, 62*(3), 295–306.
33. Murry, J. W., Jr., & Hammons, J. O. (1995). Delphi: A versatile methodology for conducting qualitative research. *The Review of Higher Education, 18*(4), 423–436.
34. Orji, I. J., Kusi-Sarpong, S., Huang, S., & Vazquez-Brust, D. (2020). Evaluating the factors that influence blockchain adoption in the freight logistics industry. *Transportation Research Part E: Logistics and Transportation Review, 141*, 102025.
35. Saaty, T. L. (2008). Decision making with the analytic hierarchy process. *International Journal of Services Sciences, 1*(1), 83–98.
36. Saberi, S., Kouhizadeh, M., & Sarkis, J. (2018). Blockchain technology: A panacea or pariah for resources conservation and recycling? *Resources, Conservation and Recycling, 130*, 80–81.
37. Schuetz, S., & Venkatesh, V. (2020). Blockchain, adoption, and financial inclusion in India: Research opportunities. *International Journal of Information Management, 52*, 101936.
38. Shafiq, A., Ahmed, M. U., & Mahmoodi, F. (2020). Impact of supply chain analytics and customer pressure for ethical conduct on socially responsible practices and performance: An exploratory study. *International Journal of Production Economics, 225*, 107571.
39. Sharma, R., Shishodia, A., Kamble, S., Gunasekaran, A., & Belhadi, A. (2020). Agriculture supply chain risks and COVID-19: Mitigation strategies and implications for the practitioners. *International Journal of Logistics Research and Applications*, 1–27.
40. Sundarakani, B., Ajaykumar, A., & Gunasekaran, A. (2021). Big data driven supply chain design and applications for blockchain: An action research using case study approach. *Omega*, 102452.
41. Sunny, J., Undralla, N., & Pillai, V. M. (2020). Supply chain transparency through blockchain-based traceability: An overview with demonstration. *Computers & Industrial Engineering*, 106895.
42. Swan, M. (2015). *Blockchain: Blueprint for a new economy*. O'Reilly Media, Inc.
43. Tapscott, A., & Tapscott, D. (2017). How blockchain is changing finance. *Harvard Business Review, 1*(9), 2–5.
44. Toufaily, E., Zalan, T., & Dhaou, S. B. (2021). A framework of blockchain technology adoption: An investigation of challenges and expected value. *Information & Management, 58*(3), 103444.
45. Tsiulin, S., Reinau, K. H., Hilmola, O.-P., Goryaev, N., & Karam, A. (2020). Blockchain-based applications in shipping and port management: A literature review towards defining key conceptual frameworks. *Review of International Business and Strategy*.
46. Venkatesh, V. G., Kang, K., Wang, B., Zhong, R. Y., & Zhang, A. (2020). System architecture for blockchain based transparency of supply chain social sustainability. *Robotics and Computer-Integrated Manufacturing, 63*, 101896.
47. Verma, A., & Singhal, N. (2018). A computing methodology for evaluating supply chain competitiveness. *Materials Today: Proceedings, 5*(2), 4183–4191.
48. Wamba, S. F., Queiroz, M. M., & Trinchera, L. (2020). Dynamics between blockchain adoption determinants and supply chain performance: An empirical investigation. *International Journal of Production Economics, 229*, 107791.
49. Wang, G., Gunasekaran, A., Ngai, E. W., & Papadopoulos, T. (2016). Big data analytics in logistics and supply chain management: Certain investigations for research and applications. *International Journal of Production Economics, 176*, 98–110.
50. Wang, Y., Singgih, M., Wang, J., & Rit, M. (2019). Making sense of blockchain technology: How will it transform supply chains? *International Journal of Production Economics, 211*, 221–236.
51. Wang, Y., & Kogan, A. (2018). Designing confidentiality-preserving Blockchain-based transaction processing systems. *International Journal of Accounting Information Systems, 30*, 1–18.

52. Xu, Y., Zhang, C., Zeng, Q., Wang, G., Ren, J., & Zhang, Y. (2020). Blockchain-enabled accountability mechanism against information leakage in vertical industry services. *IEEE Transactions on Network Science and Engineering.*
53. Yang, C.-S. (2019). Maritime shipping digitalization: Blockchain-based technology applications, future improvements, and intention to use. *Transportation Research Part E: Logistics and Transportation Review, 131*, 108–117.
54. Yuen, T. H. (2020). PAChain: Private, authenticated & auditable consortium blockchain and its implementation. *Future Generation Computer Systems, 112*, 913–929.
55. Zheng, Z., Xie, S., Dai, H.-N., Chen, X., & Wang, H. (2018). Blockchain challenges and opportunities: A survey. *International Journal of Web and Grid Services, 14*(4), 352–375.

Rohit Sharma Dr. Rohit Sharma is working as an Assistant Professor in Operations and Supply Chain Management at Jaipuria Institute of Management, Noida. His research interests include industry 4.0 technologies, operations management, and agricultural operations management. His work has been published in refereed journals such as Computers and Operations Research, Computers in Industry, International Journal of Logistics: Research and Applications, Journal of Enterprise Information Management, and International Journal of Information Management.

Anjali Shishodia Dr. Anjali Shishodia is working as an Assistant Professor in Operations and Supply Chain Management at LM Thapar School of Management, Dera Bassi, Mohali, Punjab. Her research interests include operations management, project management, and project driven supply chains. Her work has been published in refereed journals such as Production Planning and Control, Computers and Industrial Engineering, International Journal of Logistics: Research and Applications, and Benchmarking: an International Journal.

The manufacturer's authorised representative in the EU is Springer Nature Customer Service Centre GmbH, Europaplatz 3, 69115 Heidelberg, Germany. If you have any concerns regarding our products, please contact ProductSafety@springernature.com

Printed and bound by CPI Group (UK) Ltd, Croydon, CR0 4YY
28/09/2026
02224073-0012